PRINCIPLES OF
CHEMICAL REACTOR
ANALYSIS AND DESIGN

PRINCIPLES OF CHEMICAL REACTOR ANALYSIS AND DESIGN

New Tools for Industrial Chemical Reactor Operations

Second Edition

UZI MANN
Texas Tech University

A JOHN WILEY & SONS, INC., PUBLICATION

For general information on our other products and services or for technical support, please contact our
Customer Care Department within the United States at (800) 762-2974, outside the United States at
(317) 572-3993 or fax (317) 572-4002.

Wiley also publishes its books in variety of electronic formats. Some content that appears in print may not be
available in electronic formats. For more information about Wiley products, visit our web site at www.wiley.com.

Library of Congress Cataloging-in-Publication Data:

Mann, Uzi
   Principles of chemical reactor analysis and design : new tools for industrial chemical reactor
operations / Uzi Mann, M.D. Morris, advisory editor—2nd ed.
      p. cm.
   Includes index.
   ISBN 978-0-471-26180-3 (cloth)
1. Chemical reactors—Design and construction.   I. Title.

TP157.M268 2008
660′.2832—dc22

                                                                            2008044359

Printed in the United States of America

10 9 8 7 6 5 4 3 2 1

In memory of my sister, Meira Lavie

To Helen, and to David, Amy, and Joel

"Discovery consists of looking at the same thing as everyone else and thinking something different."

Albert Szent-Györgyi
Nobel Laureate, 1937

CONTENTS

Preface   xi

Notation   xv

1   Overview of Chemical Reaction Engineering   1

   1.1   Classification of Chemical Reactions, 2
   1.2   Classification of Chemical Reactors, 3
   1.3   Phenomena and Concepts, 8
      1.3.1   Stoichiometry, 8
      1.3.2   Chemical Kinetics, 9
      1.3.3   Transport Effects, 9
      1.3.4   Global Rate Expression, 14
      1.3.5   Species Balance Equation and Reactor
           Design Equation, 14
      1.3.6   Energy Balance Equation, 15
      1.3.7   Momentum Balance Equation, 15
   1.4   Common Practices, 15
      1.4.1   Experimental Reactors, 16
      1.4.2   Selection of Reactor Configuration, 16
      1.4.3   Selection of Operating Conditions, 18
      1.4.4   Operational Considerations, 18
      1.4.5   Scaleup, 19
      1.4.6   Diagnostic Methods, 20
   1.5   Industrial Reactors, 20
   1.6   Summary, 21
   References, 22

2   **Stoichiometry**   25

2.1   Four Contexts of Chemical Reaction, 25
2.2   Chemical Formulas and Stoichiometric Coefficients, 26
2.3   Extent of a Chemical Reaction, 28
2.4   Independent and Dependent Chemical Reactions, 39
2.5   Characterization of the Reactor Feed, 47
    2.5.1   Limiting Reactant, 48
    2.5.2   Excess Reactant, 49
2.6   Characterization of Reactor Performance, 54
    2.6.1   Reactant Conversion, 54
    2.6.2   Product Yield and Selectivity, 58
2.7   Dimensionless Extents, 64
2.8   Independent Species Composition Specifications, 68
2.9   Summary, 72
Problems, 72
Bibliography, 79

3   **Chemical Kinetics**   81

3.1   Species Formation Rates, 81
3.2   Rates of Chemical Reactions, 82
3.3   Rate Expressions of Chemical Reactions, 86
3.4   Effects of Transport Phenomena, 91
3.5   Characteristic Reaction Time, 91
3.6   Summary, 97
Problems, 97
Bibliography, 99

4   **Species Balances and Design Equations**   101

4.1   Macroscopic Species Balances—General Species-Based
    Design Equations, 102
4.2   Species-Based Design Equations of Ideal Reactors, 104
    4.2.1   Ideal Batch Reactor, 104
    4.2.2   Continuous Stirred-Tank Reactor (CSTR), 105
    4.2.3   Plug-Flow Reactor (PFR), 106
4.3   Reaction-Based Design Equations, 107
    4.3.1   Ideal Batch Reactor, 107
    4.3.2   Plug-Flow Reactor, 109
    4.3.3   Continuous Stirred-Tank Reactor (CSTR), 111
    4.3.4   Formulation Procedure, 112
4.4   Dimensionless Design Equations and
    Operating Curves, 113

4.5   Summary, 125
Problems, 126
Bibliography, 129

5   Energy Balances   **131**

5.1   Review of Thermodynamic Relations, 131
    5.1.1   Heat of Reaction, 131
    5.1.2   Effect of Temperature on Reaction
            Equilibrium Constant, 134
5.2   Energy Balances, 135
    5.2.1   Batch Reactors, 136
    5.2.2   Flow Reactors, 147
5.3   Summary, 156
Problems, 157
Bibliography, 158

6   Ideal Batch Reactor   **159**

6.1   Design Equations and Auxiliary Relations, 160
6.2   Isothermal Operations with Single Reactions, 166
    6.2.1   Constant-Volume Reactors, 167
    6.2.2   Gaseous, Variable-Volume
            Batch Reactors, 181
    6.2.3   Determination of the Reaction
            Rate Expression, 189
6.3   Isothermal Operations with Multiple Reactions, 198
6.4   Nonisothermal Operations, 216
6.5   Summary, 230
Problems, 231
Bibliography, 238

7   Plug-Flow Reactor   **239**

7.1   Design Equations and Auxiliary Relations, 240
7.2   Isothermal Operations with Single Reactions, 245
    7.2.1   Design, 246
    7.2.2   Determination of Reaction
            Rate Expression, 261
7.3   Isothermal Operations with Multiple
      Reactions, 265
7.4   Nonisothermal Operations, 281
7.5   Effects of Pressure Drop, 296
7.6   Summary, 308
Problems, 309

8   Continuous Stirred-Tank Reactor   **317**

8.1   Design Equations and Auxiliary Relations, 318
8.2   Isothermal Operations with Single Reactions, 322
      8.2.1   Design of a Single CSTR, 324
      8.2.2   Determination of the Reaction Rate
              Expression, 333
      8.2.3   Cascade of CSTRs Connected in Series, 336
8.3   Isothermal Operations with Multiple Reactions, 341
8.4   Nonisothermal Operations, 358
8.5   Summary, 370
Problems, 370

9   Other Reactor Configurations   **377**

9.1   Semibatch Reactors, 377
9.2   Plug-Flow Reactor with Distributed Feed, 400
9.3   Distillation Reactor, 416
9.4   Recycle Reactor, 425
9.5   Summary, 435
Problems, 435

10   Economic-Based Optimization   **441**

10.1   Economic-Based Performance Objective Functions, 442
10.2   Batch and Semibatch Reactors, 448
10.3   Flow Reactors, 450
10.4   Summary, 453
Problems, 453
Bibliography, 454

Appendix A   Summary of Key Relationships   **455**

**Appendix B   Microscopic Species Balances—Species Continuity
               Equations**   **465**

Appendix C   Summary of Numerical Differentiation and Integration   **469**

Index   **471**

PREFACE

I decided to write this book because I was not pleased with the way current textbooks present the subject of chemical reactor analysis and design. In my opinion, there are several deficiencies, both contextual and pedagogical, to the way this subject is now being taught. Here are the main ones:

- Reactor design is confined to simple reactions. Most textbooks focus on the design of chemical reactors with single reactions; only a brief discussion is devoted to reactors with multiple reactions. In practice, of course, engineers rarely encounter chemical reactors with single reactions.

- Two design formulations are presented; one for reactors with single reactions (where the design is expressed in terms of the conversion of a reactant), and one for reactors with multiple reactions (where the design formulation is based on writing the species balance equations for all the species that participate in the reactions). A unified design methodology that applies to all reactor operations is lacking.

- The operations of chemical reactors are expressed in terms of extensive, system-specific parameters (i.e., reactor volume, molar flow rates). In contrast, the common approach used in the design of most operations in chemical engineering is based on describing the operation in terms of dimensionless quantities. Dimensionless formulations provide an insight into the underlining phenomena that affect the operation, which are lost when the analysis is case specific.

- The analysis of chemical reactor operations is limited to simple reactor configurations (i.e., batch, tubular, CSTR), with little, if any analysis of other configurations (i.e., semibatch, tubular with side injection, distillation reactor), which are commonly used in industry to improve the yield and selectivity of the desirable product. These reactor configurations are discussed qualitatively in some textbooks, but no design equations are derived or provided.

- Most examples cover isothermal reactor operations; nonisothermal operations are sparsely discussed. In the few nonisothermal examples that are presented, usually single reactions are considered, and the dependency of the heat capacity of the reacting fluid on the temperature and composition is usually ignored. Consequently, the effect of the most important factor that affect the rates of the chemical reactions—the temperature—is not described in the most comprehensive way possible.

- In all solved examples, the heat-transfer coefficient is usually specified. But, what is not mentioned is the fact that the heat transfer can be determined only after the reactor size and geometry are specified, and the flow conditions are known. Those, of course, are not known in the initial steps of reactor design. What is needed is a method to estimate, a priori, the range of heat-transfer coefficient and then determine what reactor configuration and size provide them.

Considering those points, the current pedagogy of chemical reactor analysis and design falls short of providing students with the needed methodology and tools to address the actual technical challenges they will face in practice.

This book presents a different approach to the analysis of chemical reactor operations—reaction-based design formulation rather than the common species-based design formulation. This volume describes a unified methodology that applies to both single and multiple reactions (reactors with single reactions are merely simple special cases). The methodology is applicable to any type of chemical reactions (homogeneous, heterogeneous, catalytic) and any form of rate expression. Reactor operations are described in terms of *dimensionless* design equations that generate dimensionless *operating curves* that describe the progress of the individual chemical reactions, the composition of species, and the temperature. All parameters that affect the heat transfer are combined into a single dimensionless number that can be estimated a priori. Variations in the heat capacity of the reacting fluid are fully accounted. The methodology is applied readily to all reactor configurations (including semibatch, recycle, etc.), and it also provides a convenient framework for *economic-based* optimization of reactor operations.

One of the most difficult decisions that a textbook writer has to make is to select what material to cover and what topics to leave out. This is especially difficult in chemical reaction engineering because of the wide scope of the field and the diversity of topics that it covers. As the title indicates, this book focuses on the analysis and design of chemical reactors. The objective of the book is to present a comprehensive, unified methodology to analyze and design chemical reactors that overcomes the deficiencies of the current pedagogy. To concentrate on this objective, some topics that are commonly covered in chemical reaction engineering textbooks (chemical kinetics, catalysis, effect of diffusion, mass-transfer limitation, etc.) are not covered here. Those topics are discussed in detail in many excellent textbooks, and the reader is expected to be familiar with them. Also, advanced topics related to special reactor types (fluidized bed, trickle bed, etc.) are not covered in the text.

Students require knowledge of solving (numerically) simultaneous first-order differential equations (initial value problems) and multiple nonlinear algebraic equations. The use of mathematical software that provides numerical solutions to those types of equations (e.g., Matlab, Mathematica, Maple, Mathcad, Polymath, HiQ, etc.) is required. Numerical solutions of all the examples in the text are posted on the book web page.

The problems at the end of each chapter are categorized by their level of difficulty, indicated by a subscript next to the problem number. Subscript 1 indicates simple problems that require application of equations provided in the text. Subscript 2 indicates problems whose solutions require some more in-depth analysis and modifications of given equations. Subscript 3 indicates problems whose solutions require more comprehensive analysis and involve application of several concepts. Subscript 4 indicates problems that require the use of a mathematical software or the writing of a computer code to obtain numerical solutions.

I am indebted to many people for their encouragement and help during the development of this text. M. D. Morris was the driving force in developing this book from early conception of the idea to its completion. Stan Emets assisted in solving and checking the examples, and provided constructive criticism. My wife, Helen Mann, typed and retyped the text, in which she put not only her skills, but also her heart.

UZI MANN

NOTATION

All quantities are defined in their generic dimensions (length, time, mass or mole, energy, etc.). Symbols that appear in only one section are not listed. Numbers in parentheses indicate the equations where the symbol is defined or appears for the first time.

A	Cross-section area, area
a	Species activity coefficient
C	Molar concentration, mole/volume
CF	Correction factor of heat capacity, dimensionless (Eq. 5.2.19)
c_p	Mass-based heat capacity at constant pressure, energy/mass K
$\hat{c}_p$	Molar-based heat capacity at constant pressure, energy/mole K
D	Reactor (tube) diameter, length
DHR	Dimensionless heat of reaction, dimensionless (Eq. 5.2.23)
d_p	Particle diameter, length
E	Total energy, energy
E_a	Activation energy, energy/mole extent
e	Specific energy, energy/mass
F	Molar flow rate, mole/time
f	Conversion of a reactant, dimensionless (Eqs. 2.6.1a and 2.6.1b)
f	Friction factor, dimensionless
G	Mass velocity, mass/time area
G_j	Generation rate of species j in a flow reactor, moles j/time
g	Gravitational acceleration, length/time^2
H	Enthalpy, energy
$\hat{H}$	Molar-based specific enthalpy, energy/mole
ΔH_R	Heat of reaction, energy/mole extent

h	Mass-based specific enthalpy, energy/mass
HTN	Dimensionless heat-transfer number, dimensionless (Eq. 5.2.22)
$\mathbf{J}_j$	Molar flux of species j, mole j/(time area)
J	Total number of species
K	Equilibrium constant
KE	Kinetic energy, energy
$k, k(T)$	Reaction rate constant
k	Index of dependent reactions
L	Length, length
M	Mass, mass
MW	Molecular weight, mass/mole
m	Index of independent reactions
$\dot{m}$	Mass flow rate, mass/time
$N, N(t)$	Molar content in a reactor, moles
n	Index for chemical reactions
$\mathbf{n}$	Unit outward vector
$(n_j - n_{j_0})_i$	Moles of species j formed by the ith reaction, moles of species j
OC	Operating cost
P	Total pressure, force/area
PE	Potential energy, energy
$Q(t)$	Heat added to the reactor in time t, energy
$\dot{Q}$	Rate heat added to the reactor, energy/time
R	Gas constant, energy/temperature mole
R	Recycle ratio (Eq. 9.4.9) dimensionless
r	Volume-based rate of a chemical reaction, mole extent/time volume
(r_j)	Volume-based rate of formation of species j, mole j/time volume
$(r_j)_s$	Surface-based rate of formation of species j, mole j/time surface area of catalyst
$(r_j)_w$	Mass-based rate of formation of species j, mole j/time catalyst mass
S	Surface area, area
s_j	Stoichiometric coefficient of species j, mole j/mole extent
SC	Separation cost
T	Temperature, K or °R
t	Time, time
t_{cr}	Characteristic reaction time, time (Eq. 3.5.1)
U	Internal energy, energy
U	Heat-transfer coefficient, energy/time area K
u	Mass-based specific internal energy, energy/mass
u	Velocity, length/time
V	Volume, volume
V_R	Reactor volume, volume

Val_j	Value of species j, \$/mole
v	Volumetric flow rate, volume/time
W	Work, energy
X, $X(t)$	Extent of a chemical reaction, mole extent (Eq. 2.3.1)
$\dot{X}$	Reaction extent per unit time, mole extent/time (Eq. 2.3.10)
y	Molar fraction, dimensionless
Z	Dimensionless extent, dimensionless (Eqs. 2.7.1 and 2.7.2)
z	Vertical location, length

Greek Symbols

α	Order of the reaction with respect to species A, dimensionless
α_{km}	Multiplier factor of mth independent reaction for kth dependent reaction (Eq. 2.4.9)
β	Order of the reaction with respect to component B, dimensionless
γ	Dimensionless activation energy, $E_a/R \cdot T_0$, dimensionless (Eq. 3.3.5)
Δ	Change in the number of moles per unit extent, mole (Eq. 2.2.5)
ε	Void of packed bed, dimensionless
η	Yield, dimensionless (Eqs. 2.6.12 and 2.6.14)
θ	Dimensionless temperature, T/T_0, dimensionless
μ	Viscosity, mass/length time
ρ	Density, mass/volume
σ	Selectivity, dimensionless (Eqs. 2.6.16 and 2.6.18)
τ	Dimensionless operating time, t/t_{cr}, or space time, $V_R/v_0 t_{cr}$, dimensionless (Eqs. 4.4.3 and 4.4.8)
Φ	Particle sphericity, dimensionless

Subscripts

0	Reference state or stream
A	Limiting reactant
all	All
cr	Characteristic reaction
D	Dependent reaction
dep	Dependent
eq	Equilibrium
F	Heating (or cooling) fluid
gas	Gas phase
I	Inert
I	Independent reaction
i	The ith reaction
in	Inlet, inlet stream
inj	Injected stream

j	The jth species
k	Index number for dependent reactions
liq	Liquid phase
m	Index number for independent reactions
op	Operation
out	Outlet
R	Reactor
S	Surface
sh	Shaft work (mechanical work)
sp	Space
sys	System
tot	Total
V	Volume basis
vis	Viscous
W	Mass basis

1

OVERVIEW OF CHEMICAL REACTION ENGINEERING*

Chemical reaction engineering (CRE) is the branch of engineering that encompasses the selection, design, and operation of chemical reactors. Because of the diversity of chemical reactor applications, the wide spectrum of operating conditions, and the multitude of factors that affect reactor operations, CRE encompasses many diverse concepts, principles, and methods that cannot be covered adequately in a single volume. This chapter provides a brief overview of the phenomena encountered in the operation of chemical reactors and of the concepts and methods used to describe them.

A chemical reactor is an equipment unit in a chemical process (plant) where chemical transformations (reactions) take place to generate a desirable product at a specified production rate, using a given chemistry. The reactor configuration and its operating conditions are selected to achieve certain objectives such as maximizing the profit of the process, and minimizing the generation of pollutants, while satisfying several design and operating constraints (safety, controllability, availability of raw materials, etc.). Usually, the performance of the chemical reactor plays a pivotal role in the operation and economics of the entire process since its operation affects most other units in the process (separation units, utilities, etc.).

---

*This chapter is adopted from *Kirk-Othmer's Encyclopedia of Chemical Technology, 7th ed*, Wiley Interscience, NY (2007).

Chemical reactors should fulfill three main requirements:

1. Provide appropriate contacting of the reactants.
2. Provide the necessary reaction time for the formation of the desirable product.
3. Provide the heat-transfer capability required to maintain the specified temperature range.

In many instances these three requirements are not complimentary, and achieving one of them comes at the expense of another. Chemical reaction engineering is concerned with achieving these requirements for a wide range of operating conditions—different reacting phases (liquid, gas, solid), different reaction mechanisms (catalytic, noncatalytic), and different operating temperature and pressure (low temperature for biological reaction, high temperature for many reactions in hydrocarbon processing).

1.1   CLASSIFICATION OF CHEMICAL REACTIONS

For convenience, chemical reactions are classified in two groups:

- Homogeneous reactions—Reactions that occur in a single phase
- Heterogeneous reactions—Reactions that involve species (reactants or products) that exist in more than one phase. Heterogeneous reactions are categorized further as:
  - Fluid–fluid reactions—Chemical reactions between reactants that are in two immiscible phases (gas–liquid or liquid–liquid). The reaction occurs either at the interface or when one reactant dissolves in the other phase (which also contains the products). In many instances, the overall reaction rate depends on the interface area available, the miscibility of the reactant, and the transfer rates (e.g., diffusion) of the reactants to the interface and in the reacting phase.
  - Noncatalytic gas–solid reactions (e.g., combustion and gasification of coal, roasting of pyrites). These reactions occur on the surface of the solid. The gaseous reactant is transported to the interface, where it reacts with the solid reactant. Gaseous products are transported to the gas phase, and solid products (e.g., ash) remain in the solid. The overall reaction rate depends on the surface area available and the rate of transfer of the gaseous reactant to the solid surface.
  - Catalytic gas–solid reactions in which the reactants and products are gaseous, but the reaction takes place at the solid surface where a catalytic reagent is present. To facilitate the reaction, a large surface area is required; hence, porous particles are commonly used. The reaction takes place on the surface of the pores in the interior of the particle.

In many instances, the overall reaction rate is determined by the diffusion rate of reactants into the interior of the pore, and the diffusion of the product out of the pore.

- Catalytic gas–liquid–solid reactions—Reactants are gases and liquids, and the reaction takes place at a solid surface where a catalytic reagent is deposited (e.g., hydrogenation reactions). Normally, the liquid reactant covers the solid surface and the gaseous reactant is transferred (by diffusion) to the catalytic site.

Each of these reaction categories has its features and characteristics that should be described quantitatively.

1.2   CLASSIFICATION OF CHEMICAL REACTORS

Chemical reactors are commonly classified by the three main charateristics:

1. Mode of operation (e.g., batch, continuous, semibatch)
2. Geometric configuration (e.g., tubular, agitated tank, radial flow)
3. Contacting patterns between phases (e.g., packed bed, fluidized bed, bubble column)

In addition, reactor operations are also classified by the way their temperature (or heat transfer) is controlled. Three operational conditions are commonly used: (i) isothermal operation—the same temperatures exist throughout the reactor, (ii) adiabatic operation—no heat is transferred into or out of the reactor, and (iii) non-isothermal operation—the operation is neither isothermal nor adiabatic.

The following terms are commonly used:

- Batch reactors (Fig. 1.1*a*)—Reactants are charged into a vessel at the beginning of the operation, and products are discharged at the end of the operation. The chemical reactions take place over time. The vessel is usually agitated to provide good contacting between the reactants and to create uniform conditions (concentrations and temperature) throughout the vessel.
- Semibatch reactor (Fig. 1.1*b*)—A tank in which one reactant is charged initially and another reactant is added continuously during the operation. This mode of operation is used when it is desirable to maintain one reactant (the injected reactant) at low concentration to improve the selectivity of the desirable product and to supply (or remove) heat.
- Distillation reactor (Fig. 1.1*c*)—A batch reactor where volatile products are removed continuously from the reactor during the operation.
- Continuous reactor (flow reactors)—A vessel into which reactants are fed continuously and products are withdrawn continuously from it. The chemical

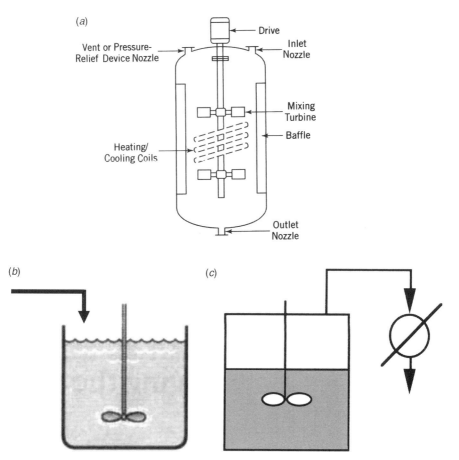

Figure 1.1   Batch operations: (*a*) batch reactor, (*b*) semibatch reactor, and (*c*) distillation reactor.

reactions take place over space (the reactor volume), and the residence time of the reacting fluid in the reactor provides the required reaction time. Common configurations of continuous reactors:

- Tubular reactor (Fig. 1.2*a*)
- Continuous stirred-tank reactor (CSTR) (Fig. 1.2*b*)
- Cascade of CSTRs (Fig. 1.2*c*)

- For multiphase reactions, the contacting patterns are used as a basis for classifying the reactors. Common configurations include:
  - Packed-bed reactor (Fig. 1.3*a*)—A vessel filled with catalytic pellets and the reacting fluid passing through the void space between them. Relatively large pellets (e.g., larger than 1 cm) are used to avoid excessive pressure drop and higher operating cost. In general, heat transfer to/from large-scale packed-bed reactors is a challenge.

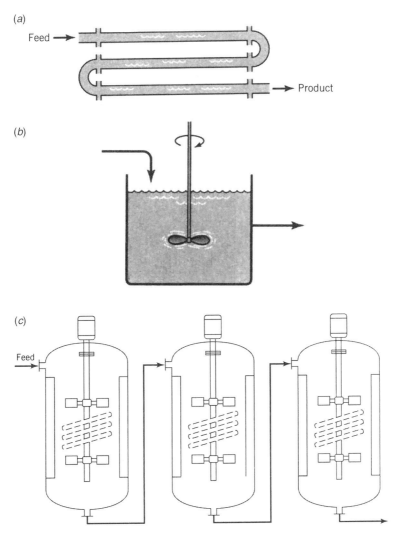

Figure 1.2 Continuous reactors: (*a*) tubular reactor, (*b*) continuous stirred-tank reactor (CSTR), and (*c*) cascade of CSTRs.

- Moving-bed reactor (Fig. 1.3*b*)—A vessel where solid particles (either reactant or catalyst) are continuously fed and withdrawn. The gas flow is maintained to allow the downward movement of the particles.
- Fluidized-bed reactor (Fig. 1.3*c*)—A vessel filled with fine particles (e.g., smaller than 500 μm) that are suspended by the upward flowing fluid. The fluidized bed provides good mixing of the particles and, consequently, a uniform temperature.
- Trickle-bed reactor—A packed bed where a liquid reactant is fed from the top, wetting catalytic pellets and a gas reactant, fed either from the top or

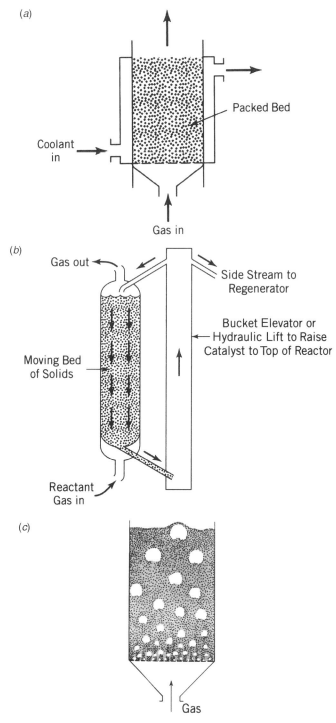

Figure 1.3 Multiphase reactors: (*a*) packed-bed reactor, (*b*) moving-bed reactor, (*c*) flui-dized-bed reactor, (*d*) bubbling column reactor, (*e*) spray reactor, and (*f*) kiln reactor.

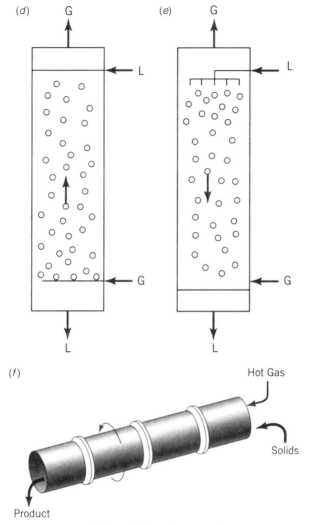

Figure 1.3 (*Continued*).

the bottom, flows through the void spaces between the pellets. The gaseous reactant must be absorbed and transported across the liquid film to the catalytic sites at the surface of the pellets.

- Bubbling column reactor (Fig. 1.3*d*)—A vessel filled with a liquid reactant and a gas reactant, fed from the bottom, moves upward in the form of bubbles. The liquid reactant is fed from the top and withdrawn from the bottom. The gaseous reactant is absorbed in the liquid reactant, and the reaction takes place in the liquid phase.
- Others [e.g., spray reactor (Fig. 1.3*e*), slurry reactor, kiln reactor (Fig. 1.3*f*), membrane reactor, etc.].

Due to the diverse applications and numerous configurations of chemical reactors, no generic design procedure exists to describe reactor operations. Rather, in each case it is necessary to identify the characteristics of the chemical reaction and the main features that the reactor should provide. Once these are identified, the appropriate physical and chemical concepts are applied to describe the selected reactor operation.

1.3 PHENOMENA AND CONCEPTS

The operation of a chemical reactor is affected by a multitude of diverse factors. In order to select, design, and operate a chemical reactor, it is necessary to identify the phenomena involved, to understand how they affect the reactor operation, and to express these effects mathematically. This section provides a brief review of the phenomena encountered in chemical reactor operations as well as the fundamental and engineering concepts that are used to describe them. Figure 1.4 shows schematically how various fundamental and engineering concepts are combined in formulating the reactor design equations.

1.3.1 Stoichiometry

Stoichiometry is an accounting system used to keep track of what species are formed (or consumed) and to calculate the composition of chemical reactors. Chapter 2 covers in detail the stoichiometric concepts and definitions used in reactor analysis.

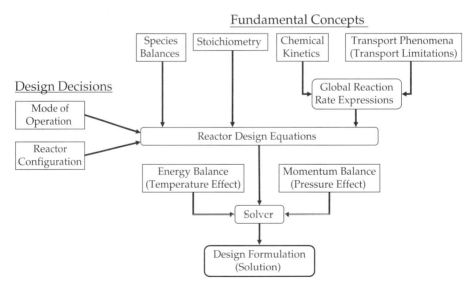

Figure 1.4 Schematic diagram of reactor design formulation.

1.3.2 Chemical Kinetics

Chemical kinetics is the branch of chemistry concerned with the rates of chemical reactions [3, 14, 19, 36–41]. Many chemical reactions involve the formation of unstable intermediate species (e.g., free radicals). Chemical kinetics is the study of the mechanisms involved in obtaining a rate expression for the chemical reaction (the reaction pathway). In most instances, the reaction rate expression is not available and should be determined experimentally. Chapter 3 covers the definitions and relations used in reactor analysis and design.

1.3.3 Transport Effects

The rate expressions obtained by chemical kinetics describe the dependency of the reaction rate on kinetic parameters related to the chemical reactions. These rate expressions are commonly referred to as the "intrinsic" rate expressions of the chemical reactions (or intrinsic kinetics). However, in many instances, the local species concentrations depend also on the rate that the species are transported in the reacting medium. Consequently, the actual reaction rate (also referred to as the *global reaction rate*) is affected by the transport rates of the reactants and products.

The effects of transport phenomena on the global reaction rate are prevalent in three general cases:

1. Fluid–solid catalytic reactions
2. Noncatalytic fluid–solid reactions
3. Fluid–fluid (liquid–liquid, gas–liquid) reactions

Incorporating the effects of species transport rates to obtain the global rates of the chemical reactions is a difficult task since it requires knowledge of the local temperature and flow patterns (hydrodynamics) and numerous physical and chemical properties (porosity, pore size and size distribution, viscosity, diffusion coefficients, thermal conductivity, etc.).

The species transfer flux to/from an interface is often described by a product of a mass-transfer coefficient, k_M, and a concentration difference between the bulk and the interface. The mass-transfer coefficient is correlated to the local flow conditions [13, 21, 26–29]. For example, in a packed bed the mass-transfer coefficient from the bulk of the fluid to the surface of a particle is obtained from a correlation of the form

$$\text{Sh} = \frac{k_M d_p}{D} = C \, \text{Re}^{0.5} \, \text{Sc}^{0.33}, \qquad (1.3.1)$$

where Sh is the Sherwood number, Re is the Reynolds number (based on the particle diameter and the superficial fluid velocity—the velocity the fluid would have if there were no particle packing), Sc is the Schmidt number, D is the diffusivity of the fluid, and C is a dimensionless constant. Similar correlations are available for mass transfer between two immiscible fluids.

In *catalytic gas–solid reactions*, the reaction takes place at catalytic sites on the surface of the solid. To obtain appreciable reaction rates, porous solids are used and the reactions take place on the surface of the pores in the interior of the particle. Hence, catalytic gas–solid reactions involve seven steps: (1) transport of the reactant from the fluid bulk to the mouth of the pore, (2) diffusion of the reactant to the interior of the pore, (3) adsorption of the reactant to the surface of the solid, (4) surface reaction at the catalytic site, (5) desorption of the product from the surface, (6) diffusion of the products to the mouth of the pore, and (7) transport of the products from the mouth of the pore to the bulk of the fluid. Steps 3–5 represent the kinetic mechanism of heterogeneous catalytic reactions. The rate of the reaction depends on the rates of these individual steps and the interactions between the catalytic site and the species, and the adsorption equilibrium constants of the various species present. A procedure, known as the Langmuir–Hishelwood–Hougen–Watson (LHHW) formulation, is used to derive and verify the reaction rate expressions for catalytic reactions [1, 3, 5, 7, 8, 14–18]. In many instances, one step is much slower than the other two steps and it determines the overall rate. This step is referred to as *the rate-limiting step*.

Often the global reaction rate of heterogeneous catalytic reactions is affected by the diffusion in the pore and the external mass-transfer rate of the reactants and the products. When the diffusion in the pores is not fast, a reactant concentration profile develops in the interior of the particle, resulting in a different reaction rate at different radial locations inside the catalytic pellet. To relate the global reaction rate to various concentration profiles that may develop, a kinetic effectiveness factor is defined [1, 3, 4, 7, 8] by

$$\left(\begin{array}{c} \text{Effectiveness} \\ \text{factor} \end{array} \right) \equiv \frac{\text{Actual reaction rate}}{\text{Reaction rate at the bulk condition}} \tag{1.3.2}$$

Hence, to express the actual reaction rate, we have to multiply the reaction rate based on the bulk condition by a correction factor, which accounts for the diffusion effects. The effectiveness factor depends on the ratio between the reaction rate and the diffusion rate and is expressed in terms of a modulus (Thiele modulus), ϕ, defined by

$$\phi^2 = \frac{\text{Characteristic reaction rate}}{\text{Characteristic diffusion rate}} \tag{1.3.3}$$

The function expressing the Thiele modulus in terms of kinetic parameters and the catalyst properties depends on the intrinsic reaction rate. For first-order reactions, the modulus is

$$\phi = L\sqrt{\frac{k}{D_{\text{eff}}}} \tag{1.3.4}$$

where k is the volume-based reaction rate constant, and D_{eff} is the effective diffusion coefficient in the particle (depending on the reactants and products, the size and size distribution of the pore, and the porosity of the pellet), and L is a characteristic length of the pellet obtained by the volume of the pellet divided by its exterior surface area. Figure 1.5 shows the relationship between the effectiveness factor and the Thiele modulus for first-order reactions. Note that for exothermic reactions the effectiveness factor may be larger than one because of the heating of the catalytic pellet. The

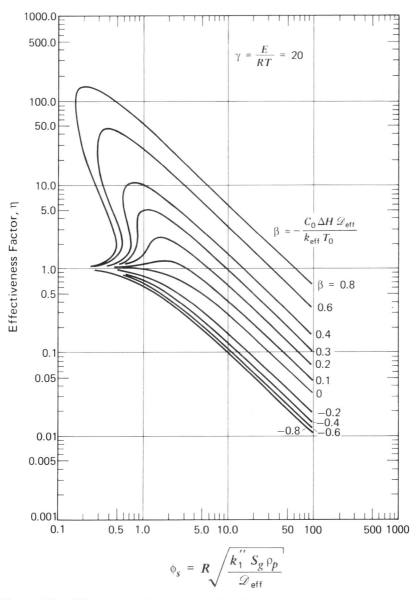

Figure 1.5 Effectiveness factor of gas-phase heterogeneous catalytic reactions.

derivation of the Thiele modulus for LHHW rate expressions is not an easy task, nor is the derivation of the relationship between the effectiveness factor and the Thiele modulus.

Noncatalytic solid–fluid reaction is a class of heterogeneous chemical reactions where one reactant is a solid and the other reactant is a fluid. The products of solid–fluid reactions may be either fluid products, solid products, or both. The rates of solid–fluid reactions depend on the phenomena affecting the transport of the fluid reactant to the surface of the solid reactant. The reaction takes place in a narrow zone that moves progressively from the outer surface of the solid particle toward the center. For convenience, noncatalytic fluid–solid reactions are divided into several categories, according to the changes that the solid particle undergoes during the reaction [1, 7, 9, 23]:

1. *Shrinking Particle*   This occurs when the particle consists entirely of the solid reactant, and the reaction does not generate any solid products. The reaction takes place on the surface of the particle, and as it proceeds, the particle shrinks, until it is consumed completely.

2. *Shrinking Core with an Ash Layer*   This occurs when one of the reaction products forms a porous layer (ash, oxide, etc.). As the reaction proceeds, a layer of ash is formed in the section of the particle that has reacted, externally to a shrinking core of the solid reactant. The fluid reactant diffuses through the ash layer, and the reaction occurs at the surface of a shrinking core until the core is consumed completely.

3. *Shrinking Core*   This occurs when the solid reactant is spread in the particle among grains of inert solid material. As the reaction proceeds, the particle remains intact, but a core containing the solid reactant is formed covered by a layer of the inert grains. The fluid reactant diffuses through the layer, and the reaction occurs at the surface of a shrinking core until the core is consumed completely.

4. *Progressive Conversion*   This occurs when the solid reactant is in a porous particle. The gaseous reactant penetrates through pores and reacts with the solid reactant (distributed throughout the particle) at all time. The concentration of the solid reactant progressively reduced until it is consumed completely. The size of the particle does not vary during the reaction.

In each of the cases described above, the global reaction rate depends on three factors: (i) the rate the fluid reactant is transported from the bulk to the outer surface of the particle, (ii) the rate the fluid reactant diffuses through the porous solid (ash or particle) to the surface of the unreacted core, and (iii) the reaction rate. The global reaction rate is usually expressed in terms of the ratios of the rates of these phenomena, as well as the dimensions of the ash layer and the unreacted core. Various mathematical models are available in the literature, providing the time needed for complete conversion of the solid reactants [1, 7, 9, 23].

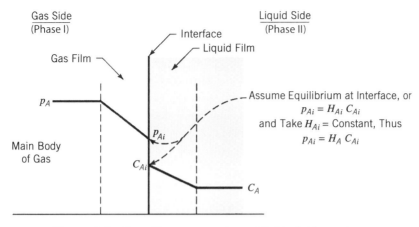

Figure 1.6   Two films presentation of fluid–fluid reactions.

Fluid–fluid reactions are reactions that occur between two reactants where each of them is in a different phase. The two phases can be either gas and liquid or two immiscible liquids. In either case, one reactant is transferred to the interface between the phases and absorbed in the other phase, where the chemical reaction takes place. The reaction and the transport of the reactant are usually described by the two-film model, shown schematically in Figure 1.6. Consider reactant A is in phase I, reactant B is in phase II, and the reaction occurs in phase II. The overall rate of the reaction depends on the following factors: (i) the rate at which reactant A is transferred to the interface, (ii) the solubility of reactant A in phase II, (iii) the diffusion rate of the reactant A in phase II, (iv) the reaction rate, and (v) the diffusion rate of reactant B in phase II. Different situations may develop, depending on the relative magnitude of these factors, and on the form of the rate expression of the chemical reaction. To discern the effect of reactant transport and the reaction rate, a reaction modulus is usually used. Commonly, the transport flux of reactant A in phase II is described in two ways: (i) by a diffusion equation (Fick's law) and/or (ii) a mass-transfer coefficient (transport through a film resistance) [7, 9]. The dimensionless modulus is called the Hatta number (sometimes it is also referred to as the Damkohler number), and it is defined by

$$\text{Ha}^2 = \frac{\text{Maximum reaction rate in the film}}{\text{Maximum transport rate through the film}} \qquad (1.3.5)$$

For second-order reactions (first order with respect to each reactant), the Hatta number is calculated in one of two ways, depending on the available parameters [7, 9]:

$$\text{Ha} = L\sqrt{\frac{kC_B}{D}} = \sqrt{\frac{kC_B D}{k_{\text{AII}}}} \qquad (1.3.6)$$

where k is the reaction rate constant, D is the diffusion coefficient of reactant A in phase II, L is a characteristic length (usually the film thickness), and $k_{A_{II}}$ is the mass-transfer coefficient of reactant A across the film. Fluid–fluid reaction are characterized by the value of the Hatta number. When Ha > 2, the reaction is fast and takes place only in the film near the interface. When $0.2 <$ Ha < 2, the reaction is slow enough such that reactant A diffuses to the bulk of phase II. When Ha < 0.2, the reaction is slow and takes place throughout phase II [7, 9, 22, 24, 25].

1.3.4 Global Rate Expression

The global rate expression is a mathematical function that expresses the *actual* rate of a chemical reaction per unit volume of the reactor, accounting for *all* the phenomena and mechanisms that take place. Knowledge of the global reaction rate is essential for designing and operating chemical reactors. For most homogeneous chemical reactions, the global rate is the same as the intrinsic kinetic rate. However, for many heterogeneous chemical reactions, a priori determination of the global reaction rate is extremely difficult.

The global reaction rate depends on three factors; (i) chemical kinetics (the intrinsic reaction rate), (ii) the rates that chemical species are transported (transport limitations), and (iii) the interfacial surface per unit volume. Therefore, even when a kinetic-transport model is carefully constructed (using the concepts described above), it is necessary to determine the interfacial surface per unit volume. The interfacial surface depends on the way the two phases are contacted (droplet, bubble, or particle size) and the holdup of each phase in the reactor. All those factors depend on the flow patterns (hydrodynamics) in the reactor, and those are not known a priori. Estimating the global rate expression is one of the most challenging tasks in chemical reaction engineering.

1.3.5 Species Balance Equation and Reactor Design Equation

The genesis of the reactor design equations is the conservation of mass. Since reactor operations involve changes in species compositions, the mass balance is written for individual species, and it is expressed in terms of moles rather than mass. Species balances and the reactor design equations are discussed in detail in Chapter 4. To obtain a complete description of the reactor operation, it is necessary to know the local reaction rates at all points inside the reactor. This is a formidable task that rarely can be carried out. Instead, the reactor operation is described by idealized models that approximate the actual operation. Chapters 5–9 cover the applications of reactor design equations to several ideal reactor configurations that are commonly used.

For flow reactors, the plug-flow and the CSTR models represent two limiting cases. The former represents continuous reactor without any mixing, where the reactant concentrations decrease along the reactor. The latter represents a reactor with complete mixing where the outlet reactant concentration exists throughout the reactor. Since in practice reactors are neither plug flow nor CSTR, it is common to obtain the performance of these two ideal reactors to identify the performance boundaries of the actual reactors.

When the behavior of a reactor is not adequately described by one of the ideal-ized models, a more refined model is constructed. In such models the reactor is divided into sections, each is assumed to have its own species concentrations and temperature, with material and heat interchanged between them [6, 7, 10, 11, 43]. The volume of each zone and the interchanges are parameters determined from the reactor operating data. The advantage of such refined models is that they provide a more detailed representation of the reactor, based on actual operating data. However, their application is limited to existing reactors.

Recent advances in computerized fluid dynamics (CFD) and developments of advanced mathematical methods to solve coupled nonlinear differential equations may provide tools for phenomenological representations of reactor hydrodynamics [40–43]. High speed and reduced cost of computation and increased cost of labora-tory and pilot-plant experimentation make such tools increasingly attractive. The utility of CFD software packages in chemical reactor simulation depends on the fol-lowing factors: (i) reliability of predicting the flow patters, (ii) ease of incorporating of the chemical kinetics and adequacy of the physical and chemical representations, (iii) scale of resolution for the application and numerical accuracy of the solution algorithms, and (iv) skills of the user.

1.3.6 Energy Balance Equation

To express variations of the reactor temperature, we apply the energy balance equation (first law of thermodynamics). Chapter 5 covers in detail the derivation and application of the energy balance equation in reactor design. The applications of the energy balance equation to ideal reactor configurations are covered in Chapters 5–9.

1.3.7 Momentum Balance Equation

In most reaction operations, it is not necessary to use the momentum balance equation. For gas-phase reaction, when the pressure of the reacting fluid varies sub-stantially and it affects the reaction rates, we apply the momentum balance equation to express the pressure variation. This occurs in rare applications (e.g., long tubular reactor with high velocity). The last section of Chapter 7 covers the application of the momentum balance equation for plug-flow reactors.

1.4 COMMON PRACTICES

Inherently, the selection and design of a chemical reactor are made iteratively because, in many instances, the global reaction rates are not known a priori. In fact, the flow patterns of reacting fluid (which affect the global rates) can be esti-mated only after the reactor vessel has been specified and the operating conditions have been selected. This section provides a review of commonly used practices.

1.4.1   Experimental Reactors

Often the kinetics of the chemical reaction and whether or not the reaction rate is affected by transport limitation are not known a priori. Lab-scale experimental reac-tors are structured such that they are operated isothermally and can be described by one of three ideal reactor models (ideal batch, CSTR, and plug flow). Isothermal operation is achieved by providing a large heat-transfer surface and maintaing the reactor in a constant-temperature bath. Experiments are conducted at different initial (or inlet) reactant proportions (to determine the form of the rate expression) and at different temperatures (to determine the activation energy).

A batch experimental reactor is used for slow reactions since species compo-sitions can be readily measured with time. The determination of reaction rate expression is described in Chapter 6. A tubular (plug-flow) experimental reactor is suitable for fast reactions and high-temperature experiments. The species compo-sition at the reactor outlet is measured for different feed rates. Short packed beds are used as differential reactors to obtain instantaneous reaction rates. The reaction rate is determined from the design equation, as described in Chapter 7. An experimental CSTR is a convenient tool in determining reaction rate since the reaction rate is directly obtained from the design equation, as discussed in Chapter 8.

The rate expressions of catalytic heterogenous reactions are generally carried out in flow reactors. When a packed-bed reactor is used (Fig. 1.7a), it is necessary to acertain that a plug-flow behavior is maintained. This is achieved by sufficiently high velocity, and having a tube-to-particle diameter ratio of at least 10 (to avoid bypassing near the wall, where the void fraction is higher than in the bed). The tube diameter should not be too large to avoid radial gradient of temperature and concentrations. A spinning basket reactor (Fig. 1.7b) is a useful tool for determin-ing the reaction rate of heterogeneous catalytic reactions and the effectiveness factor. At sufficiently high rotation speeds, the external transport rate (between the bulk to the surface of the catalytic pellet) does not affect the overall reaction rate. The effectiveness factor is determined by conducting a series of experiments with different pellet diameters.

When the heat of reaction is not known, experiments are conducted on a well-stirred calorimeter (either batch or continuous). The adiabatic temperature change is measured and the heat of reaction is determined from the energy balance equation.

1.4.2   Selection of Reactor Configuration

The first step in the design of a chemical reactor is the selection of the operating mode—batch or continuous. The selection is made on the basis of both economic

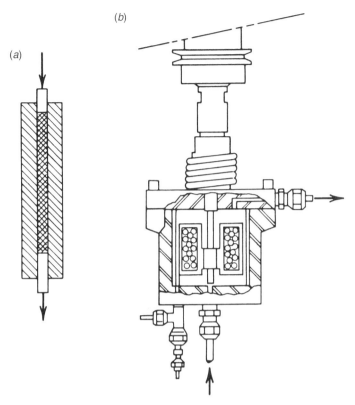

Figure 1.7 Experimental reactors for gas-phase catalytic reactions: (*a*) packed-bed and (*b*) spinning basket.

and operational considerations. Batch operations are suitable for small quantity production of high-value products, and for producing multiple products with the same equipment. Batch reactors are also used when the reacting fluid is very viscous (e.g., in the manufacture of polymer resins). Batch operations require downtime between batches for charging, discharging, and cleaning. Another drawback of this operation mode is variations among batches. Batch reactors have relative low capital investment, but their operating cost is relatively high. Continuous reactor operations are suitable for large-volume production and provide good product uniformity. Continuous reactors require relatively high capital investment, but their operating expense is relatively low.

Next, it is necessary to identify the dominating factors that affect the chemical reactions and select the most suitable reactor configuration. For homogeneous chemical reactions, one of three factors often dominates: (i) equilibrium limitation of the desirable reaction, (ii) the formation of undesirable products (by side reactions), and (iii) the amount of heat that should be transferred. For example, if a low concentration of the reactants suppresses the formation of the undesirable product, a CSTR is preferred over a tubular reactor even though a larger reactor volume is needed. When high heat-transfer rate is required, a tubular reactor with relatively small diameter (providing high surface-to-volume ratio) is used.

For heterogeneous catalytic reactions, the size of the catalyst pellets is usually the dominating factor. Packed beds with large-diameter pellets have low pressure drop (and low operating cost), but the large pellets exhibit high pore diffusion limitation and require a larger reactor. Often, the pellet size is selected on the basis of economic considerations balancing between the capital cost and the operating cost. When fine catalytic particles are required, a fluidized bed is used. In fluidized-bed reactors the reacting fluid mixed extensively, and a portion of it passes through the reactor in large bubbles with little contact with the catalytic particles. Consequently, a larger reactor volume is needed. In many noncatalytic gas–solid reactions, the feeding and movement of the solid reactant is dominating. For fluid–fluid reactions, contacting between the reactants (the interfacial area per unit volume) dominates.

1.4.3 Selection of Operating Conditions

Once the reactor type and configuration have been selected, the reactor operating conditions should be selected. For example, should the reactor be operated such that high conversion of the reactant is achieved, or should it be operated at lower conversion (with higher recycle of the unconverted reactant). The selection of the reactor operating conditions is done on the basis of an optimization objective function (e.g., maximizing profit, maximizing product yield or selectivity, minimizing generation of pollutants), as discussed in Chapter 10. When an economic criterion is used, the performance of the entire process (i.e., the reactor, separation system, utilities) is considered rather than the performance of the reactor alone.

1.4.4 Operational Considerations

Considerations should be given to assure that the reactor is operational (i.e., startup and shutdown), controllable, and does not create any safety hazards. Also, chemical reactors can operate at multiple conditions (the design and energy balance equations have multiple solutions), some of them may be unstable. Such situation is illustrated in Figure 1.8, which shows the heat generation and heat removal curves of CSTR with an exothermic reaction [2, 3]. The intersections of the two curves represent plausible operating conditions. Operating point b is unstable since any upset in the operating conditions will result in the reactor operating at point a or c.

Safe operation is a paramount concern in chemical reactor operations. Runaway reactions occur when the heat generated by the chemical reactions exceed the heat that can be removed from the reactor. The surplus heat increases the temperature of the reacting fluid, causing the reaction rates to increase further (heat generation increases exponentially with temperature while the rate of heat transfer increases linearly). Runaway reactions lead to rapid rise in the temperature and pressure,

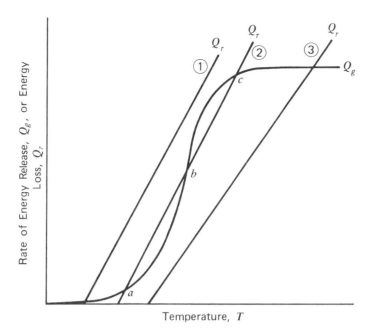

Figure 1.8   Heat generation and removal in CSTR.

which if not relieved may cause an explosion. Experience has shown that the following factors are prevalent in accidents involving chemical reactors: (i) inadequate temperature control, (ii) inadequate agitation, (iii) little knowledge of the reaction chemistry and thermochemistry, and (iv) raw material quality [12, 20].

1.4.5   Scaleup

The objective of scaleup is to design industrial-sized reactors on the basis of experimental data obtained from lab-scale reactors. A reliable scaleup requires insight of the phenomena and mechanisms that affect the performance of the reactor operation. Once these factors are identified and quantified, the task is to establish similar conditions in the industrial-size reactor. The difficulty arises from the fact that not all the factors can be maintained similar simultaneously upon scaleup [37]. For example, it is often impossible to maintain similar flow conditions (e.g., Reynolds number) and the same surface heat-transfer area per unit volume. Good understanding of the phenomena and mechanisms can enable the designer to account for the different conditions. Unfortunately, in many instances, considerable uncertainties exist with regard to the mechanisms and the magnitude of the parameters. As a result, an experimental investigation on a pilot-scale reactor is conducted to improve the reliability of the design of an industrial-scale reactor.

In many processes that apply to an agitated tank, the main task is to maintain sufficient mixing during scaleup. Considerable information is available in the literature on scaling up of agitated tanks [30–36].

1.4.6 Diagnostic Methods

In practice, especially in large-scale reactors, plug-flow or complete mixing are rarely achieved, and it is desirable to quantify the deviation from those idealized flow conditions. Also, when a chemical reactor does not perform at the expected level, it is necessary to identify the reason. A diagnostic method that is applied in such situations is based on measuring the residence time distribution (RTD) in the reactor. An inert tracer is injected at the reactor inlet, and its concentration at the reactor outlet is measured with time. By comparing the outlet concentration curve to the inlet concentration curve, the RTD curve of the reacting fluid in the reactor can be constructed [1, 7, 10, 43].

The measured mean residence time and shape of the RTD curve provide valuable information on the flow of the reacting fluid in the reactor. Based on fundamental physical concepts, the mean residence time is the quotient of the volume of the reacting fluid in the reactor and its volumetric flow rate:

$$\bar{t} = \frac{V}{v} \tag{1.4.1}$$

Since mean residence time and the volumetric flow rate are known, value of V (the "active volume" of the fluid in the reactor) is readily calculated. If the calculated value of V is smaller than the reactor volume, it indicates that a stagnant zone (not available to the flowing fluid) exists in the reactor. In heterogeneous fluid–fluid reactors, measuring the mean residence time of each fluid provides the holdup of each in the reactor. Comparing the RTD curve to that of CSTR and plug-flow reactor provides an indication on the deviations of the actual flow patterns from those of idealized flows.

It is important to recognize the limitations of the RTD method. Residence time distribution does not discern between a reacting fluid that is mixed on the molecular level (micromixing) and one that flows in segregated blobs. Also, the same RTD is obtained when the reacting fluid is mixed near the entrance or near the exit. Both of these factors affect the chemical reactions and the performance of the reactor.

1.5 INDUSTRIAL REACTORS

Figure 1.9 shows the structures of two industrial reactors used to produce two of the largest volume chemicals. Figure 1.9a shows the internals of an ammonia converter, and Figure 1.9b shows schematically the fluidized-bed catalytic cracking (FCC) reactor that converts heavy petroleum crude to lighter hydrocarbon cuts.

The ammonia synthesis reaction is an exothermic reversible reaction that is carried out in a packed-bed of catalytic pellets. The removal of the heat generated by the reaction is the dominating factor in the design of the reactor. Since the

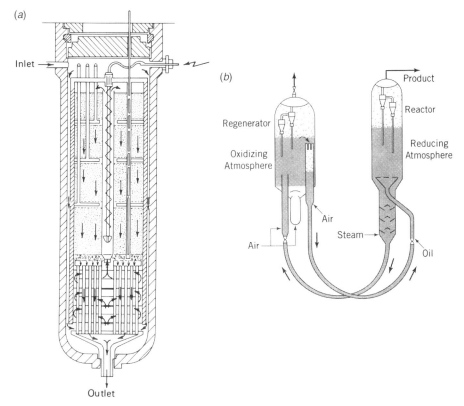

Figure 1.9   Industrial reactors: (*a*) ammonia reactor and (*b*) fluidized-bed catalytic cracking reactor.

reaction is reversible, an increase in the temperature limits the conversion (equilibrium limited). The intricate reactor was designed to provide efficient heat transfer and to use some of the heat of reaction to heat the feed.

The FCC unit consists of two fluidized-bed reactors, one used as the cracking reactor and the second as a catalyst regenerator. During the cracking reaction, carbon builds up on the catalytic particles, causing deactivation. To regenerate the catalytic particles, they are transported to a regeneration unit where the carbon is burned with air. The heat generated during the combustion of the carbon is carried with the hot particles to the cracking reactor; hence, the circulating particles provide the heat for the endothermic cracking reaction.

1.6   SUMMARY

The objective of this chapter was to describe the diversity and complexity of chemical reactor operations and to provide an overview of the phenomena encountered, and the concepts used to describe them. The chapter attempts to convey three points:

1. The difficulty in obtaining a global reaction rate expression—an expression that accounts for both intrinsic kinetics and transport effects.
2. The inherent difficulty in designing chemical reactors, which is due to two factors: (i) Global reaction rates depend on the local flow conditions, which are not known a priori, and (ii) even when the global reaction rate expressions are known, solving the reactor design equations is a formidable task that rarely can be performed in the design exercise.
3. Approximate engineering approaches are successfully employed (using ideal reactor models) to estimate the reactor operations and guesstimate the limits of its performance.

This volume describes a methodology to describe the operations ideal chemical reactor.

REFERENCES

Additional discussions of chemical reactor design can be found in:
1. G. F. Froment and K. B. Bischoff, *Chemical Reactor Analysis and Design*, 2nd ed., Wiley, New York, 1990.
2. K. G. Denbigh and J. C. R. Turner, *Chemical Reactor Theory*, Cambridge University Press, New York, 1971.
3. C. G. Hill, *An Introduction to Chemical Engineering Kinetics and Reactor Design*, Wiley, New York, 1977.
4. L. D. Schmidt, *The Engineering of Chemical Reactions*, 2nd ed., Oxford University Press, New York, 2005.
5. M. E. Davis and R. J. Davis, *Fundamentals of Chemical Reaction Engineering*, McGraw-Hill, New York, 2003.
6. K. R. Westerterp, W. P. M. van Swaaij, and A. A. C. M. Beenackers, *Chemical Reactor Design and Operation*, 2nd ed., Wiley, New York, 1988.
7. O. Levenspiel, *Chemical Reaction Engineering*, 3rd ed., Wiley, New York, 1999.
8. H. S. Fogler, *Elements of Chemical Reaction Engineering*, 4th ed., Prentice-Hall, Englewood Cliffs, NJ, 2005.
9. R. W. Missen, C. A. Mims, and B. A. Saville, *Introduction to Chemical Reaction Engineering and Kinetics*, Wiley, New York, 1999.
10. E. B. Nueman, *Chemical Reactor Design, Optimization, and Scale-up*, McGraw-Hill, New York, 2002.
11. M. Baerns, H. Hofmann, and A. Renken, *Chemische Reaktionstechnik*, Thieme, Stuttgart, 1999.
12. A. K. Coker, *Modeling of Chemical Kinetics and Reactor Design*, Butterworth-Heinemann, Boston, 2001.

13. L. K. Doraiswany and M. M. Sharma, *Heterogeneous Reactions, Analyses, Examples and Reactor Design, Vol. 2, Fluid–Fluid Solid Reactions*, Wiley, New York, 1984.

More detailed treatments of chemical kinetics, reaction mechanisms, catalysis, and the theoretical basis for the rate expression can be found in:

14. R. M. Masel, *Chemical Kinetics and Catalysis*, Wiley Interscience, Hoboken, NJ, 2001.

15. R. M. Masel, *Principles of Adsorption and Reaction on Solid Surfaces*, Wiley, New York, 1996.

16. C. H. Bartholomew and R. J. Farrauto, *Fundamentals of Industrial Catalytic Processes*, Wiley Interscience, Hoboken, NJ, 2006.

17. B. C. Gates, *Catalytic Chemistry*, Wiley, New York, 1992.

18. B. C. Gates, J. R. Katzer, and G. C. A. Schuit, *Chemistry of Catalytic Processes*, McGraw-Hill, New York, 1979.

19. K. J. Laidler, *Chemical Kinetics*, 3rd ed., Harper & Row, New York, 1987.

20. J. Barton and R. Rodgers, eds., *Chemical Reaction Hazards*, Institution of Chemical Engineers, Rugby, Warwickshire, UK, 1993.

More detailed treatments of multi-phase reactions can be found in:

21. H. H. Lee, *Heterogeneous Reactor Design*, Butterworth, Boston, 1985.

22. P. V. Danckwerts, *Gas-Liquid Reactions*, McGraw-Hill, New York, 1970.

23. J. Szekeley, J. W. Evans, and H. Y. Sohn, *Gas-Solid Reactions*, Academic, New York, 1976.

24. Y. T. Shah, *Gas–Liquid–Solid Reactor Design*, McGraw-Hill, New York, 1979.

25. L. S. Fan and K. Tsuchiya, *Bubble Wake Dynamics in Liquids and Liquid–Solid Suspensions*, Butterworth-Heinemann, Boston, 1990.

More detailed treatments of transport phenomena and their effects on chemical reactions can be found in:

26. R. B. Bird, W. E. Stewart, and E. N. Lightfoot, *Transport Phenomena*, 2nd ed., Wiley, Hoboken, NJ, 2002.

27. J. R. Welty, C. E. Wicks, and R. E. Wilson, *Fundamentals of Momentum Heat and Mass Transfer*, 2nd ed., Wiley, New York, 1984.

28. G. Astarita, *Mass Transfer with Chemical Reaction*, Elsevier, Amsterdam, 1967.

29. L. A. Belfiore, *Transport Phenomena for Chemical Reactor Design*, Wiley Intersience, Hoboken, NJ, 2003.

More detailed treatments of mixing technology and scale-up can be found in:

30. J. Y. Oldshue, *Fluid Mixing Technology*, McGraw-Hill, New York, 1983.

31. M. Zlokrnic, *Stirring: Theory and Practice*, Wiley-VCH, Hoboken, NJ, 2001.

32. E. L. Paul, V. Atiemo-Oberg, and S. M. Kresta, eds., *Handbook of Industrial Mixing; Science and Practice*, Wiley Interscience, Hoboken, NJ, 2003.

33. N. Harnby, M. F. Edward, and A. W. Nienow, *Mixing in the Process Industry*, Butterworth-Heinemann, Boston, 1997.

34. G. B. Tatterson, *Fluid Mixing and Gas Dispersion in Agitated Tanks*, McGraw-Hill, New York, 1991.

35. J. J. Ulbrecht, *Mixing of Liquids by Mechanical Agitation*, Taylor & Francis, Boston, 1985.

36. A. Bisio and R. L. Kabel, *Scaleup of Chemical Processes*, Wiley, New York, 1985.

More detailed treatments of fluidization technology can be found in:

37. D. Kunii and O. Levenspiel, *Fluidization Engineering*, 2nd ed., Butterworth-Heinemann, Boston, 1991.

38. J. G. Yates, *Fundamentals of Fluidized-Bed Chemical Processes*, Butterworth, London, New York, 1983.

39. J. F. Davidson and D. Harrison, *Fluidised Particles*, Cambridge University Press, New York, 1963.

More detailed treatments of computational fluid dynamics and modeling of low systems can be found in:

40. V. V. Ranade, *Computational Flow Modeling for Chemical Reactor Engineering (Process Systems Engineering)*, Academic, New York, 2001.

41. R. V. A. Oliemans, ed., *Computational Fluid Dynamics for the Petrochemical Process Industry*, Kluwer Academic, The Netherlands, 1991.

42. R. J. Kee, *Chemically Reacting Flow*, Wiley Interscience, Hoboken, NJ, 2003.

43. C. Y. Wen and L. T. Fan, *Models for Flow Systems and Chemical Reactors*, Marcel Dekker, New York, 1975.

2

STOICHIOMETRY

This chapter covers a tool used extensively in the analysis of processes involving chemical reactions—stoichiometry. Literally, stoichiometry means measurement of elements. In practice, stoichiometry is an accounting system that provides a framework to describe chemical transformations. Stoichiometry keeps track of the amount of species that are being formed and consumed and enables us to calculate the composition of chemical reactors. This chapter describes a systematic stoichiometric methodology that enables us to handle any reacting system with multiple chemical reactions. Specifically, the methodology indicates how many chemical reactions should be considered to determine all the state quantities of the reactor and the number of design equations necessary to describe the reactor operation. It also indicates what set of chemical reactions is most suitable for the design formulation of chemical reactors. The stoichiometric methodology presented in this chapter is used throughout the text, and therefore a good knowledge of its key definitions and structure is essential.

2.1 FOUR CONTEXTS OF CHEMICAL REACTION

Before we begin the discussion of stoichiometry, it is important to recognize that the term *chemical reaction* is used in four contexts:

- As a chemical formula
- As stoichiometric relation between the species

Principles of Chemical Reactor Analysis and Design, Second Edition. By Uzi Mann
Copyright © 2009 John Wiley & Sons, Inc.

- As a presentation of the pathway of the chemical transformation
- As an elementary reaction

The *chemical formula* is essentially the selection of a framework (or a "basis") for the calculation of chemical transformations. It is discussed in details in Section 2.2.

A *stoichiometric relation* is merely a representation of the proportion between chemical species. To illustrate, consider the chlorination of methane to produce trichloro methane. When methane and chlorine are contacted in a reactor, the following chemical reactions occur:

$$\text{Reaction 1:} \quad CH_4 + Cl_2 \quad \longrightarrow \quad CH_3Cl + HCl \tag{2.1.1}$$

$$\text{Reaction 2:} \quad CH_3Cl + Cl_2 \quad \longrightarrow \quad CH_2Cl_2 + HCl \tag{2.1.2}$$

$$\text{Reaction 3:} \quad CH_2Cl_2 + Cl_2 \quad \longrightarrow \quad CHCl_3 + HCl \tag{2.1.3}$$

$$\text{Reaction 4:} \quad CHCl_3 + Cl_2 \quad \longrightarrow \quad CCl_4 + HCl \tag{2.1.4}$$

To relate the amount of trichloro methane (the desirable product) generated to the reactants (methane and chlorine), we add the first three reactions to obtain the following stoichiometric relation:

$$CH_4 + 3Cl_2 \quad \longrightarrow \quad CHCl_3 + 3HCl \tag{2.1.5}$$

Reaction 2.1.5 does not take place, rather it merely provides a relation between the reactants and the desired product. Usually, stoichiometric relations are used to determine the limiting reactant and the yield of the desirable products.

Reaction pathways represent the routes by which the chemical species are formed and consumed. For example, in the methane chlorination above, the four reactions indicate the pathways by which the various species are generated or consumed. In practice, each of these reactions may involve the formation and destruction of intermediates and unstable species (e.g., free radicals). The rates of the reaction pathways should be known in order to determine the rate of formation (or depletion) of the chemical species in the reactor.

An *elementary reaction* is a representation of the interactions between the species on the molecular level. Elementary reactions are used to describe the mechanisms by which species are formed. The term *elementary* is used because each reactant is assumed to be in its most simple form.

2.2   CHEMICAL FORMULAS AND STOICHIOMETRIC COEFFICIENTS

The first step in analyzing any engineering problem is to define the system and select the framework for solving the problem. This step is commonly referred to as selecting a *basis* for the calculation. A similar step is required when dealing with processes involving chemical reactions. A chemical transformation can be expressed in many forms called chemical formulas, and we have to select one of them as the framework (or basis) for the analysis. To illustrate this point, consider, for example, the reaction between oxygen and carbon monoxide to form carbon dioxide. This chemical transformation can be described by one of many chemical formulas, such as

$$CO + \tfrac{1}{2}O_2 \quad \longrightarrow \quad CO_2 \tag{2.2.1}$$

$$2CO + O_2 \quad \longrightarrow \quad 2CO_2 \tag{2.2.2}$$

or, if you wish to do so,

$$20CO + 10O_2 \quad \longrightarrow \quad 20CO_2 \tag{2.2.3}$$

Each chemical formula represents a given amount of mass. For example, Reaction 2.2.1 represents 44 *units* of mass, Reaction 2.2.2 represents 88 *units* of mass, and Reaction 2.2.3 represents 880 *units* of mass. We select one of these chemical formulas and relate all relevant quantities (heat of reaction, rate of reaction, etc.) to it.

We adopt common conventions concerning chemical reactions. Each chemical reaction has an arrow indicating the direction of the chemical transformation and clearly defines the reactants and products of the reaction. Species to the left of the arrow are called *reactants*, and those on the right are called *products*. Reversible reactions are treated as two distinct reactions, one forward and one backward. The arrow also serves as an equality sign for the total *mass* represented by the chemical formula. Thus, each chemical reaction should be balanced; an unbalanced chemical reaction violates the conservation of mass principle.

Once the specific chemical formula is selected, the stoichiometric coefficients of the individual species are defined as follows: For each product species, the stoichiometric coefficient is identical to the coefficient of that species in the chemical formula. For each reactant, the stoichiometric coefficient is the negative value of the coefficient of that species in the chemical reaction. If a species does not participate in the reaction, its stoichiometric coefficient is zero.

Consider the general chemical reaction

$$aA + bB \quad \longrightarrow \quad cC + dD \tag{2.2.4}$$

The species' stoichiometric coefficients are: $s_A = -a$, $s_B = -b$, $s_C = c$, $s_D = d$, and, for any inert species I, $s_I = 0$. By defining the stoichiometric coefficients in this manner, chemical reactions are expressed as homogeneous algebraic equations. For example, Reaction 2.2.4 is expressed as

$$-aA - bB + cC + dD = 0$$

The advantage of doing so will become clear later when we consider multiple simultaneous chemical reactions.

Each chemical reaction is characterized by the sum of its stoichiometric coefficients,

$$\Delta = \sum_{j}^{J} s_j = s_A + s_B + \cdots \tag{2.2.5}$$

where J indicates the number of species. The parameter Δ indicates the change in the number of moles per "unit" of the chemical formula selected. For example, for Reaction 2.2.4, $\Delta = (-a) + (-b) + c + d$. As we will see later, this parameter is extremely useful in the design formulation of chemical reactors because it enables us to express the changes in the total number of moles in the reactor.

The following points concerning stoichiometry are worth noting:

1. The stoichiometric coefficient of species j, s_j, is dimensionless, expressed in (moles of j)/(moles of reaction extent), as discussed in Section 2.3.
2. The mathematical condition for a balanced chemical reaction is

$$\sum_{j}^{J} s_j(\mathrm{MW}_j) = 0 \qquad j = A, B, \ldots \tag{2.2.6}$$

   where s_j and MW_j are, respectively, the stoichiometric coefficient and the molecular mass of species j.
3. Usually, for convenience, we write chemical equations such that either the largest species' coefficient is one (as in Reaction 2.2.1) or the smallest species' coefficient is the smallest integer (as in Reaction 2.2.2).
4. The numerical values of the species' coefficients of the reaction depend on the specific chemical formula selected. However, the ratio of any two coefficients is the same, regardless of the chemical formula used.

2.3 EXTENT OF A CHEMICAL REACTION

Once the chemical formula is selected, the progress of the chemical reaction can be quantified. It is convenient to express the progress of the chemical reaction in terms of the chemical formula selected. To do so we define a quantity called the reaction extent.

Definition The extent of a chemical reaction is one unit of the chemical formula selected.

The extent represents a certain amount of mass, and its units are mole extent. For example, for Reaction 2.2.1, one extent means that *one mole* of CO reacts with *half a mole* of O_2 to form *one mole* of CO_2 and it represents 44 *units* of mass.

The extent of a chemical reaction is determined from the number of moles of any species, say species j, formed (or depleted) by the reaction

$$X \equiv \frac{\text{Moles of species } j \text{ formed by the reaction}}{\text{Stoichiometric coefficient of species } j} = \frac{n_j - n_{j_0}}{s_j} \qquad (2.3.1)$$

For example, for Reaction 2.2.4, the extent is calculated by one of the following relations:

$$X = \frac{n_A - n_{A_0}}{-a} = \frac{n_B - n_{B_0}}{-b} = \frac{n_C - n_{C_0}}{c} = \frac{n_D - n_{D_0}}{d}$$

The following points concerning the extent of a chemical reaction are worth noting:

1. The extent is a calculated quantity, and in order to obtain it we have to derive relations between the extent and measurable quantities (see Example 2.3). However, when the chemical extent is known, the amount of all species can be readily calculated.
2. The heat of reaction of a chemical reaction is expressed in terms of energy per mole extent.

Next we derive relationships between the species composition in chemical reactors to the chemical reactions taking place in them. For convenience, we distinguish between two modes of reactor operations: batch operation (batch reactors) and steady continuous operation (flow reactors), shown schematically in Figure 2.1. In batch reactors, reactants are charged into the reactor and, after a certain period of time, the products are discharged from the reactor; hence, the chemical reactions take place over time. In steady-flow reactors, reactants are continuously fed into the reactor, and products are continuously withdrawn from the reactor outlet; hence, the chemical reactions take place over space.

First, consider a batch reactor where n_R simultaneous chemical reactions take place and focus attention on species j. The total number of moles of species j in

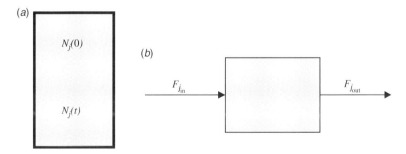

Figure 2.1   Modes of reactor operations: (*a*) batch reactor and (*b*) flow reactor.

the reactor at time t, $N_j(t)$, is related to the number of moles of species j formed by each of the individual chemical reactions

$$N_j(t) - N_j(0) = (n_j - n_{j_0})_1 + \cdots + (n_j - n_{j_0})_i + \cdots + (n_j - n_{j_0})_{n_R} \quad (2.3.2)$$

where $(n_j - n_{j_0})_i$ is the number of moles of species j formed by the ith chemical reaction in time t. Note that species j may be formed in some reactions and consumed in others. As will be discussed in Section 2.4, to determine the species composition (and, in general, all other state quantities), only a set of *independent* reactions should be considered, and *not* all the chemical reactions that take place. Hence, using Eq. 2.3.1, Eq. 2.3.2 reduces to

$$N_j(t) = N_j(0) + \sum_{m}^{n_I} (s_j)_m X_m(t) \quad j = A, B, \ldots \quad (2.3.3)$$

where m is an index for independent reactions, $(s_j)_m$ is the stoichiometric coefficient of species j in the mth independent reaction, $X_m(t)$ is the extent of the mth independent reaction at time t, and n_I is the number of independent reactions. Equation 2.3.3 relates the species composition in a batch reactor to the extents of the independent chemical reactions.

To relate the total number of moles in the reactor at time t, $N_{tot}(t)$ to the extents, we write Eq. 2.2.3 for each species in the reactor and sum the relations, and then collect terms by the individual reaction extents,

$$N_{tot}(t) = N_{tot}(0) + \sum_{m}^{n_I} \Delta_m X_m(t) \quad (2.3.4)$$

where Δ_m is the change in the number of moles per unit extent of the mth independent chemical reaction, defined by Eq. 2.2.5, and $N_{tot}(0)$ is the total number of moles initially in the reactor.

When a *single* chemical reaction takes place, Eq. 2.3.3 reduces to

$$N_j(t) = N_j(0) + s_j X(t) \quad j = A, B, \ldots \quad (2.3.5)$$

Writing Eq. 2.3.5 for any two species, say A and j,

$$N_j(t) = N_j(0) + \frac{s_j}{s_A}[N_A(t) - N_A(0)] \quad (2.3.6)$$

Equation 2.3.6 provides an algebraic relation between the number of moles of any two species in the reactor at time t without calculating the extent itself. To determine the total number of moles in a batch reactor at time t, $N_{tot}(t)$, use Eq. 2.3.4, which reduces to

$$N_{tot}(t) = N_{tot}(0) + \Delta X(t) \tag{2.3.7}$$

To determine $N_{tot}(t)$ without calculating the extent explicitly, take the summation of Eq. 2.3.6 over all species,

$$N_{tot}(t) = N_{tot}(0) + \frac{\Delta}{s_A}[N_A(t) - N_A(0)] \tag{2.3.8}$$

Consider now a *steady*-flow reactor and conduct a species balance over the reactor. At steady state, the molar flow rate of species j at the reactor outlet is equal to the molar flow rate of species j at the reactor inlet plus the rate species j is being generated inside the reactor by the reaction, G_j,

$$F_{j_{out}} = F_{j_{in}} + G_j \qquad j = A, B, \ldots \tag{2.3.9}$$

The generation term, G_j, is discussed in detail in Chapter 4; here, we focus on its relation to the extents of the chemical reactions. The rate species j is being generated by each reaction is expressed in a relation similar to Eq. 2.3.2. Differentiating Eq. 2.3.3 with respect to time, the rate of generation of species j is

$$G_j = \sum_m^{n_I} (s_j)_m \dot{X}_m \qquad j = A, B, \ldots \tag{2.3.10}$$

where $\dot{X}_m$ is the extent of the mth independent reaction converted per unit time in the reactor. Substituting Eq. 2.3.10 into Eq. 2.3.9,

$$F_{j_{out}} = F_{j_{in}} + \sum_m^{n_I} (s_j)_m \dot{X}_m \qquad j = A, B, \ldots \tag{2.3.11}$$

Summing Eq. 2.3.11 over all species, the total molar flow rate at the reactor outlet relates to the extents of the reactions by

$$F_{tot_{out}} = F_{tot_{in}} + \sum_m^{n_I} \Delta_m \dot{X}_m \tag{2.3.12}$$

where Δ_m is the change in the number of moles per unit extent of the mth independent chemical reaction, defined by Eq. 2.2.5.

For steady-flow reactors with a *single* chemical reaction, Eq. 2.3.11 reduces to

$$F_{j_{out}} = F_{j_{in}} + s_j \dot{X} \qquad j = A, B, \ldots \tag{2.3.13}$$

and Eq. 2.3.12 reduces to

$$F_{tot_{out}} = F_{tot_{in}} + \Delta \dot{X} \tag{2.3.14}$$

We can write Eq. 2.3.11 for any two species, say A and j, to obtain

$$F_{j_{out}} = F_{j_{in}} + \frac{s_j}{s_A}(F_{A_{out}} - F_{A_{in}}) \quad j = B, C, \ldots \tag{2.3.15}$$

Taking the summation of Eq. 2.3.15 over all species,

$$F_{tot_{out}} = F_{tot_{in}} + \frac{\Delta}{s_A}(F_{A_{out}} - F_{A_{in}}) \tag{2.3.16}$$

Equations 2.3.15 and 2.3.16 enable us to determine, respectively, the molar flow rate of any species and the total molar flow rate at the reactor outlet in terms of the molar flow rate of any other species without calculating the extent itself.

Example 2.1   A batch reactor contains 4 mol of CO and 1 mol of O_2. Calculate the extent of the reaction and the composition of the reactor when the reaction goes to completion.
a. Use Reaction 2.2.1 and determine the species compositions from the extent.
b. Use Reaction 2.2.2 and determine the species compositions without calculating the extent.

Solution

a. We select the chemical formula

$$CO + \tfrac{1}{2}O_2 \longrightarrow CO_2$$

and its stoichiometric coefficients are

$$s_{CO} = -1 \qquad s_{O_2} = -\tfrac{1}{2} \qquad s_{CO_2} = 1 \qquad \Delta = -\tfrac{1}{2}$$

The reaction reaches completion when one of the reactants is depleted. Assuming that CO is depleted—hence, $N_{CO}(t) = 0$, and we write Eq. 2.3.5 to determine the extent of the reaction,

$$X(t) = \frac{N_{CO}(t) - N_{CO}(0)}{s_{CO}} = \frac{0 - 4.0}{-1} = 4 \, \text{mol extent}$$

Now that we know the extent of the chemical reaction, we can calculate the moles of O_2 and CO_2 in the reactor at the end of the operation by Eq. 2.3.5:

$$N_{O_2}(t) = N_{O_2}(0) + s_{O_2}X(t) = 1 + (-0.5)(4) = -1 \, mol$$

$$N_{CO_2}(t) = N_{CO_2}(0) + s_{CO_2}X(t) = 0 + (1)(4) = 4 \, mol$$

Since the number of moles of any species cannot be negative, obtaining such values is an indication that there is an error in the calculation, or that it is based on incorrect information. The error here is, of course, the assumption that the CO is depleted before the oxygen, while actually oxygen is depleted first. Hence, to determine the extent at completion, we write Eq. 2.3.5 for oxygen with $N_{O_2}(t) = 0$,

$$X(t) = \frac{N_{O_2}(t) - N_{O_2}(0)}{s_{O_2}} = \frac{0 - 1.0}{-0.5} = 2 \, mol \, extent$$

We calculate the moles of CO and CO_2 in the reactor at the end of the operation by Eq. 2.3.5:

$$N_{CO}(t) = N_{CO}(0) + s_{CO}X(t) = 4 + (-1)(2) = 2 \, mol$$

$$N_{CO_2}(t) = N_{CO_2}(0) + s_{CO_2}X(t) = 0 + (1)(2) = 2 \, mol$$

The total number of moles in the reactor at time t is

$$N_{tot}(t) = N_{O_2}(t) + N_{CO}(t) + N_{CO_2}(t) = 0 + 2 + 2 = 4 \, mol$$

We can also calculate $N_{tot}(t)$ by Eq. 2.3.7:

$$N_{tot}(t) = N_{tot}(0) + \Delta X(t) = 5 + \left(-\frac{1}{2}\right)(2) = 4 \, mol$$

b. In this case, we select the chemical formula

$$2\,CO + O_2 \longrightarrow 2\,CO_2$$

and its stoichiometric coefficients are

$$s_{CO} = -2 \qquad s_{O_2} = -1 \qquad s_{CO_2} = 2 \qquad \Delta = -1$$

Using the given initial composition, $N_{O_2}(0) = 1 \, mol$, and that at completion $N_{O_2}(t) = 0$, we write Eq. 2.3.2 for oxygen with $N_{O_2}(t) = 0$,

$$X(t) = \frac{N_{O_2}(t) - N_{O_2}(0)}{s_{O_2}} = \frac{0 - 1.0}{-1} = 1 \, mol \, extent$$

To determine the moles of CO and CO_2 without calculating the extent, Eq. 2.3.6 is used:

$$N_{CO}(t) = N_{CO}(0) + \frac{s_{CO}}{s_{O_2}}[N_{O_2}(t) - N_{O_2}(0)] = 4 + \frac{-2}{-1}(0-1) = 2\,\text{mol}$$

$$N_{CO_2}(t) = N_{CO_2}(0) + \frac{s_{CO_2}}{s_{O_2}}[N_{O_2}(t) - N_{O_2}(0)] = 0 + \frac{2}{-1}(0-1) = 2\,\text{mol}$$

Note that the final reactor composition does not depend on the specific chemical formula selected. Also, note that the value of the extent in part (b) is half the value of that in part (a).

Example 2.2 Ethylene oxide is produced by catalytic oxidation on a silver catalyst according to the following chemical reaction:

$$C_2H_4 + \tfrac{1}{2}O_2 \longrightarrow C_2H_4O$$

A gaseous stream consisting of 60% C_2H_4, 30% O_2, and 10% N_2 (by mole) is fed at a rate of 40 mol/min into a flow reactor operating at steady state. If the mole fraction of oxygen in the reactor effluent stream is 0.08, calculate the production rate of ethylene oxide.

Solution We select the written chemical reaction as the chemical formula; hence, the stoichiometric coefficients are

$$s_{C_2H_4} = -1 \qquad s_{O_2} = -\tfrac{1}{2} \qquad s_{C_2H_4O} = 1 \qquad s_{N_2} = 0 \qquad \Delta = -\tfrac{1}{2}$$

To determine the extent of the chemical reaction, we use the given composition of the oxygen in the exit stream. First, we use Eq. 2.3.13 to express the oxygen molar flow rate at the reactor exit:

$$(F_{O_2})_{\text{out}} = (F_{O_2})_{\text{in}} + s_{O_2}\dot{X} = (0.3)(40) + \left(-\tfrac{1}{2}\right)\dot{X} \tag{a}$$

Next, we use Eq. 2.3.14 to express the total molar flow rate of the effluent stream:

$$(F_{\text{tot}})_{\text{out}} = (F_{\text{tot}})_{\text{in}} + \Delta\dot{X} = 40 + \left(-\tfrac{1}{2}\right)\dot{X} \tag{b}$$

Thus, the oxygen mole fraction in the exit stream is

$$(y_{O_2})_{\text{out}} = \frac{(F_{O_2})_{\text{out}}}{(F_{\text{tot}})_{\text{out}}} = \frac{(0.3)(40) + \left(-\tfrac{1}{2}\right)\dot{X}}{40 + \left(-\tfrac{1}{2}\right)\dot{X}} = 0.08 \tag{c}$$

Solving (c), $\dot{X} = 19.13 \, \text{mol/min}$. Now that the value of $\dot{X}$ is known, Eq. 2.3.13 is used to calculate the production rate of ethylene oxide:

$$(F_{C_2H_4O})_{out} = (F_{C_2H_4O})_{in} + s_{C_2H_4O}\dot{X} = 0 + (1)(19.13) = 19.13 \, \text{mol/min}$$

Example 2.3   An aqueous solution with a concentration of 0.8 mol/L of reactant A is fed into a flow reactor at a rate of 150 L/min. The following chemical reaction takes place in the reactor:

$$2A \ \longrightarrow \ B$$

Conductivity cells are used to determine the compositions of the inlet and outlet streams. The conductivity reading at the inlet is 140 units and at the outlet is 90 units. If the conductivity is proportional to the sum of the concentrations of A and B, determine:

a. The extent of the reaction
b. The production rate of B

Solution   This example illustrates how the reaction extent can be determined from a measurable quantity. The stoichiometric coefficients are

$$s_A = -2 \qquad s_B = 1 \qquad \Delta = -1$$

a. To determine the extent, we have to derive a relationship between the extent and the conductivity. Since the conductivity is proportional to $(C_A + C_B)$, we can write

$$\frac{\lambda_{out}}{\lambda_{in}} = \frac{C_{A_{out}} + C_{B_{out}}}{C_{A_{in}} + C_{B_{in}}} = \frac{90}{140} \tag{a}$$

The species concentrations are related to the molar flow rates and the volumetric flow rates by

$$C_A = \frac{F_A}{v} \qquad C_B = \frac{F_B}{v} \tag{b}$$

For liquid-phase reactions, the density is constant. Using Eq. 2.3.12 and noting that $C_{B_{in}} = 0$, the molar flow rates of species A and B at the reactor outlet are

$$F_{A_{out}} = F_{A_{in}} + s_A\dot{X} = vC_{A_{in}} + s_A\dot{X} = 120 - 2\dot{X} \tag{c}$$

$$F_{B_{out}} = F_{B_{in}} + s_B\dot{X} = vC_{B_{in}} + s_B\dot{X} = \dot{X} \tag{d}$$

Substituting (c) and (d) into (b) and the latter into (a)

$$\frac{\lambda_{\text{out}}}{\lambda_{\text{in}}} = \frac{F_{A_{\text{in}}} - \dot{X}}{F_{A_{\text{in}}}} = \frac{120 - \dot{X}}{120} = \frac{90}{140} \tag{e}$$

or

$$\dot{X} = 120\left(1 - \frac{90}{140}\right) = 42.86 \, \text{mol/min} \tag{f}$$

b. Using (d), the production rate of product B is 42.86 mol/min.

Example 2.4 The following two simultaneous chemical reactions take place in a batch reactor:

$$\text{Reaction 1:} \quad CO + \tfrac{1}{2}O_2 \longrightarrow CO_2 \tag{a}$$

$$\text{Reaction 2:} \quad C + \tfrac{1}{2}O_2 \longrightarrow CO \tag{b}$$

Initially, the reactor contains 4 mol of CO, 4 mol of O_2, and 2 mol of C. At the end of the operation, the reactor contains 2 mol of CO and 2 mol of O_2. Determine:

a. The composition of the reactor at the end of the operation
b. The portion of O_2 that reacts in each reaction

Solution Since each chemical reaction has a species that does not participate in the other, the two reactions are independent. The stoichiometric coefficients of the species in the respective reactions are

$$(s_{CO})_1 = -1 \quad (s_{O_2})_1 = -\tfrac{1}{2} \quad (s_{CO_2})_1 = 1 \quad (s_C)_1 = 0 \quad \Delta_1 = -\tfrac{1}{2}$$

$$(s_{CO})_2 = 1 \quad (s_{O_2})_2 = -\tfrac{1}{2} \quad (s_{CO_2})_2 = 0 \quad (s_C)_2 = -1 \quad \Delta_2 = -\tfrac{1}{2}$$

a. To determine the extents of the chemical reactions, we first write Eq. 2.3.3 for CO:

$$N_{CO}(t) = N_{CO}(0) + (s_{CO})_1 X_1(t) + (s_{CO})_2 X_2(t) = 2 \, \text{mol}$$

Substituting the numerical values for the stoichiometric coefficients and the initial composition, we obtain

$$-X_1(t) + X_2(t) = -2 \tag{c}$$

Next we write Eq. 2.3.3 for O_2:

$$N_{O_2}(t) = N_{O_2}(0) + (s_{O_2})_1 X_1(t) + (s_{O_2})_2 X_2(t) = 2\,\text{mol}$$

and obtain

$$X_1(t) + X_2(t) = 4 \qquad\qquad (d)$$

Solving (c) and (d), $X_1(t) = 3\,\text{mol}$ and $X_2(t) = 1\,\text{mol}$. Now that the extents of the two chemical reactions are known, we can calculate the amount of C and CO in the reactor. Using Eq. 2.3.3,

$$N_C(t) = N_C(0) + (s_C)_1 X_1(t) + (s_C)_2 X_2(t) = 1\,\text{mol}$$

$$N_{CO_2}(t) = N_{CO_2}(0) + (s_{CO_2})_1 X_1(t) + (s_{CO_2})_2 X_2(t) = 3\,\text{mol}$$

The total number of moles at time t is

$$N_{tot}(t) = N_{CO}(t) + N_{O_2}(t) + N_C(t) + N_{CO_2}(t) = 8\,\text{mol}$$

It can be calculated also by using Eq. 2.3.4,

$$N_{tot}(t) = 10.0 + \left(-\tfrac{1}{2}\right)(3.0) + \left(-\tfrac{1}{2}\right)(1.0) = 8.0\,\text{mol}$$

b. To determine the number of moles of oxygen reacted by each chemical reaction we use Eq. 2.3.1:

$$(n_{O_2} - n_{O_2 0})_1 = (s_{O_2})_1 X_1(t) = \left(-\tfrac{1}{2}\right)(3.0) = -1.5\,\text{mol}$$

$$(n_{O_2} - n_{O_2 0})_2 = (s_{O_2})_2 X_2(t) = \left(-\tfrac{1}{2}\right)(1.0) = -0.5\,\text{mol}$$

The minus sign indicates that oxygen is being consumed in the chemical reactions. Thus, 75% of the oxygen is consumed by Reaction 1 and 25% by Reaction 2.

Example 2.5 Wafers for integrated circuits are made of pure silicon that is produced by reacting raw silicon with HCl to form silicon trichloride, $SiHCl_3$. The silicon trichloride is reduced later with hydrogen to provide pure silicon. At the reactor operating conditions, silicon tetrachloride, $SiCl_4$, is also formed. Molten raw silicon is fed into a flow reactor at a rate of 80 lbmol/h, and gaseous HCl is fed in proportion of 4 mol HCl per mole silicon. The reactor operates at 1250°C and the following reactions take place:

$$\text{Reaction 1:} \quad Si + 3HCl \longrightarrow SiHCl_3 + H_2 \qquad (a)$$

$$\text{Reaction 2:} \quad Si + 4HCl \longrightarrow SiCl_4 + 2H_2 \qquad (b)$$

If all the silicon is consumed, and the effluent stream contains 40% H_2 (by mole), determine:

a. The production rate of $SiHCl_3$

b. The amount of HCl reacted

Solution Since each chemical reaction has a species that does not appear in the other, the two reactions are independent, and their stoichiometric coefficients are

$$(s_{Si})_1 = -1 \quad (s_{HCl})_1 = -3 \quad (s_{SiHCl_3})_1 = 1 \quad (s_{SiCl_4})_1 = 0 \quad (s_{H_2})_1 = 1 \quad \Delta_1 = -2$$

$$(s_{Si})_2 = -1 \quad (s_{HCl})_2 = -4 \quad (s_{SiHCl_3})_2 = 0 \quad (s_{SiCl_4})_2 = 1 \quad (s_{H_2})_2 = 2 \quad \Delta_2 = -2$$

a. We select the inlet stream as a basis for the calculation; thus, $(F_{Si})_{in} = 80$ lbmol/h and $(F_{HCl})_{in} = 320$ lbmol/h, and $(F_{tot})_{in} = 400$ lbmol/h. Using Eq. 2.3.11, the production rate of $SiHCl_3$ is

$$(F_{SiHCl_3})_{out} = 0 + (s_{SiHCl_3})_1 \dot{X}_1 + (s_{SiHCl_3})_2 \dot{X}_2 = \dot{X}_1 \qquad (c)$$

To obtain the extents of the reactions, we first use the fact that all the silicon is consumed in the reactor; hence, $(F_{Si})_{out} = 0$. Using Eq. 2.3.11,

$$(F_{Si})_{out} = 80 + (-1)\dot{X}_1 + (-1)\dot{X}_2$$

or

$$\dot{X}_1 + \dot{X}_2 = 80 \text{ lbmol/h} \qquad (d)$$

Next we use the given composition of H_2 in the outlet stream:

$$(y_{H_2})_{out} = \frac{(F_{H_2})_{out}}{(F_{tot})_{out}} = 0.4 \qquad (e)$$

Using Eq. 2.3.11,

$$(F_{H_2})_{out} = 0 + \dot{X}_1 + 2\dot{X}_2 \qquad (f)$$

and using Eq. 2.3.12,

$$(F_{tot})_{out} = 400 - 2\dot{X}_1 - 2\dot{X}_2 \qquad (g)$$

Substituting (f) and (g) into (e),

$$(1.8)\dot{X}_1 + (2.8)\dot{X}_2 = 160 \text{ lbmol/min} \qquad (h)$$

Solving (h) and (d), $\dot{X}_1 = 64$ lbmol/h, $\dot{X}_2 = 16$ lbmol/h, and from (c), the production rate of $SiHCl_3$ is $(F_{SiHCl_3})_{out} = 64$ lbmol/h.

b. Using Eq. 2.3.11, the flow rate of HCl at the reactor exit is

$$(F_{HCl})_{out} = 320 + (-3)(64) + (-4)(16) = 64 \, lbmol/h \qquad (i)$$

The amount of HCl reacted in the reactor is

$$(F_{HCl})_{in} - (F_{HCl})_{out} = 320 - 64 = 256 \, lbmol/h \qquad (j)$$

2.4 INDEPENDENT AND DEPENDENT CHEMICAL REACTIONS

In the preceding section we noted that when multiple chemical reactions take place, only a set of independent chemical reactions should be considered to determine the species composition. As indicated, the summations in Eq. 2.3.3 and Eq. 2.3.11 are taken over a set of *independent* chemical reactions and not over *all* the reactions that take place. This point deserves a closer examination.

The concept of independent reactions, or, more accurately, independent stoichiometric relations, is an important concept in stoichiometry and reactor analysis. The number of independent reactions indicates the smallest number of stoichiometric relations needed to describe the chemical transformations that take place and to determine all the *state* quantities of a chemical reactor (species composition, temperature, enthalpy, etc.). As will be seen later, the number of independent reactions also indicates the smallest number of design equations needed to describe the reactor operation. Since state quantities are independent of the path, we can select different sets of independent reactions to determine the change from one state to another. Below, we discuss the roles of independent and dependent reactions in describing reactor operations. We also describe a procedure to determine the number of independent reactions and how to identify a set of independent reactions.

To develop insight into the concept of independent reactions, consider the following reversible chemical reaction:

$$CO + H_2O \rightleftharpoons CO_2 + H_2 \qquad (2.4.1)$$

According to the stoichiometric methodology adopted here, reversible reactions are represented as two distinct reactions, a forward and a reverse. Hence, Reaction 2.4.1 is described by two chemical reactions: a forward reaction,

$$CO + H_2O \longrightarrow CO_2 + H_2 \qquad (2.4.1a)$$

whose stoichiometric coefficients are

$$s_{CO} = -1 \qquad s_{H_2O} = -1 \qquad s_{CO_2} = 1 \qquad s_{H_2} = 1$$

and a backward reaction,

$$CO_2 + H_2 \longrightarrow CO + H_2O \qquad (2.4.1b)$$

whose stoichiometric coefficients are

$$s_{CO} = 1 \qquad s_{H_2O} = 1 \qquad s_{CO_2} = -1 \qquad s_{H_2} = -1$$

Although two chemical reactions take place here, both of them provide the same information on the proportions among the individual species. This is because Reaction 2.4.1b is the reverse of Reaction 2.4.1a, and its stoichiometric coefficients have the negative values of those of Reaction 2.4.1a. In mathematical terms, we say that the two reactions are linearly dependent. Hence, only one chemical reaction (stoichiometric relation) is needed to determine the species compositions.

For more complex reaction systems, the dependency between the chemical reactions may be due to a linear combination of two (or more) independent reactions. Consider, for example, the following simultaneous chemical reactions:

$$C + \tfrac{1}{2}O_2 \;\longrightarrow\; CO \tag{2.4.2a}$$

$$CO + \tfrac{1}{2}O_2 \;\longrightarrow\; CO_2 \tag{2.4.2b}$$

$$C + O_2 \;\longrightarrow\; CO_2 \tag{2.4.2c}$$

A close examination of these reactions reveals that Reaction 2.4.2c is the sum of Reaction 2.4.2a and Reaction 2.4.2b. Hence, there are only two independent chemical reactions, and, to determine any state quantity, we have to consider only a set of two independent reactions. In this case, any two reactions among the three form a set of independent reactions.

To determine the number of independent reactions in a set of multiple chemical reactions, we construct the matrix of the stoichiometric coefficients for the chemical reactions. We designate a row for each chemical reaction and a column for each chemical species and write the stoichiometric coefficient of each species in the respective reaction in the corresponding matrix element. The order that the reactions (rows) or the species (columns) are assigned in the matrix is not important. However, to avoid forming ill-behaved matrices, it is prudent to consider the important (dominant) reactions first and write the species in the order they appear in the reactions. Once a column is assigned to a specific species, it should not be changed. For example, for Reaction 2.4.1, the matrix of stoichiometric coefficients is constructed by listing the two individual reactions as they are given and the species in the order they appear in the respective reactions,

$$\begin{array}{cccc} CO & H_2O & CO_2 & H_2 \end{array}$$
$$\begin{bmatrix} -1 & -1 & 1 & 1 \\ 1 & 1 & -1 & -1 \end{bmatrix} \tag{2.4.3}$$

For Reactions 2.4.2a, 2.4.2b, and 2.4.2c, the stoichiometric matrix is

$$\begin{array}{cccc} C & O_2 & CO & CO_2 \end{array}$$
$$\begin{bmatrix} -1 & -0.5 & 1 & 0 \\ 0 & -0.5 & -1 & 1 \\ -1 & -1 & 0 & 1 \end{bmatrix} \tag{2.4.4}$$

The fact that chemical reactions are expressed as linear homogeneous equations allows us to exploit the properties of such equations and to use the associated algebraic tools. Specifically, we use elementary row operations to reduce the stoichiometric matrix to a reduced form, using Gaussian elimination. A reduced matrix is defined as a matrix where all the elements below the diagonal (elements 1,1; 2,2; 3,3; etc.) are zero. The number of nonzero rows in the reduced matrix indicates the number of independent chemical reactions. (A zero row is defined as a row in which all elements are zero.) The nonzero rows in the reduced matrix represent one set of independent chemical reactions (i.e., stoichiometric relations) for the system.

Elementary row operations are mathematical operations that can be performed on individual equations in a system of linear equations without changing the solution of the system. There are three elementary row operations: (i) interchanging any two rows, (ii) multiplying a row by a nonzero constant, and (iii) adding a scalar-multiplied row to another row. To reduce a matrix to a diagonal form, the three elementary row operations are applied to modify the matrix such that all the entries below the diagonal are zeros. First, we check if the first diagonal element (the element in the first row) is nonzero. If it is nonzero, we use the three elementary row operations to convert all the elements in the first column below it to zero. For example, in stoichiometric Matrix 2.4.3, the diagonal element in the first row is nonzero, and, to eliminate the entry below it, we add the first row to the second row and obtain

$$\begin{vmatrix} -1 & -1 & 1 & 1 \\ 0 & 0 & 0 & 0 \end{vmatrix} \qquad (2.4.5)$$

In general, when the diagonal element is zero, we replace it (if possible) with a nonzero element by interchanging the row with a lower row. We repeat the procedure for the diagonal element in the second column, then the third column, and so on until we obtain a reduced matrix. In Matrix 2.4.5, all the elements below the diagonal are zero, therefore, it is a reduced matrix. Since it has only one nonzero row, the system has one independent chemical reaction. Note that unlike a conventional Gaussian elimination procedure, the elements on the diagonal are not converted to 1, since by multiplying a row by a negative constant we change the corresponding chemical reaction.

To determine the number of independent reactions among Reactions 2.4.2a, 2.4.2b, and 2.4.2c, we reduce Matrix 2.4.4. Since the diagonal element in the first row is nonzero, we leave the first row unchanged and eliminate the nonzero elements in the first column below it. Similarly, since the first element in the second row is zero, leave the second row unchanged. To eliminate the nonzero element in the first column of the third row, subtract the first row from the third row and obtain the following matrix:

$$\begin{bmatrix} -1 & -0.5 & 1 & 0 \\ 0 & -0.5 & -1 & 1 \\ 0 & -0.5 & -1 & 1 \end{bmatrix} \qquad (2.4.6)$$

Now check the diagonal element of the second column. Since it is nonzero, leave the second row unchanged and eliminate the nonzero elements in the second column below it. To eliminate the nonzero element in the second column of the third row, subtract the second row from the third row and obtain the following matrix:

$$\begin{bmatrix} -1 & -0.5 & 1 & 0 \\ 0 & -0.5 & -1 & 1 \\ 0 & 0 & 0 & 0 \end{bmatrix} \qquad (2.4.7)$$

In this matrix, all the elements below the diagonal are zero; hence, it is a reduced matrix. Since it has two nonzero rows, there are two independent reactions.

Once a reduced stoichiometric matrix is obtained, we can identify one set of independent chemical reactions from the nonzero rows of the reduced matrix. In the case of Matrix 2.4.5, the nonzero row represents Reaction 2.4.1a. In the case of Matrix 2.4.7, the two nonzero rows represent Reactions 2.4.2a and 2.4.2b. Note that the set of independent reactions is not unique, and other sets can be generated by replacing one or more reactions in the original set by a linear combination of some or all reactions in the original set. For example, Reaction 2.4.1b, the reverse of Reaction 2.4.1a, can serve as the independent reaction of Reactions 2.4.1a and 2.4.1b. In Matrix 2.4.7, the second row can be replaced by the sum of the first and second row to obtain Reactions 2.4.2a and 2.4.2c as a set of independent reactions. In fact, for this case, any two reactions of the original three form a set of independent reactions. In principle, the set of independent reactions may include a reaction that does not actually take place in the reactor, yet we can use the set to calculate the reactor composition and other state variables. In practice, to simplify the calculations, we select a convenient set of independent reactions. In Chapter 4, a methodology for identifying the most suitable set of independent chemical reactions for designing chemical reactors is described.

As will be discussed later, the rates at which chemical species are being formed (or depleted) depend on *all* the chemical reactions that actually take place in the reactor (reaction pathways). Hence, to design chemical reactors with multiple reactions, we consider all the chemical reactions that are taking place, including the *dependent* reactions. Therefore, it is necessary to express the dependent reactions in terms of the independent reactions. Next, we describe how to do so.

Once a set of independent reactions is selected, write each of the dependent reactions as a linear combination of the independent reactions. The total number of reactions is the sum of the independent and dependent reactions:

$$n_R = n_I + n_D \qquad (2.4.8)$$

Let index m denote the mth independent reaction and index k the kth dependent reaction. For example, for Reactions 2.4.1a and 2.4.1b, there are two reactions but one independent reaction. By selecting Reaction 2.4.1a as the independent reaction, $m = 1$, and the index of the dependent reaction is $k = 2$. For Reactions 2.4.2a, 2.4.2b, and 2.4.2c, the set consists of three reactions, but among them only two are independent reactions. If Reactions 2.4.2a and 2.4.2b are selected as the independent reactions, and Reaction 2.4.2c as the dependent reaction, then, $m = 1, 2$, and $k = 3$.

To determine the relationships between the dependent and independent reactions, let α_{km} denote the scalar factor relating the kth dependent reaction to the mth independent reaction. Thus, α_{km} is the multiplier of the mth independent reaction to obtain the kth independent reaction. To determine the numerical values of the α_{km} factors, we conduct species balances for each dependent reaction. For the kth dependent reaction, the following set of linear equations should be satisfied for each species:

$$\sum_{m}^{n_I} \alpha_{km}(s_j)_m = (s_j)_k \quad j = A, B, \ldots \tag{2.4.9}$$

Hence, Eq. 2.4.9 provides a set of linear equations whose unknowns are α_{km}'s. As we will see later, these factors play an important role in formulating the design equations for chemical reactors with multiple reactions.

To illustrate how the α_{km} factors are determined, consider, for example, Reactions 2.4.1a and 2.4.1b. Selecting Reaction 2.4.1a as the independent reaction ($m = 1$) and Reaction 2.4.1b as the dependent reaction ($k = 2$), write Eq. 2.4.9 for Reaction 2.4.1b and take $j = CO$; thus,

$$\alpha_{21}(-1) = 1$$

and $\alpha_{21} = -1$. Indeed, Reaction 2.4.1b is the reverse of Reaction 2.4.1a and is obtained by multiplying the latter by -1. To determine the relationships between dependent and independent reactions among Reactions 2.4.2a, 2.4.2b, and 2.4.2c, we select Reactions 2.4.2a and 2.4.2b as the independent reactions and Reaction 2.4.2c as the dependent reaction. Hence, $m = 1, 2$ and $k = 3$. Write Eq. 2.4.9 for Reaction 2.4.2c for two species. For $j = C$,

$$\alpha_{31}(-1) + \alpha_{32}(0) = -1$$

and $\alpha_{31} = 1$. For $j = CO$,

$$\alpha_{31}(1) + \alpha_{32}(-1) = 0$$

and $\alpha_{32} = 1$. Hence, Reaction 2.4.2c is the sum of Reactions 2.4.2a and 2.4.2b,

$$\text{Reaction } 3 = \text{Reaction } 1 + \text{Reaction } 2$$

Example 2.6 Carbon disulfide is a common solvent that is produced by reacting sulfur vapor with methane. It is produced in a steady-flow reactor where the following chemical reactions take place:

$$
\begin{array}{lll}
\text{Reaction 1:} & CH_4 + 2S & \longrightarrow \quad CS_2 + 2H_2 \\
\text{Reaction 2:} & CH_4 + 4S & \longrightarrow \quad CS_2 + 2H_2S \\
\text{Reaction 3:} & CH_4 + 2H_2S & \longrightarrow \quad CS_2 + 4H_2
\end{array}
$$

Methane is fed to the reactor at a rate of 80 mol/min and vapor sulfur at a rate of 400 mol/min. The fraction of the methane converted in the reactor is 80%, and the hydrogen mole fraction in the product stream is 10.5%.

a. Identify a set of independent reactions.

b. Determine the production rate of carbon disulfide.

c. Determine the composition of the outlet stream.

d. Express each of the dependent reactions in terms of the independent reactions.

Solution

a. To determine the number of independent reactions, first construct a matrix of stoichiometric coefficients for the given reactions.

$$
\begin{array}{ccccc}
CH_4 & S & CS_2 & H_2 & H_2S
\end{array}
$$

$$
\begin{bmatrix}
-1 & -2 & 1 & 2 & 0 \\
-1 & -4 & 1 & 0 & 2 \\
-1 & 0 & 1 & 4 & -2
\end{bmatrix}
\tag{a}
$$

The diagonal element in the first column is nonzero, so the first row is left unchanged. To eliminate the nonzero elements below the diagonal in the first column, we subtract the first row from the second row and the first row from the third row. Matrix (a) reduces to

$$
\begin{bmatrix}
-1 & -2 & 1 & 2 & 0 \\
0 & -2 & 0 & -2 & 2 \\
0 & 2 & 0 & 2 & -2
\end{bmatrix}
\tag{b}
$$

Now all the elements in the first column below the diagonal element are zero, and we proceed to the second column. The diagonal element in the second column (in the second row) is nonzero, so the second row is left unchanged. To eliminate the nonzero elements below the diagonal in the second column in Matrix (b), we add the second row to the third row. To simplify the reduced matrix, we divide the second row by 2, and the matrix reduces to

$$\begin{bmatrix} -1 & -2 & 1 & 2 & 0 \\ 0 & -1 & 0 & -1 & 1 \\ 0 & 0 & 0 & 0 & 0 \end{bmatrix} \tag{c}$$

Since all the elements under the diagonal in Matrix (c) are zero, this matrix is a *reduced* matrix. Since Matrix (c) has two nonzero rows, there are two independent chemical reactions. In this case, we can select any two reactions of the original system as a set of independent reactions. The nonzero rows in Matrix (c) represent another set of independent reactions:

Reaction 1: $CH_4 + 2S \longrightarrow CS_2 + 2H_2$
Reaction 4: $H_2 + S \longrightarrow H_2S$

Note that the second row in Matrix (c) represents a chemical reaction (or more accurately a stoichiometric relation) that is not among the original reactions, and we denote it as Reaction 4. Hence, in this case, there are four plausible stoichiometric relations among the species, but only two of them are independent. Also, only Reactions 1, 2, and 3 actually take place in the reactor (i.e., reaction pathways).

b. To determine the species flow rates at the reactor outlet, a set of two independent reactions is selected and its extents calculated. We select Reactions 1 and 4 as the set of independent reactions; Hence, the indices of the independent reactions are $m = 1, 4$, and the indices of the dependent reactions are $k = 2, 3$. The stoichiometric coefficients of the two independent reactions are

$$(s_{CH_4})_1 = -1 \quad (s_S)_1 = -2 \quad (s_{CS_2})_1 = 1 \quad (s_{H_2})_1 = 2 \quad (s_{H_2S})_1 = 0 \quad \Delta_1 = 0$$

$$(s_{CH_4})_4 = 0 \quad (s_S)_4 = -1 \quad (s_{CS_2})_4 = 0 \quad (s_{H_2})_4 = -1 \quad (s_{H_2S})_4 = 1 \quad \Delta_4 = -1$$

To determine the required quantities, we express the molar flow rates of the individual species in terms of the extents of Reactions 1 and 4. Selecting the given inlet flow rates as a basis for the calculation,

$$(F_{CH_4})_{in} = 80 \, \text{mol/min} \quad (F_S)_{in} = 400 \, \text{mol/min} \quad (F_{tot})_{in} = 480 \, \text{mol/min}$$

To express the molar flow rates of the individual species in the product stream, we use Eq. 2.3.11 for the individual species:

$$(F_{CH_4})_{out} = 80 + (-1)\dot{X}_1 + (0)\dot{X}_4 = 80 - \dot{X}_1 \tag{d}$$

$$(F_S)_{out} = 400 + (-2)\dot{X}_1 + (-1)\dot{X}_4 = 400 - 2\dot{X}_1 - \dot{X}_4 \tag{e}$$

$$(F_{CS_2})_{out} = 0 + (1)\dot{X}_1 + (0)\dot{X}_4 = \dot{X}_1 \tag{f}$$

$$(F_{H_2})_{out} = 0 + (2)\dot{X}_1 + (-1)\dot{X}_4 = 2\dot{X}_1 - \dot{X}_4 \tag{g}$$

$$(F_{H_2S})_{out} = 0 + (0)\dot{X}_1 + (1)\dot{X}_4 = \dot{X}_4 \tag{h}$$

and, using Eq. 2.3.12,

$$(F_{tot})_{out} = 480 + (0)\dot{X}_1 + (-1)\dot{X}_4 = 480 - \dot{X}_4 \tag{i}$$

Expressing the fraction of CH_4 that is converted,

$$\frac{(F_{CH_4})_{in} - (F_{CH_4})_{out}}{(F_{CH_4})_{in}} = \frac{80 - (80 - \dot{X}_1)}{80} = 0.8$$

and we obtain $\dot{X}_1 = 64\,kmol/min$. Using (g) and (i), the fraction of H_2 in the outlet stream is

$$(y_{H_2})_{out} = \frac{(F_{H_2})_{out}}{(F_{tot})_{out}} = \frac{2\dot{X}_1 - \dot{X}_4}{480 - \dot{X}_4} = 0.105 \tag{j}$$

Substituting $\dot{X}_1 = 64\,mol/min$ into (j), $\dot{X}_4 = 86.7\,mol/min$. Now that the extents of the two independent reactions are known, we calculate the species molar flow rates of the effluent stream using (d) through (h):

$$(F_{CH_4})_{out} = 16 \qquad (F_S)_{out} = 185.3 \qquad (F_{CS_2})_{out} = 64$$

$$(F_{H_2})_{out} = 41.3 \qquad (F_{H_2S})_{out} = 86.7 \qquad (F_{tot})_{out} = 393.3\,mol/min$$

The production rate of CS_2 is 64 mol/min.

c. Now that the molar flow rates of the individual species in the outlet stream are known, the stream's composition can be calculated using (j). The respective mole fractions are

$$(y_{CH_4})_{out} = 0.041 \qquad (y_S)_{out} = 0.471 \qquad (y_{CS_2})_{out} = 0.163$$

$$(y_{H_2})_{out} = 0.105 \qquad (y_{H_2S})_{out} = 0.220$$

The reader is challenged to verify that the same outlet species molar flow rates are obtained when a different set of independent reactions is selected, say Reactions 1 and 2.

d. To determine the relations between the two dependent reactions (Reactions 2 and 3) and the two independent reactions (Reactions 1 and 4), write Eq. 2.4.9 for each dependent reaction. We start with the first dependent reaction $(k = 2)$:

$$\sum_m^{n_I} \alpha_{2m}(s_j)_m = \alpha_{21}(s_j)_1 + \alpha_{24}(s_j)_4 = (s_j)_2 \tag{k}$$

Write (k) for $j = CH_4$:

$$\alpha_{21}(-1) + \alpha_{24}(0) = -1 \tag{l}$$

and obtain $\alpha_{21} = 1$. Next, we write (k) for $j = H_2S$:

$$\alpha_{21}(0) + \alpha_{24}(1) = 2 \tag{m}$$

and obtain $\alpha_{24} = 2$. Thus, Reaction 2 relates to the two independent reactions by

$$\text{Reaction 2} = \text{Reaction 1} + 2(\text{Reaction 4})$$

Consider now the second dependent reaction and write Eq. 2.4.9 for $k = 3$:

$$\sum_m^{n_I} \alpha_{3m}(s_j)_m = \alpha_{31}(s_j)_1 + \alpha_{34}(s_j)_4 = (s_j)_3 \tag{n}$$

First, we write (n) for $j = CH_4$:

$$\alpha_{31}(-1) + \alpha_{34}(0) = -1 \tag{o}$$

and obtain $\alpha_{31} = 1$. Then we write (n) for $j = H_2S$:

$$\alpha_{31}(0) + \alpha_{34}(1) = -2 \tag{p}$$

and obtain $\alpha_{34} = -2$. Thus, Reaction 3 relates to the two independent reactions by

$$\text{Reaction 3} = \text{Reaction 1} - 2(\text{Reaction 4})$$

The previous example illustrates that the composition of a reactor can be determined by using a set of independent chemical reactions (stoichiometric relations) that includes one or more reactions that do not actually take place in the reactor.

2.5 CHARACTERIZATION OF THE REACTOR FEED

As indicated by Section 2.3, to determine the content of a batch reactor, or the outlet composition of a flow reactor, the composition of the initial state, or the inlet stream should be specified. So far, the initial contents of a batch reactor $N_j(0)$, or the inlet stream of a flow reactor, $F_{j_{in}}$, have been specified. However, in some instances it is convenient to characterize the reactor feed in terms of stoichiometric parameters of the chemical reactions that take place in the reactor. Also, as illustrated in Example 2.1, it is useful to identify the limiting reactant. This section covers the common quantities used to characterize the reactor feed.

2.5.1   Limiting Reactant

Let A and B be two reactants of a chemical reaction taking place in a batch reactor. Using Eq. 2.3.6, at any time t, the molar contents of the two reactants are related by

$$\frac{N_A(t) - N_A(0)}{s_A} = \frac{N_B(t) - N_B(0)}{s_B}$$

If reactant A is indeed the limiting reactant, $N_A(t)$ vanishes before $N_B(t)$, therefore, the following relation should be satisfied:

$$\left|\frac{N_A(0)}{s_A}\right| \leq \left|\frac{N_B(0)}{s_B}\right| \tag{2.5.1a}$$

Note that because the stoichiometric coefficients of reactants are negative, absolute values are used. Similarly, for steady-flow reactors, using Eq. 2.3.15, the species molar flow rates of the two reactants are related by

$$\frac{F_{A_\text{out}} - F_{A_\text{in}}}{s_A} = \frac{F_{B_\text{out}} - F_{B_\text{in}}}{s_B}$$

and if A is the limiting reactant, the following condition should be satisfied:

$$\left|\frac{F_{A_\text{in}}}{s_A}\right| \leq \left|\frac{F_{B_\text{in}}}{s_B}\right| \tag{2.5.1b}$$

When reactants A and B are in stoichiometric proportion, Eqs. 2.5.1a and 2.5.1b reduce, respectively, to

$$\frac{N_A(0)}{s_A} = \frac{N_B(0)}{s_B} \quad \text{or} \quad \frac{F_{A_\text{in}}}{s_A} = \frac{F_{B_\text{in}}}{s_B} \tag{2.5.2}$$

Equation 2.5.2 is the mathematical condition for stoichiometric proportion of the reactants in the reactor feed.

When multiple chemical reactions take place, the chemical reaction whose stoichiometric coefficients are used in Eqs. 2.5.1a and 2.5.1b is the stoichiometric relation that ties the reactants fed to the desirable product. This is illustrated in Example 2.11.

The procedure for identifying the limiting reactant of a chemical reaction is quite simple. For each reactant calculate

$$\left|\frac{N_j(0)}{s_j}\right| \quad \text{or} \quad \left|\frac{F_{j_\text{in}}}{s_j}\right| \qquad j = A, B, \ldots \tag{2.5.3}$$

The reactant with the smallest value is the limiting reactant. Since very rarely a chemical reaction has more than three reactants, the procedure is rather short. When the reactants are in stoichiometric proportion, any reactant can be considered as the limiting reactant. In the remainder of the text, the subscript A denotes the limiting reactant; other species, either reactants or products, are labeled by other letters.

2.5.2 Excess Reactant

Excess reactant is a quantity indicating the surplus amount of a reactant over the stoichiometric amount. It is usually used in combustion reactions to indicate the surplus amount of oxygen provided, which characterizes the combustion conditions. The excess amount of reactant B is defined by

$$\begin{pmatrix} \text{Excess} \\ \text{reactant} \\ \text{B} \end{pmatrix} \equiv \frac{\text{Amount of } B \text{ fed} - \text{Stoichiometric amount of } B \text{ needed}}{\text{Stoichiometric amount of } B \text{ needed}} \qquad (2.5.4)$$

The stoichiometric amount of reactant B is determined for the specific chemical reaction or reactions under consideration. For combustion reactions, the convention is to select the chemical reactions that provide complete oxidation of all the fuel components to their highest oxidation level (all carbon atoms to CO_2, all sulfur atoms to SO_2, etc.). Hence, although other chemical reactions may take place during the operation, generating CO and other products, the excess oxygen is defined and calculated on the basis of *complete* oxidation reactions.

In practice, combustion operations involve the introduction of external source of oxygen, usually in the form of air. Also, in most instances, combustion involves multiple reactions. To characterize combustion operations, a quantity called "excess air" is defined by

$$\left\{ \begin{matrix} \text{Excess} \\ \text{air} \end{matrix} \right\} \equiv \frac{\text{External air supplied} - \text{Stoichiometric external air required}}{\text{Stoichiometric external air required}} \qquad (2.5.5)$$

Excess air is commonly used to characterize the operating conditions of burners. Note that in Eq. 2.5.5 only *external* oxygen is considered (with the exclusion of oxygen that might be in the fuel) and that the stoichiometric amount required is determined on the basis of all complete combustion reactions that might take place.

Example 2.7 The gas-phase reaction

$$2CO + O_2 \longrightarrow 2CO_2$$

takes place in a closed, constant-volume batch reactor at isothermal conditions. Initially, the reactor contains 1.5 kmol of CO, 1 kmol of O_2, 1 kmol of CO_2, and 0.5 kmol of N_2, and its pressure is 5 atm. At time t, the reactor pressure is $P(t) = 4.5$ atm. Assume ideal-gas behavior.

a. Identify the limiting reactant.

b. Determine the excess amount of the other reactant.

c. Determine the reactor pressure when the reaction goes to completion.

d. Determine the extent of the reaction at time t.

e. Determine the reactor content (in moles) at time t.

Solution   We select the given chemical reaction as the chemical formula, and the stoichiometric coefficients are

$$s_{CO} = -2 \qquad s_{O_2} = -1 \qquad s_{CO_2} = 2 \qquad s_{N_2} = 0 \qquad \Delta = -1$$

a. To identify the limiting reactant, apply Eq. 2.5.3 for each reactant:

$$\left|\frac{N_{CO}(0)}{s_{CO}}\right| = \left|\frac{1.5}{-2}\right| = 0.75 \qquad \left|\frac{N_{O_2}(0)}{s_{O_2}}\right| = \left|\frac{1.0}{-1}\right| = 1.0 \tag{a}$$

Hence, CO is the limiting reactant.

b. Now that the limiting reactant is identified, the stoichiometric amount of O_2 is readily determined using Eq. 2.5.2:

$$(N_{O_2})_{stoich} = \frac{s_{O_2}}{s_{CO}} N_{CO}(0) = \frac{-1}{-2}(1.5\,kmol) = 0.75\,kmol \tag{b}$$

Using Eq. 2.5.4, the excess O_2 is

$$\left\{\begin{matrix} Excess \\ O_2 \end{matrix}\right\} \equiv \frac{(1\,kmol) - (0.75\,kmol)}{0.75\,kmol} = 0.33 \tag{c}$$

c. At completion, $N_{CO}(t) = 0$, and, using Eq. 2.3.5, the extent of the reaction is

$$X(t) = \frac{N_{CO}(t) - N_{CO}(0)}{s_{CO}} = \frac{0 - 1.5}{-2} = 0.75\,kmol \tag{d}$$

To determine the pressure, we derive a relationship between the pressure and the extent of reaction. For ideal-gas behavior,

$$P(t) = \frac{RT}{V} N_{tot}(t) \tag{e}$$

where $N_{tot}(t)$ is given by Eq. 2.3.7. Hence, for isothermal operation,

$$\frac{P(t)}{P(0)} = \frac{N_{tot}(t)}{N_{tot}(0)} = \frac{N_{tot}(0) + \Delta X(t)}{N_{tot}(0)} \tag{f}$$

and, in this case, and, at $t = 0$, $N_{tot}(0) = 4\,kmol$, so

$$\frac{P(t)}{P(0)} = \frac{4 - X(t)}{4} \tag{g}$$

The reactor pressure at completion is

$$P = (5 \text{ atm}) \frac{4 - 0.75}{4} = 4.06 \text{ atm}$$

d. Rearranging (f), the reaction extent when $P = 4.5$ atm is

$$X(t) = \left[1 - \frac{P(t)}{P(0)}\right] N_{\text{tot}}(0) = \left(1 - \frac{4.5}{5}\right)(4) = 0.4 \text{ kmol} \qquad \text{(h)}$$

e. Now that the extent at time t is known, we can determine the amounts of each species, using Eq. 2.3.5:

$$N_{CO}(t) = N_{CO}(0) + s_{CO} X(t) = 1.5 + (-2)(0.4) = 0.7 \text{ kmol}$$

$$N_{O_2}(t) = N_{O_2}(0) + s_{O_2} X(t) = 1.0 + (-1)(0.4) = 0.6 \text{ kmol}$$

$$N_{CO_2}(t) = N_{CO_2}(0) + s_{CO_2} X(t) = 1.0 + (2)(0.4) = 1.8 \text{ kmol}$$

$$N_{N_2}(t) = N_{N_2}(0) + s_{N_2} X(t) = 0.5 + (0)(0.4) = 0.5 \text{ kmol}$$

and, using Eq. 2.3.7,

$$N_{\text{tot}}(t) = N_{\text{tot}}(0) + \Delta X(t) = 4.0 + (-1)(0.4) = 3.6 \text{ kmol}$$

Example 2.8 A gaseous fuel consisting of 72% CH_4, 24% C_2H_6, 3% N_2, and 1% O_2 (mole percent) is fed into a combustion chamber at a rate of 10 g-mol/min. A stream of external air is mixed with the fuel, and the following chemical reactions are believed to take place in the combustion chamber:

Reaction 1: $CH_4 + 2O_2 \longrightarrow CO_2 + 2H_2O$

Reaction 2: $2CH_4 + 3O_2 \longrightarrow 2CO + 4H_2O$

Reaction 3: $2C_2H_6 + 7O_2 \longrightarrow 4CO_2 + 6H_2O$

Reaction 4: $2C_2H_6 + 5O_2 \longrightarrow 4CO + 6H_2O$

Reaction 5: $2CO + O_2 \longrightarrow 2CO_2$

An analysis of the effluent stream indicates that all the ethane has been converted and that, on a dry basis, its composition is 83.96% N_2, 7.05% CO_2, 0.18% CO, and 8.69% O_2 (mole percent). Determine:

a. The flow rate of the air fed to the reactor

b. The excess air

Solution   First, we determine the number of independent reactions. We construct a matrix of stoichiometric coefficients for the given reactions:

$$\begin{array}{cccccc} CH_4 & O_2 & CO_2 & H_2O & CO & C_2H_6 \end{array}$$

$$\begin{bmatrix} -1 & -2 & 1 & 2 & 0 & 0 \\ -2 & -3 & 0 & 4 & 2 & 0 \\ 0 & -7 & 4 & 6 & 0 & -2 \\ 0 & -5 & 0 & 6 & 4 & -2 \\ 0 & -1 & 2 & 0 & -2 & 0 \end{bmatrix} \tag{a}$$

Applying Gaussian elimination, Matrix (a) reduces to

$$\begin{bmatrix} -1 & -2 & 1 & 2 & 0 & 0 \\ 0 & 1 & -2 & 0 & 2 & 0 \\ 0 & 0 & -5 & 3 & 7 & -1 \\ 0 & 0 & 0 & 0 & 0 & 0 \\ 0 & 0 & 0 & 0 & 0 & 0 \end{bmatrix} \tag{b}$$

Since Matrix (b) has three nonzero rows, there are three independent chemical reactions. We select Reactions 1, 3, and 5 as a set of independent reactions, and their stoichiometric coefficients are

$$(s_{CH_4})_1 = -1 \quad (s_{O_2})_1 = -2 \quad (s_{CO_2})_1 = 1 \quad (s_{H_2O})_1 = 2 \quad (s_{CO})_1 = 0 \quad (s_{C_2H_6})_1 = 0$$

$$(s_{CH_4})_3 = 0 \quad (s_{O_2})_3 = -7 \quad (s_{CO_2})_3 = 4 \quad (s_{H_2O})_3 = 6 \quad (s_{CO})_3 = 0 \quad (s_{C_2H_6})_3 = 0$$

$$(s_{CH_4})_5 = 0 \quad (s_{O_2})_5 = -1 \quad (s_{CO_2})_5 = 2 \quad (s_{H_2O})_5 = 0 \quad (s_{CO})_5 = -2 \quad (s_{C_2H_6})_5 = 2$$

$$\Delta_1 = 0 \quad \Delta_3 = 1 \quad \Delta_5 = -1$$

We select the fuel feed as a basis for the calculation ($F_1 = 10\,mol/min$) and denote the fed air stream by F_2. The inlet stream is $F_1 + F_2$.

a. First we want to determine the extents of the three independent reactions (i.e., Reactions 1, 3, and 5), using the given dry-basis compositions. Using Eq. 2.3.12, the total molar flow rate of the flue gas is

$$(F_{tot})_{out} = (10 + F_2) + \dot{X}_3 - \dot{X}_5 \tag{c}$$

Using Eq. 2.3.11, the molar flow rate of H_2O in the flue gas is

$$(F_{H_2O})_{out} = (0) + 2\dot{X}_1 + 6\dot{X}_3 \tag{d}$$

Combining (c) and (d), the total molar flow rate of the dry flue gas is

$$(F_{tot})_{dry} = (F_{tot})_{out} - (F_{H_2O})_{out} = 10 + F_2 - 2\dot{X}_1 - 5\dot{X}_3 - \dot{X}_5 \tag{e}$$

Using Eqs. 2.3.11 and (e) for each of the given dry compositions, we obtain

$$y_{N_2} = \frac{0.3 + 0.79F_2}{10 + F_2 - 2\dot{X}_1 - 5\dot{X}_3 - \dot{X}_5} = 0.8399 \qquad (f)$$

$$y_{O_2} = \frac{0.1 + 0.21F_2 - 2\dot{X}_1 - 7\dot{X}_3 - \dot{X}_5}{10 + F_2 - 2\dot{X}_1 - 5\dot{X}_3 - \dot{X}_5} = 0.0865 \qquad (g)$$

$$y_{CO_2} = \frac{\dot{X}_1 + 4\dot{X}_3 + 2\dot{X}_5}{10 + F_2 - 2\dot{X}_1 - 5\dot{X}_3 - \dot{X}_5} = 0.0708 \qquad (h)$$

$$y_{CO} = \frac{-2\dot{X}_5}{10 + F_2 - 2\dot{X}_1 - 5\dot{X}_3 - \dot{X}_5} = 0.0018 \qquad (i)$$

Solving (f), (g), (h), and (i), we obtain $F_2 = 172.9\,\text{mol/min}$, $\dot{X}_1 = 7.03\,\text{mol/min}$, $\dot{X}_3 = 1.2\,\text{mol/min}$, and $\dot{X}_5 = -0.147\,\text{mol/min}$. The molar feed rate of the air stream is $172.9\,\text{mol/min}$. The amount of external oxygen fed is $0.21F_2 = 36.31\,\text{mol/min}$. The total amount of oxygen fed to the combustor is $0.01(10) + 36.31 = 36.41\,\text{mol/min}$.

b. To determine the excess amount of external oxygen, we first have to determine the stoichiometric amount needed. For methane and ethane, the stoichiometric amount is determined on the basis of the reactions for *complete* combustion:

$$\text{Reaction 1:} \quad CH_4 + 2O_2 \longrightarrow CO_2 + 2H_2O \qquad (j)$$

$$\text{Reaction 3:} \quad 2C_2H_6 + 7O_2 \longrightarrow 4CO_2 + 6H_2O \qquad (k)$$

Note that we do not consider Reaction 5 because CO is not a species in the fuel. Using Eq. 2.5.2, the stoichiometric amount of oxygen needed for complete combustion is

$$(F_{O_2})_{\text{stoich}} = \frac{(s_{O_2})_1}{(s_{CH_4})_1}(F_{CH_4})_{\text{in}} + \frac{(s_{O_2})_3}{(s_{C_2H_6})_3}(F_{C_2H_6})_{\text{in}} \qquad (l)$$

The total stoichiometric amount of oxygen needed is $22.8\,\text{mol/min}$, but $0.1\,\text{mol/min}$ is fed in the fuel stream itself. Hence, the stoichiometric amount of external oxygen needed is $22.7\,\text{mol/min}$. Using Eq. 2.5.5, the excess air is

$$\left\{ \begin{array}{c} \text{Excess} \\ \text{air} \end{array} \right\} = \left\{ \begin{array}{c} \text{Excess} \\ \text{external } O_2 \end{array} \right\} = \frac{36.41 - 22.7}{22.7} - 0.604 \qquad (m)$$

2.6 CHARACTERIZATION OF REACTOR PERFORMANCE

In the preceding sections, the stoichiometric relationships used to quantify the operation of chemical reactors were expressed in terms of extensive quantities (moles, molar flow rates, reaction extents, etc.) whose numerical values depend on the basis selected for the calculation. In most applications, it is convenient to define intensive dimensionless quantities that characterize the operation of chemical reactors and provide quick measures of the reactor performance. In this section, we define and discuss some common stoichiometric quantities used in reactor analysis.

2.6.1 Reactant Conversion

The conversion is defined as the fraction of a reactant that has been consumed. For batch reactors, the conversion of reactant A at time t, $f_A(t)$, is defined by

$$f_A(t) \equiv \frac{\text{Moles of reactant A consumed in time } t}{\text{Moles of reactant A charged to the reactor}} = \frac{N_A(0) - N_A(t)}{N_A(0)} \quad (2.6.1a)$$

For flow reactors operating at steady state, the conversion of reactant A in the reactor is defined by

$$f_{A\,\text{out}} \equiv \frac{\text{Rate reactant A is consumed in the system}}{\text{Rate reactant A is fed to the system}} = \frac{F_{A_\text{in}} - F_{A_\text{out}}}{F_{A_\text{in}}} \quad (2.6.1b)$$

Three points concerning the conversion should be noted:

1. The conversion is defined only for *reactants*, and, by definition, its value is between 0 and 1.
2. The conversion is related to the composition (or flow rate) of a reactant, and it is not defined on the basis of any specific chemical reaction. When multiple chemical reactions take place, a reactant may be consumed in several chemical reactions. However, if reactant A is produced by any independent chemical reaction, its conversion is not defined.
3. The conversion depends on the initial state selected, $N_A(0)$ (for batch reactors) and on the boundaries of the system, "in" and "out" (for flow reactors)—see Example 2.9.

When a *single* chemical reaction takes place, the conversion of a reactant relates to the extent of the reaction. For batch reactors, from Eq. 2.3.5,

$$N_A(0) - N_A(t) = -s_A X_1(t)$$

and, substituting in Eq. 2.6.1a, the relationship between the conversion of reactant A and the reaction extent is

$$f_A(t) = -\frac{s_A}{N_A(0)} X(t) \quad (2.6.2)$$

To express the number of moles of any species in terms of the conversion of reactant A, substitute $X(t)$ from Eq. 2.6.2 into Eq. 2.3.6 and obtain

$$N_j(t) = N_j(0) - \frac{s_j}{s_A} N_A(0) f_A(t) \tag{2.6.3}$$

To express the total number of moles in the reactor in terms of the conversion of reactant A, substitute $X(t)$ from Eq. 2.3.5 into Eq. 2.3.7 and obtain

$$N_{tot}(t) = N_{tot}(0) - \frac{\Delta}{s_A} N_A(0) f_A(t) \tag{2.6.4}$$

where Δ, defined by Eq. 2.2.5, denotes the change in the number of moles per unit extent.

To obtain a relationship between the conversion and the reaction extent in steady-flow reactors with *single* chemical reactions, we write Eq. 2.3.13 for reactant A and substitute it in Eq. 2.6.1b,

$$f_{A_{out}} = - \frac{s_A}{F_{A_{in}}} \dot{X}_{out} \tag{2.6.5}$$

To express the molar flow rate of any species at the reactor outlet in terms of the conversion of reactant A, substitute Eq. 2.6.5 into Eq. 2.3.13 and obtain

$$F_{j_{out}} = F_{j_{in}} - \frac{s_j}{s_A} F_{A_{in}} f_{A_{out}} \tag{2.6.6}$$

To relate the total molar flow rate at the reactor exit to the conversion, substitute Eq. 2.6.6 into Eq. 2.3.14 and obtain

$$F_{tot_{out}} = F_{tot_{in}} - \frac{\Delta}{s_A} F_{A_{in}} f_{A_{out}} \tag{2.6.7}$$

When species A is a reactant in several chemical reactions, the term $N_A(0) - N_A(t)$ in Eq. 2.6.1a should account for the consumption of reactant A by all the *independent* reactions. Using Eq. 2.3.3 for reactant A,

$$f_A(t) \equiv \frac{N_A(0) - N_A(t)}{N_A(0)} = - \frac{1}{N_A(0)} \sum_{m}^{n_I} (s_A)_m X_m(t)$$

or

$$f_A(t) = \sum_{m}^{n_I} f_{A_m}(t) \tag{2.6.8}$$

where $f_{A_m}(t)$ is the conversion of reactant A by the mth independent reaction, defined by

$$f_{A_m}(t) \equiv - \frac{(s_A)_m}{N_A(0)} X_m(t) \tag{2.6.9}$$

Similarly, for steady-flow reactors

$$f_A = \sum_m^{n_I} f_{A_m} \qquad (2.6.10)$$

where f_{A_m} is the conversion of A by the mth independent reaction defined by

$$f_{A_m} \equiv -\frac{(s_A)_m}{F_{A_{in}}} \dot{X}_m \qquad (2.6.11)$$

Example 2.9 Ammonia is produced in a continuous catalytic reactor according to the reaction

$$N_2 + 3H_2 \longrightarrow 2NH_3$$

A synthesis gas stream (stream 1) consisting of 24.5% N_2, 73.5% H_2, and 2% argon is fed at a rate of 60 mol/min. At the operating conditions, the conversion per pass in the reactor is 12%. To enhance the operation, a portion of the reactor effluent stream (stream 3) is recycled and combined with the fresh synthesis gas as illustrated in Figure E2.9.1. If the mole fraction of the argon in the product stream (stream 4) is 3%, determine:

a. The ammonia production rate
b. The overall nitrogen conversion in the process
c. The recycle ratio (F_5/F_4)

Solution The stoichiometric coefficients of the chemical reaction are

$$s_{N_2} = -1 \qquad s_{H_2} = -3 \qquad s_{NH_3} = 2 \qquad s_{Ar} = 0 \qquad \Delta = -2$$

Note that the argon is an inert species, and its stoichiometric coefficient is zero.

a. Selecting the entire process as the system (the inlet is stream 1 and the outlet is stream 4); hence, $(F_{tot})_{in} = F_1 = 60$ mol/min and $(F_{tot})_{out} = F_4$. Writing an argon balance using Eq. 2.3.3,

$$(F_{Ar})_4 = (F_{Ar})_1 + s_{Ar}\dot{X}$$

$$(0.03)(F_{tot})_4 = (0.02)(F_{tot})_1 = (0.02)(60 \text{ mol/min}) \qquad \text{(a)}$$

Solving (a), $(F_{tot})_4 = 40$ mol/min. Using Eq. 2.3.14 to relate the total molar flow rate of the outlet stream to the extent,

$$(F_{tot})_4 = (F_{tot})_1 + \Delta\dot{X}$$

$$40 = 60 + (-2)\dot{X} \qquad \text{(b)}$$

and $\dot{X} = 10$ mol/min. Now that the extent of the reaction is known, the production rate of ammonia can be calculated by using Eq. 2.3.13,

$$(F_{NH_3})_4 = (F_{NH_3})_1 + s_{NH_3}\dot{X} = 0 + (2)(10) = 20 \text{ mol/min} \qquad \text{(c)}$$

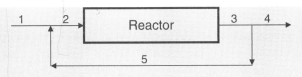

Figure E2.9.1 Recycle of ammonia reactor.

b. To calculate the nitrogen molar flow rate at the outlet stream, we write Eq. 2.3.13 for nitrogen:

$$(F_{N_2})_4 = (F_{N_2})_1 + s_{N_2}\dot{X} = (0.245)(60) + (-1)(10) = 4.70 \, \text{mol/min} \quad \text{(d)}$$

The overall nitrogen conversion in the process, defined by Eq. 2.6.1b, is

$$f_{N_2} \equiv \frac{(F_{N_2})_1 - (F_{N_2})_4}{(F_{N_2})_1} = \frac{14.7 - 4.7}{14.7} = 0.680 \quad \text{(e)}$$

Note that the same conversion is obtained by using Eq. 2.6.5.

c. To calculate the recycle ratio, we select the reactor itself as the system (the inlet stream is stream 2 and the outlet stream is stream 3). Using the given conversion per pass,

$$f_{N_2} \equiv \frac{(F_{N_2})_2 - (F_{N_2})_3}{(F_{N_2})_2} = 0.12$$

or

$$(F_{N_2})_3 = 0.880(F_{N_2})_2 \quad \text{(f)}$$

Using Eq. 2.3.13 to express $(F_{N_2})_3$ and $(F_{N_2})_2$ in terms of the extent,

$$(F_{N_2})_3 = (F_{N_2})_2 + s_{N_2}\dot{X} = (F_{N_2})_2 + (-1)(10) \quad \text{(g)}$$

and substituting (f),

$$(0.880)(F_{N_2})_2 = (F_{N_2})_2 + (-1)(10) \quad \text{(h)}$$

Solving (h), $(F_{N_2})_2 = 83.33 \, \text{mol/min}$, and, from (f), $(F_{N_2})_3 = 73.33 \, \text{mol/min}$. Now, writing a nitrogen balance over the mixing point,

$$(F_{N_2})_2 = (F_{N_2})_1 + (F_{N_2})_5 = 0.880(F_{N_2})_2$$
$$83.33 = (0.245)(60) + (F_{N_2})_5 \quad \text{(i)}$$

Solving (i), $(F_{N_2})_5 = 68.63 \, \text{mol/min}$. Thus, the recycle ratio is

$$R \equiv \frac{(F_{N_2})_5}{(F_{N_2})_4} = \frac{68.63}{4.70} = 14.6$$

2.6.2   Product Yield and Selectivity

When several simultaneous chemical reactions take place producing both desired and undesired products, it is convenient to define parameters that indicate what portion of the reactant was converted to valuable products. Below, we define and discuss two quantities that are commonly used: yield and selectivity of the desirable product.

Yield is a measure of the portion of a reactant converted to the desired product by the *desirable* chemical reaction. It indicates the amount of the desirable product, species V, produced relative to the amount of V that could have been produced if only the desirable reaction took place. The yield is defined such that its value is between zero and one.

For *batch reactors*, the yield of product V at time t is

$$\eta_V(t) \equiv \left(\begin{array}{c} \text{Stoichiometric} \\ \text{factor} \end{array} \right) \frac{\text{Moles of product V formed in time } t}{\text{Moles of reactant A initially in the reactor}}$$

and in mathematical terms

$$\eta_V(t) \equiv -\left(\frac{s_A}{s_V} \right)_{\text{des}} \frac{N_V(t) - N_V(0)}{N_A(0)} \tag{2.6.12}$$

where s_A and s_V are, respectively, the stoichiometric coefficients of A and V in the *desirable* chemical reaction. Using stoichiometric relations (Eq. 2.3.3), the yield relates to the extents of the independent chemical reactions by

$$\eta_V(t) = -\left(\frac{s_A}{s_V} \right)_{\text{des}} \frac{1}{N_A(0)} \left[\sum_{m}^{n_I} (s_V)_m X_m(t) \right] \tag{2.6.13}$$

For *flow reactors*, the yield of product V at the reactor outlet is

$$\eta_{V_{\text{out}}} \equiv \left(\begin{array}{c} \text{Stoichiometric} \\ \text{factor} \end{array} \right) \frac{\text{Rate product V is formed in the reactor}}{\text{Rate of reactant A is fed into the reactor}}$$

and in mathematical terms

$$\eta_{V_{\text{out}}} \equiv -\left(\frac{s_A}{s_V} \right)_{\text{des}} \frac{F_{V_{\text{out}}} - F_{V_{\text{in}}}}{F_{A_{\text{in}}}} \tag{2.6.14}$$

Using Eq. 2.3.11, the yield relates to the extents of the independent chemical reactions by

$$\eta_{V_{\text{out}}} = -\left(\frac{s_A}{s_V} \right)_{\text{des}} \frac{1}{F_{A_{\text{in}}}} \left[\sum_{m}^{n_I} (s_V)_m \dot{X}_m \right] \tag{2.6.15}$$

Product selectivity indicates the amount of product V produced relative to the theoretical amount of V that could be produced if all reactant A *consumed* were reacted by the desirable chemical reaction. For batch reactors, the selectivity of product V at time t is defined by

$$\sigma_V(t) \equiv \left(\begin{array}{c}\text{Stoichiometric}\\\text{factor}\end{array}\right)\frac{\text{Moles of product V formed in time } t}{\text{Moles of reactant A consumed in time } t}$$

In mathematical terms

$$\sigma_V(t) \equiv -\left(\frac{s_A}{s_V}\right)_{\text{des}}\frac{N_V(t) - N_V(0)}{N_A(0) - N_A(t)} \tag{2.6.16}$$

Using Eq. 2.3.3,

$$\sigma_V(t) = -\left(\frac{s_A}{s_V}\right)_{\text{des}}\frac{\displaystyle\sum_{m}^{n_I}(s_V)_m X_m(t)}{\displaystyle\sum_{m}^{n_I}(s_A)_m X_m(t)} \tag{2.6.17}$$

For steady-flow reactors, the selectivity of product V is defined by

$$\sigma_V(t) \equiv \left(\begin{array}{c}\text{Stoichiometric}\\\text{factor}\end{array}\right)\frac{\text{Rate Product V is formed in the reactor}}{\text{Rate Reactant A is consumed in the reactor}}$$

In mathematical terms

$$\sigma_V \equiv -\left(\frac{s_A}{s_V}\right)_{\text{des}}\frac{F_{V_{\text{out}}} - F_{V_{\text{in}}}}{F_{A_{\text{in}}} - F_{A_{\text{out}}}} \tag{2.6.18}$$

Using Eq. 2.3.11, the yield relates to the extents of the independent chemical reactions:

$$\sigma_V = -\left(\frac{s_A}{s_V}\right)_{\text{des}}\frac{\sum_{m}^{n_I}(s_V)_m \dot{X}_m}{\sum_{m}^{n_I}(s_A)_m \dot{X}_m} \tag{2.6.19}$$

Several points should be noted:

1. Both the yield and the selectivity are defined such that their numerical values are between 0 and 1.
2. The subscript "des" in the definitions refers to the chemical formula that ties reactant A to the product V (the desirable reaction). In some cases (e.g., in sequential reactions) the desirable reaction does not actually take place, but rather it is merely stoichiometric relations (see Example 2.11).
3. There is a simple relationship between the yield and the selectivity. Using the conversion definition, Eq. 2.6.1a or 2.6.1b,

$$\eta_V(t) = \sigma_V(t)f_A(t) \quad \text{or} \quad \eta_V = \sigma_V f_A \tag{2.6.20}$$

Example 2.10 Ethylene oxide is produced in a catalytic steady-flow reactor. A feed consisting of 70% C_2H_4 and 30% O_2 (mole percent) is fed into the reactor at a rate of 100 mol/s. The following chemical reactions take place in the reactor:

$$\text{Reaction 1:} \quad 2C_2H_4 + O_2 \longrightarrow 2C_2H_4O$$

$$\text{Reaction 2:} \quad C_2H_4 + 3O_2 \longrightarrow 2CO_2 + 2H_2O$$

$$\text{Reaction 3:} \quad C_2H_4 + 2O_2 \longrightarrow 2CO + 2H_2O$$

An analysis of the exit stream indicates that its composition is 41.17% C_2H_4, 37.65% C_2H_4O, and 7.06% O_2 (mole percent). Determine:

a. The conversions of ethylene and oxygen
b. The yield of ethylene oxide
c. The selectivity of ethylene oxide

Solution Since each chemical reaction has a species that does not participate in the other two, the three reactions are independent, and their stoichiometric coefficients are

$$(s_{C_2H_4})_1 = -2 \quad (s_{O_2})_1 = -1 \quad (s_{C_2H_4O})_1 = 2 \quad (s_{H_2O})_1 = 0 \quad (s_{CO_2})_1 = 0 \quad (s_{CO})_1 = 0$$
$$(s_{C_2H_4})_2 = -1 \quad (s_{O_2})_2 = -3 \quad (s_{C_2H_4O})_2 = 0 \quad (s_{H_2O})_2 = 2 \quad (s_{CO_2})_2 = 2 \quad (s_{CO})_2 = 0$$
$$(s_{C_2H_4})_3 = -1 \quad (s_{O_2})_3 = -2 \quad (s_{C_2H_4O})_3 = 0 \quad (s_{H_2O})_3 = 2 \quad (s_{CO_2})_3 = 0 \quad (s_{CO})_3 = 2$$
$$\Delta_1 = -1 \quad \Delta_2 = 0 \quad \Delta_3 = 1$$

To calculate the required quantities, first determine the extents of the reactions. Using Eqs. 2.3.11 and 2.3.12, the molar fraction of species j in the outlet stream is

$$y_j = \frac{F_{j_{in}} + \sum_{m}^{n_I} (s_j)_m \dot{X}_m}{F_{tot_{in}} + \sum_{m}^{n_I} \Delta_m \dot{X}_m} \tag{a}$$

Writing (a) for C_2H_4, C_2H_4O, and O_2,

$$y_{C_2H_4} = \frac{70 - 2\dot{X}_1 - \dot{X}_2 - \dot{X}_3}{100 - \dot{X}_1 + \dot{X}_3} = 0.4117 \tag{b}$$

$$y_{C_2H_4O} = \frac{2\dot{X}_1}{100 - \dot{X}_1 + \dot{X}_3} = 0.3765 \tag{c}$$

$$y_{O_2} = \frac{30 - \dot{X}_1 - 3\dot{X}_2 - 2\dot{X}_3}{100 - \dot{X}_1 + \dot{X}_3} = 0.0706 \tag{d}$$

Solving (b), (c), and (d), $\dot{X}_1 = 16$, $\dot{X}_2 = 2$, and $\dot{X}_3 = 1\,\mathrm{mol/s}$. Using Eq. 2.3.11, the molar flow rates of the respective species are

$$F_{C_2H_4} = 70 - 2(16) + (-2)1 = 35\,\mathrm{mol/s} \tag{e}$$

$$F_{C_2H_4O} = 2(16) = 32\,\mathrm{mol/s} \tag{f}$$

$$F_{O_2} = 30 + (-1)16 + (-3)2 + (-2)1 = 6\,\mathrm{mol/s} \tag{g}$$

$$F_{CO_2} = 2(2) = 4\,\mathrm{mol/s} \tag{h}$$

a. Using the conversion definition (Eq. 2.6.1b) together with (e) and (g), the conversions of C_2H_4 and O_2 are, respectively,

$$f_{C_2H_4} = \frac{70 - 35}{70} = 0.50 \tag{i}$$

$$f_{O_2} = \frac{30 - 6}{30} = 0.80 \tag{j}$$

b. The desirable reaction is Reaction 1 and the limiting reactant is O_2. Hence, using Eq. 2.6.14, the yield of ethylene oxide (on the basis of the oxygen fed) is

$$\eta_{C_2H_4O} = -\left(\frac{-1}{2}\right)\frac{32}{30} = 0.533 \tag{k}$$

c. Using Eq. 2.6.18, the selectivity of ethylene oxide is

$$\sigma_{C_2H_4O} = -\left(\frac{-1}{2}\right)\frac{32}{30 - 6} = 0.667 \tag{l}$$

Example 2.11 Dichloromethane is formed by reacting methane and chlorine. The following reactions take place in the reactor:

Reaction 1: $CH_4 + Cl_2 \longrightarrow CH_3Cl + HCl$

Reaction 2: $CH_3Cl + Cl_2 \longrightarrow CH_2Cl_2 + HCl$

Reaction 3: $CH_2Cl_2 + Cl_2 \longrightarrow CHCl_3 + HCl$

Reaction 4: $CHCl_3 + Cl_2 \longrightarrow CCl_4 + HCl$

A feed consisting of 40% methane and 60% chlorine is fed to a steady-flow reactor at a rate of 100 mol/min. The desirable product is dichloromethane. At the outlet, the molar flow rate of methane is 10 mol/min and the stream contains three times monochloromethane as dichloromethane, and the content of the latter is twice that of trichloromethane. Determine:

a. The limiting reactant
b. The conversions of methane and chlorine
c. The yield of dichloromethane
d. The selectivity of dichloromethane.

Solution   The stoichiometric relation between the desirable product (dichloromethane) and the reactants is obtained by adding Reaction 1 and Reaction 2:

$$CH_4 + 2Cl_2 \longrightarrow CH_2Cl_2 + 2HCl$$

This stoichiometric relation represents the desirable reaction, and its stoichiometric coefficients are

$$s_{CH_4} = -1 \qquad s_{Cl_2} = -2 \qquad s_{CH_3Cl} = 1 \qquad s_{HCl} = 2 \qquad \Delta = 0$$

a. To identify the limiting reactant, use Eq. 2.5.3 for the two reactants:

$$\left| \frac{F_{CH_{4_{in}}}}{s_{CH_4}} \right| = \left| \frac{40}{-1} \right| = 40 \quad \text{and} \quad \left| \frac{F_{Cl_{2_{in}}}}{s_{Cl_2}} \right| = \left| \frac{60}{-2} \right| = 30$$

Hence, the limiting reactant is chlorine.

b. To determine the conversions, we have to calculate the extents of the given four independent reactions. The stoichiometric coefficients of the species in the four reactions are

$$(s_{CH_4})_1 = -1 \qquad (s_{CH_4})_2 = 0 \qquad (s_{CH_4})_3 = 0 \qquad (s_{CH_4})_4 = 0$$

$$(s_{Cl_2})_1 = -1 \qquad (s_{Cl_2})_2 = -1 \qquad (s_{Cl_2})_3 = -1 \qquad (s_{Cl_2})_4 = -1$$

$$(s_{CH_3Cl})_1 = 1 \qquad (s_{CH_3Cl})_2 = -1 \qquad (s_{CH_3Cl})_3 = 0 \qquad (s_{CH_3Cl})_4 = 0$$

$$(s_{CH_2Cl_2})_1 = 0 \qquad (s_{CH_2Cl_2})_2 = 1 \qquad (s_{CH_2Cl_2})_3 = -1 \qquad (s_{CH_2Cl_2})_4 = 0$$

$$(s_{CHCl_3})_1 = 0 \qquad (s_{CHCl_3})_2 = 0 \qquad (s_{CHCl_3})_3 = 1 \qquad (s_{CHCl_3})_4 = -1$$

$$(s_{CCl_4})_1 = 0 \qquad (s_{CCl_4})_2 = 0 \qquad (s_{CCl_4})_3 = 0 \qquad (s_{CCl_4})_4 = 1$$

Using Eq. 2.3.11, the species molar flow rates at the reactor outlet are

$$(F_{CH_4})_{out} = 40 - \dot{X}_1$$

$$(F_{Cl_2})_{out} = 60 - \dot{X}_1 - \dot{X}_2 - \dot{X}_3 - \dot{X}_4$$

$$(F_{HCl})_{out} = \dot{X}_1 + \dot{X}_2 + \dot{X}_3 + \dot{X}_4$$

$$(F_{CH_3Cl})_{out} = \dot{X}_1 - \dot{X}_2$$

$$(F_{CH_2Cl_2})_{out} = 0 + \dot{X}_2 - \dot{X}_3$$

$$(F_{CHCl_3})_{out} = 0 + \dot{X}_3 - \dot{X}_4$$

$$(F_{CCl_4})_{out} = 0 + \dot{X}_4$$

Using the given information,

$$40 - \dot{X}_1 = 10$$

$$\frac{(F_{CH_3Cl})_{out}}{(F_{CH_2Cl_2})_{out}} = \frac{\dot{X}_1 - \dot{X}_2}{\dot{X}_2 - \dot{X}_3} = 0.5$$

$$\frac{(F_{CH_2Cl_2})_{out}}{(F_{CHCl_3})_{out}} = \frac{\dot{X}_2 - \dot{X}_3}{\dot{X}_3 - \dot{X}_4} = 2$$

$$\frac{(F_{CHCl_3})_{out}}{(F_{CCl_4})_{out}} = \frac{\dot{X}_3 - \dot{X}_4}{\dot{X}_4} = 2$$

The solution of these four equations is

$$\dot{X}_1 = 30 \qquad \dot{X}_2 = 22 \qquad \dot{X}_3 = 6 \qquad \dot{X}_4 = 2\,\text{mol/min}$$

The species flow rates at the reactor outlet are

$$(F_{CH_3Cl})_{out} = 8 \quad (F_{CH_2Cl_2})_{out} = 16 \quad (F_{CHCl_3})_{out} = 4 \quad (F_{CCl_4})_{out} = 2\,\text{mol/min}$$

$$(F_{Cl_2})_{out} = 0 \quad (F_{HCl})_{out} = 60\,\text{mol/min}$$

Using Eq. 2.6.1b, the conversions of the two reactants are

$$f_{CH_4} \equiv \frac{(F_{CH_4})_{in} - (F_{CH_4})_{out}}{(F_{CH_4})_{in}} = \frac{40 - 10}{40} = 0.75$$

$$f_{Cl_2} \equiv \frac{(F_{Cl_2})_{in} - (F_{Cl_2})_{out}}{(F_{Cl_2})_{in}} = \frac{60 - 0}{60} = 1.0$$

c. Using Eq. 2.6.15, the yield of the dichloromethane (with respect to the methane) is

$$\eta_{CH_2Cl_2} = -\left(\frac{1}{-1}\right)\left(\frac{16 - 0}{40}\right) = 0.4$$

d. Using Eq. 2.6.18, the selectivity of the dichloromethane (with respect to the methane) is

$$\eta_{CH_2Cl_2} = -\left(\frac{1}{-1}\right)\left(\frac{16-0}{40-30}\right) = 0.533$$

2.7 DIMENSIONLESS EXTENTS

The stoichiometric relations derived so far provide a glimpse at the key role the reaction extents play in the analysis of chemical reactors. Whenever the extents of the independent reactions are known, the reactor composition and all other stated variables (temperature, enthalpy, etc.) can be determined. Unfortunately, the extent has two deficiencies:

- It is not a measurable quantity and, consequently, must be related to other measurable quantities (concentrations, pressure, etc.).
- It is an extensive quantity depending on the amount of reactants initially in the reactor or on the inlet flow rate into the reactor.

While the use of calculated quantities may seem, at first, cumbersome and even counterproductive, it actually simplifies the analysis of chemical reactors with multiple reactions. In fact, calculated quantities such as enthalpy and free energy are commonly used in thermodynamics resulting in simplified expressions. Here too, by using the extents of independent reactions, we formulate the design of chemical reactors by the smallest number of design equations.

To characterize the generic behavior of chemical reactors, it is preferred to describe their operations in terms of intensive dimensionless quantities. To convert the reaction extents to intensive quantities, dimensionless extents are defined. For batch reactors, the dimensionless extent, Z_m, of the mth independent reaction is defined by

$$Z_m \equiv \frac{\text{Extent of the } m\text{th independent reaction}}{\text{Total number of moles of reference state}} = \frac{X_m}{(N_{tot})_0} \qquad (2.7.1)$$

where $(N_{tot})_0$ is the total number of moles of a conveniently selected reference state. The selection of the reference state will be discussed below, but in most applications the initial state of the reactor is taken as the reference state.

For flow reactors, the dimensionless extent, Z_m, is defined by

$$Z_m = \frac{\text{Extent per time of the } m\text{th independent reaction}}{\text{Total molar flow rate of reference stream}} = \frac{\dot{X}_m}{(F_{tot})_0} \qquad (2.7.2)$$

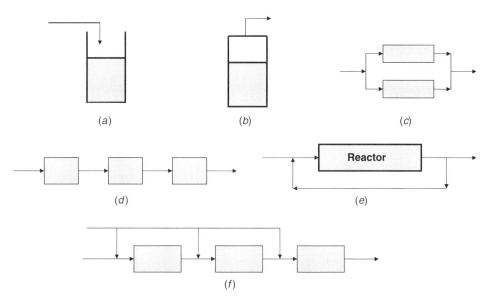

Figure 2.2 Various reactor Configurations: (a) semibatch, (b) distillation reactor, (c) split feed, (d) cascade, (e) recycle reactor, and (f) side injection.

where $(F_{\text{tot}})_0$ is the total molar flow rate of a conveniently selected reference stream. These definitions should apply for all reactor configurations; some are shown in Figure 2.2, The selection of the reference stream is discussed below. Note that the numerical values of dimensionless extents depend on the reference state or reference stream as well as on the chemical formulas used to represent the chemical transformations.

For batch reactors, using Eqs. 2.7.1 and 2.3.3, the species composition reduces to

$$N_j(t) = (N_{\text{tot}})_0 \left\{ \left[\frac{N_{\text{tot}}(0)}{(N_{\text{tot}})_0} \right] y_j(0) + \sum_m^{n_I} (s_j)_m Z_m(t) \right\} \qquad (2.7.3)$$

where $y_j(0) = N_j(0)/N_{\text{tot}}(0)$ is the initial molar fraction of species j in the reactor. When the initial state is selected as the reference state, Eq. 2.7.3 reduces to

$$N_j(t) = (N_{\text{tot}})_0 \left[y_{j_0} + \sum_m^{n_I} (s_j)_m Z_m(t) \right] \qquad (2.7.4)$$

Similarly, using Eq. 2.3.4, the total molar content in the reactor

$$N_{\text{tot}}(t) = (N_{\text{tot}})_0 \left\{ \left[\frac{N_{\text{tot}}(0)}{(N_{\text{tot}})_0} \right] + \sum_m^{n_I} \Delta_m Z_m(t) \right\} \qquad (2.7.5)$$

When the initial state is selected as the reference state, Eq. 2.7.5 reduces to

$$N_{tot}(t) = (N_{tot})_0 \left[1 + \sum_{m}^{n_I} \Delta_m Z_m(t) \right] \tag{2.7.6}$$

For steady-flow reactors, using Eq. 2.7.2, Eq. 2.3.11 reduces to

$$F_j = (F_{tot})_0 \left\{ \left[\frac{(F_{tot})_{in}}{(F_{tot})_0} \right] y_{j_{in}} + \sum_{m}^{n_I} (s_j)_m Z_m \right\} \tag{2.7.7}$$

where $(F_{tot})_{in}$ is the total molar flow rate of the inlet stream and $y_{j_{in}} = F_{j_{in}}/(F_{tot})_{in}$ is the molar fraction of species j in the inlet stream. When the inlet stream is selected as the reference stream, Eq. 2.7.7 reduces to

$$F_j = (F_{tot})_0 \left[y_{j_0} + \sum_{m}^{n_I} (s_j)_m Z_m \right] \tag{2.7.8}$$

Similarly, using Eq. 2.3.12, the total molar flow rate in the reactor

$$F_{tot} = (F_{tot})_0 \left\{ \left[\frac{(F_{tot})_{in}}{(F_{tot})_0} \right] + \sum_{m}^{n_I} \Delta_m Z_m \right\} \tag{2.7.9}$$

When the inlet stream is selected as the reference stream, Eq. 2.7.9 reduces to

$$F_{tot} = (F_{tot})_0 \left[1 + \sum_{m}^{n_I} \Delta_m Z_m \right] \tag{2.7.10}$$

Example 2.12   The gas-phase catalytic oxidation of ammonia is investigated in an isothermal batch reactor. The following reactions take place in the reactor:

Reaction 1: $4NH_3 + 5O_2 \longrightarrow 4NO + 6H_2O$

Reaction 2: $4NH_3 + 3O_2 \longrightarrow 2N_2 + 6H_2O$

Reaction 3: $2NO + O_2 \longrightarrow 2NO_2$

Reaction 4: $4NH_3 + 6NO \longrightarrow 5N_2 + 6H_2O$

Initially, the reactor contains 4 mol of NH_3 and 6 mol of O_2, and its pressure is 2 atm. At time t, the reactor pressure is 2.12 atm, and an analysis of the reactor content indicates that mole fraction of NH_3 is 0.07547, and of N_2 is 0.1132. Using the dimensionless extents, determine the reactor composition.

Solution First we have to determine the number of independent reactions, and to select a set of independent reactions. We construct a matrix of stoichiometric coefficients for the given reactions:

$$
\begin{array}{cccccc}
NH_3 & O_2 & NO & H_2O & N_2 & NO_2
\end{array}
$$

$$
\begin{bmatrix}
-4 & -5 & 4 & 6 & 0 & 0 \\
-4 & -3 & 0 & 6 & 2 & 0 \\
0 & -1 & -2 & 0 & 0 & 2 \\
-4 & 0 & -6 & 6 & 5 & 0
\end{bmatrix} \tag{a}
$$

We conduct a Gaussian elimination procedure and reduce Matrix (a) to

$$
\begin{bmatrix}
-4 & -5 & 4 & 6 & 0 & 0 \\
0 & 1 & -2 & 0 & 1 & 0 \\
0 & 0 & -4 & 0 & 1 & 2 \\
0 & 0 & 0 & 0 & 0 & 0
\end{bmatrix} \tag{b}
$$

Matrix (b) is a reduced matrix, and, since it has three nonzero rows, there are three independent chemical reactions. The nonzero rows in Matrix (b) provide the following set of independent reactions (stoichiometric relations):

Reaction 1: $\quad 4NH_3 + 5O_2 \longrightarrow 4NO + 6H_2O$

Reaction 5: $\quad 2NO \longrightarrow O_2 + N_2$

Reaction 6: $\quad 4NO \longrightarrow N_2 + 2NO_2$

Note that the second and third rows in Matrix (b) represent two chemical reactions that are not among the original reactions, and we denote them as Reactions 5 and 6. The stoichiometric coefficients of these three independent chemical reactions are

$(s_{NH_3})_1 = -4 \quad (s_{O_2})_1 = -5 \quad (s_{NO})_1 = 4 \quad (s_{H_2O})_1 = 6 \quad (s_{N_2})_1 = 0$

$(s_{NO_2})_1 = 0 \quad \Delta_1 = 1$

$(s_{NH_3})_5 = 0 \quad (s_{O_2})_5 = 1 \quad (s_{NO})_5 = -2 \quad (s_{H_2O})_5 = 0 \quad (s_{N_2})_5 = 1$

$(s_{NO_2})_5 = 0 \quad \Delta_5 = 0$

$(s_{NH_3})_6 = 0 \quad (s_{O_2})_6 = 0 \quad (s_{NO})_6 = -4 \quad (s_{H_2O})_6 = 0 \quad (s_{N_2})_6 = 1$

$(s_{NO_2})_6 = 2 \quad \Delta_6 = -1$

To determine the species compositions, we express them in terms of the extents of these three independent reactions, Z_1, Z_5, and Z_6. Selecting the initial state as the reference state, and using Eq. 2.7.4,

$$N_{NH_3}(t) = N_{tot}(0)\lfloor y_{NH_3}(0) + (s_{NH_3})_1 Z_1 + (s_{NH_3})_5 Z_5 + (s_{NH_3})_6 Z_6 \rfloor$$

$$= 10(0.4 - 4Z_1) \tag{c}$$

$$N_{N_2}(t) = N_{tot}(0)\lfloor y_{N_2}(0) + (s_{N_2})_1 Z_1 + (s_{N_2})_5 Z_5 + (s_{N_2})_6 Z_6 \rfloor = 10(5Z_1 + Z_6) \tag{d}$$

Using, Eq. 2.7.6,

$$N_{tot}(t) = N_{tot}(0)[1 + \Delta_1 Z_1 + \Delta_5 Z_5 + \Delta_6 Z_6] = 10(1 + Z_1 - Z_6) \tag{e}$$

Using the given data, and assuming ideal gas behavior,

$$\frac{P(t)}{P(0)} = \frac{N_{tot}(t)}{N_{tot}(0)} = 1 + Z_1 - Z_6 = 1.06 \tag{f}$$

$$y_{NH_3}(t) = \frac{0.4 - 4Z_1}{1 + Z_1 - Z_6} = 0.07547 \tag{g}$$

$$y_{N_2}(t) = \frac{5Z_5 + Z_6}{1 + Z_1 - Z_6} = 0.1132 \tag{h}$$

Solving Eqs. (f), (g), and (h), $Z_1 = 0.08$; $Z_5 = 0.10$; $Z_6 = 0.02$. Using Eqs. 2.7.4 and Eq. 2.7.6, the compositions of the other species are

$$y_{NO}(t) = \frac{4Z_1 - 2Z_5 - 4Z_6}{1 + Z_1 - Z_6} = 0.0377$$

$$y_{H_2O}(t) = \frac{6Z_1}{1 + Z_1 - Z_6} = 0.4528$$

$$y_{O_2}(t) = \frac{0.6 - 5Z_1 + Z_5}{1 + Z_1 - Z_6} = 0.283$$

$$y_{NO_2}(t) = \frac{2Z_6}{1 + Z_1 - Z_6} = 0.0377$$

2.8 INDEPENDENT SPECIES COMPOSITION SPECIFICATIONS

In the preceding section, we discussed how to determine the number of independent chemical reactions and how to select a set of independent reactions. The number of independent reactions indicates the number of equations that we should solve to determine the composition of the reactor. To solve these equations, the specified species conditions should provide independent information. Below, we describe a method to determine what sets of species compositions provide independent information.

Let us examine more closely the nature of the problem. Consider first a chemical reactor where the following two reactions take place:

$$N_2 + 3H_2 \longrightarrow 2NH_3$$

$$2NO + O_2 \longrightarrow 2NO_2$$

Since each chemical reaction has at least one species that does not participate in the other, both reactions are independent. Hence, we need to specify two species compositions to determine the extents of the two reactions. At first glance, it seems that we can specify the composition of any two species. However, a closer examination of the reactions reveals that none of the species participates in both reactions. By specifying the amount of two species from the same chemical reaction, we cannot determine the extents of the second reaction and the reactor composition. Hence, we have to specify two independent compositions such that each relates to a distinct independent chemical reaction. Similar situations may arise when we deal with more complex sets of chemical reactions, where the identification of independent compositions is not so obvious. Consider the hydrogenation of toluene where the following two reactions take place:

$$C_6H_5CH_3 + H_2 \longrightarrow C_6H_6 + CH_4 \qquad (2.8.1)$$

$$2C_6H_5CH_3 + H_2 \longrightarrow (C_6H_5)_2 + 2CH_4 \qquad (2.8.2)$$

These two chemical reactions are independent, yet if we specify the amounts of toluene and methane, we cannot determine the composition of the other species. The reason is that, while the two reactions are independent, we cannot select any two species compositions to provide information on all other species. In this case, compositions of the toluene and methane do not provide us with a relationship on the amount of diphenyl and hydrogen. This becomes evident if we multiply Reaction 2.8.1 by 2 and subtract Reaction 2.8.2:

$$(C_6H_5)_2 + H_2 \longrightarrow 2C_6H_6 \qquad (2.8.3)$$

It is clear that specifications of the amounts of toluene and methane do not provide any information on the amounts of diphenyl and hydrogen.

Independent composition specifications depend on the relationships among individual species and the chemical reactions taking place in the reactor, but they are invariant of the specific set of independent reactions selected. Since the set of independent reactions generated by the reduced matrix of the Gaussian elimination consists of chemical reactions with the least number of species, we apply the reduced matrix to identify independent species composition specifications. To determine the extents of the independent chemical reactions, we usually specify either the amount of individual species (moles of molar flow rates) or a quantity related to all the independent reactions (total number of moles, pressure of the system, etc.) together with compositions of certain species. Note that specifications of species mole fractions contain information on the total number of moles or, for flow reactors, on the molar flow rate.

To identify independent species composition specifications, we adopt the following procedure:

1. Construct a matrix of stoichiometric coefficients for the given reactions, and then reduce it to the reduced matrix (see Section 2.4) using Gaussian elimination.
2. Select a species (a nonzero stoichiometric coefficient) to relate to the first reaction, and then remove the column and the row from the reduced matrix.
3. Repeat step (2) for the remaining reactions until we complete the matrix. The species that we selected provide independent species compositions.
4. If the given data relates to the total number of moles rather than individual species (e.g., pressure), we add another column to the reduced matrix, containing the Δ factors (defined by Eq. 2.2.5) for each reaction in the reduced matrix. We treat the Δ column as the measurable quantity that relates to the sum of species.

For example, for Reactions 2.8.1 and 2.8.2, the reduced matrix is

$$
\begin{array}{ccccc}
\text{T} & \text{H} & \text{B} & \text{M} & \text{D} \\
\begin{bmatrix} -1 & -1 & 1 & 1 & 0 \\ 0 & 1 & -2 & 0 & 1 \end{bmatrix}
\end{array}
\qquad (2.8.4)
$$

The two rows in Matrix 2.8.4 correspond to Reactions 2.8.1 and 2.8.3, respectively. If we select methane (M) to relate to Reaction 2.8.3, we now remove the fourth column and the first row. We can now select either H, B, or D to relate to Reaction 2.8.3. Adding the Δ factor column to Matrix 2.8.4,

$$
\begin{array}{cccccc}
\text{T} & \text{H} & \text{B} & \text{M} & \text{D} & \Delta \\
\begin{bmatrix} -1 & -1 & 1 & 1 & 0 & 0 \\ 0 & 1 & -2 & 0 & 1 & 0 \end{bmatrix}
\end{array}
\qquad (2.8.5)
$$

Matrix 2.8.5 indicates that information on the total molar content in not useful in this case.

Example 2.13   Consider the chemical reactions of Example 2.12:

Reaction 1: $4NH_3 + 5O_2 \longrightarrow 4NO + 6H_2O$

Reaction 2: $4NH_3 + 3O_2 \longrightarrow 2\,N_2 + 6H_2O$

Reaction 3: $NO + O_2 \longrightarrow 2NO_2$

Reaction 4: $4NH_3 + 6NO \longrightarrow 5\,N_2 + 6H_2O$

Identify a set of species compositions that can be specified to determine the extents when one of the measurable quantities is the pressure.

Solution   The reduced matrix for this case was derived in Example 2.12. Adding to it the column of the Δ factors,

$$
\begin{array}{ccccccc}
NH_3 & O_2 & NO & H_2O & N_2 & NO_2 & \Delta
\end{array}
$$

$$
\begin{bmatrix}
-4 & -5 & 4 & 6 & 0 & 0 & 1 \\
0 & 1 & -2 & 0 & 1 & 0 & 0 \\
0 & 0 & -4 & 0 & 1 & 2 & -1
\end{bmatrix}
\tag{a}
$$

We select NH_3 to relate to the first reaction and remove the first row and first column:

$$
\begin{array}{cccccc}
O_2 & NO & H_2O & N_2 & NO_2 & \Delta
\end{array}
$$

$$
\begin{bmatrix}
1 & -2 & 0 & 1 & 0 & 0 \\
0 & -4 & 0 & 1 & 2 & -1
\end{bmatrix}
\tag{b}
$$

We select the total pressure to relate to the last reaction and remove the last row and last column:

$$
\begin{array}{ccccc}
O_2 & NO & H_2O & N_2 & NO_2
\end{array}
$$

$$
\begin{bmatrix} 1 & -2 & 0 & 1 & 0 \end{bmatrix}
\tag{c}
$$

From Matrix (c), we can select either O_2, NO, or N_2 to relate to the last reaction. Hence, we can specify the amount NH_3 and of any one of these species together with the total pressure to determine the composition of the reactor. The reader is challenged to check that when the amounts of NH_3, H_2O, and the pressure are specified, we cannot determine the reactor composition.

2.9 SUMMARY

In this chapter, we described a methodology to express the reactor composition in terms of the chemical reactions taking place. We covered the following topics:

1. Selection of a chemical reaction as a basis for the calculation and defining its stoichiometric coefficients
2. Extent of a chemical reaction and how to calculate it
3. Definition of conversion in batch and flow reactors and its relation to the extent of a chemical reaction
4. A procedure to identify the limiting reactant of a chemical reaction
5. Relation between the composition of a reactor and the extents of independent reactions
6. A method to determine the number of independent chemical reactions and how to select a set of independent reactions
7. A method to express dependent chemical reactions in terms of the independent reactions
8. The role of independent and dependent reactions in reactor calculations

For convenience, Table A.1 in Appendix A provides a summary of the definitions and stoichiometric relations derived in this chapter.

PROBLEMS*

2.1₂ Equation 2.2.6 is the mathematical condition for a balanced chemical reaction. Show that the total mass is conserved when this relation is applied. Show that for a batch reactor $M(0) = M(t)$. Show that for steady-flow reactor, the mass flow rates at the reactor inlet and outlet are identical.

2.2₂ The gas-phase decomposition reaction

$$C_2H_6 \longrightarrow C_2H_2 + 2H_2$$

is being investigated in a batch, constant-volume isothermal reactor. The reactor is charged with 20 mol of ethane, and the initial pressure is 2 atm.

---

*Subscript 1 indicates simple problems that require application of equations provided in the text. Subscript 2 indicates problems whose solutions require some more in-depth analysis and modifications of given equations. Subscript 3 indicates problems whose solutions require more comprehensive analysis and involve application of several concepts. Subscript 4 indicates problems that require the use of a mathematical software or the writing of a computer code to obtain numerical solutions.

When the reaction is stopped, the pressure of the reactor is 5 atm. Assuming ideal-gas behavior, calculate:

a. The extent of the reaction

b. The conversion of ethane

c. The partial pressure of H_2 and C_2H_2 at the end of the reaction

2.3$_2$ Repeat Problem 2.2 where a mixture of 50% ethane and 50% hydrogen is charged into the reactor. The initial pressure is 2 atm, and the final pressure is 3 atm.

2.4$_2$ The gas-phase decomposition reaction

$$C_2H_6 \longrightarrow C_2H_4 + H_2$$

is being investigated in an isobaric, batch reactor operated isothermally. Initially, 10 mol of ethane (pure) are charged into the reactor. If the final volume of the reactor is 80% larger than the initial volume, calculate:

a. The conversion

b. The reaction extent

c. The mole fraction of H_2 at the end of the operation
Assume ideal-gas behavior.

2.5$_2$ The gas-phase reaction

$$A \longrightarrow 2R + P$$

takes place in a constant-volume batch reactor. A thermal conductivity detector is used to determine the progress of the reaction. The conductivity reading is proportional to the sum of the concentrations of A and R. At the beginning of the operation, 2 kmol of A and 1 kmol of P are charged into the reactor and the conductivity reading is 120 (arbitrary units.) At time t, the conductivity reading is 180. Calculate:

a. The conversion of reactant A at time t

b. The composition of the reactor at time t

2.6$_2$ In many organic substitution reactions, the product generated by the reaction is prone to additional substitution. A semibatch reactor was used to produce monochlorobenzene by reacting benzene with chlorine. The reactor was charged with 20 mol of liquid benzene. A stream of gaseous chlorine bubbled through the liquid, and the chlorine not reacted upon was recycled. During the operation, monochlorobenzene reacted with the chlorine to produce dichlorobenzene, and the dichlorobenzene reacted with the chlorine to produce trichlorobenzene:

$$C_6H_6(l) + Cl_2(g) \longrightarrow C_6H_5Cl(l) + HCl(g)$$

$$C_6H_5Cl(l) + Cl_2(g) \longrightarrow C_6H_4Cl_2(l) + HCl(g)$$

$$C_6H_4Cl_2(l) + Cl_2(g) \longrightarrow C_6H_3Cl_3(l) + HCl(g)$$

At the end of the operation, the reactor contained 11 mol of benzene and a product mixture in which the amount of monochlorobenzene was three times the amount of dichlorobenzene, and the amount of the latter was twice the amount of trichlorobenzene. The total amount of chlorine fed during the operation was 40 mol. Find:

a. The conversion of benzene

b. The composition of the reactor (liquid contents)

c. The amount of HCl produced

d. The conversion of chlorine

2.7₂ The dimerization reaction

$$2A \longrightarrow R$$

is taking place in a liquid solution. The progress of the reaction is monitored by an infrared (IR) analyzer whose signal is adjusted such that the percent of the IR absorbed by the solution is proportional to the sum of the concentrations of A and R. Initially, the solution contains 5 kmol of A and no R, and the reading of the analyzer is 85%. At time t, the analyzer reading is 60%. Calculate:

a. The extent of reaction at time t

b. The conversion at time t

c. The amount of R in the solution at time t

2.8₂ A gaseous fuel consisting of a mixture of methane (CH_4) and ethane (C_2H_6) is fed into a burner in a proportion of 1 mol of fuel per 20 mol of air. The following reactions are believed to take place in the reactor:

$$CH_4 + 2O_2 \longrightarrow CO_2 + 2H_2O$$

$$2C_2H_6 + 7O_2 \longrightarrow 4CO_2 + 6H_2O$$

An analysis of the flue gas (dry basis) indicates that it consists of 83.7% N_2, 7.01% CO_2, 9.15% O_2, and 0.14% methane (percent mole). Calculate:

a. The composition of the fuel

b. The conversion of oxygen

c. The conversion of methane

d. The conversion of ethane

e. The excess amount of air used

2.9₁ Ethylene oxide is produced by a catalytic reaction of ethylene and oxygen at controlled conditions. However, side reactions cannot be completely prevented, and it is believed that the following reactions take place:

$$2C_2H_4 + O_2 \longrightarrow 2C_2H_4O$$

$$C_2H_4 + 3O_2 \longrightarrow 2CO_2 + 2H_2O$$

$$C_2H_4 + 2O_2 \longrightarrow 2CO + 2H_2O$$

$$2CO + O_2 \longrightarrow 2CO_2$$

$$CO + H_2O \longrightarrow CO_2 + H_2$$

a. Determine the number of independent reactions.

b. Choose a set of independent reactions among the given reactions above.

c. Express the dependent reactions in terms of the independent reactions.

2.10₂ Ammonium nitrate is a raw material used in the manufacture of agricultural chemicals and explosives. The following reactions are believed to take place in the production of ammonium nitrate:

$$4NH_3 + 5O_2 \longrightarrow 4NO + 6H_2O$$

$$2NO + O_2 \longrightarrow 2NO_2$$

$$4NH_3 + 7O_2 \longrightarrow 4NO_2 + 6H_2O$$

$$3NO_2 + H_2O \longrightarrow 2HNO_3 + NO$$

$$NH_3 + HNO_3 \longrightarrow NH_4NO_3$$

a. Determine the number of independent reactions.

b. Choose a set of independent reactions among the reactions above.

c. Express the dependent reactions in terms of the independent reactions.

2.11₁ In the reforming of methane, the following chemical reactions may occur:

$$CH_4 + 2H_2O \longrightarrow CO_2 + 4H_2$$

$$CH_4 + 2O_2 \longrightarrow CO_2 + 2H_2O$$

$$2CH_4 + O_2 \longrightarrow 2CH_3OH$$

$$2CH_4 + 3O_2 \longrightarrow 2CO + 4H_2O$$

$$CO_2 + 3H_2 \longrightarrow CH_3OH + H_2O$$

$$2CH_3OH + 3O_2 \longrightarrow 2CO_2 + 4H_2O$$

$$2CO + O_2 \longrightarrow 2CO_2$$

$$CO + H_2O \longrightarrow CO_2 + H_2$$

a. Determine the number of independent reactions.

b. Choose a set of independent reactions from among the reactions above.

c. Express the dependent reactions in terms of the independent reactions.

2.12₁ Consider the classic mechanism of the reaction between hydrogen and bromine to form hydrogen bromide (the asterisks indicate free radicals):

$$Br_2 \longrightarrow 2Br^*$$

$$Br^* + H_2 \longrightarrow HBr + H^*$$

$$H^* + Br_2 \longrightarrow HBr + Br^*$$

$$H^* + HBr \longrightarrow H_2 \longrightarrow Br^*$$

$$2Br^* \longrightarrow Br_2$$

a. Determine the number of independent reactions.

b. Identify a set of independent reactions from the reactions above.

c. Express the dependent reactions in terms of the independent reactions.

2.13₃ Methanol is being produced according to the reaction

$$CO + 2H_2 \longrightarrow CH_3OH$$

A synthesis gas stream consisting of 67.1% H_2, 32.5% CO, and 0.4% CH_4 (by mole) is fed into the process described below at a rate of 100 mol/h. The effluent stream from the reactor is fed into a separator (Fig. P2.13) where the methanol is completely removed, and the unconverted reactants are recycled to the reactor. To avoid the buildup of methane, a portion of the recycled stream is purged. At present operating conditions, the CO

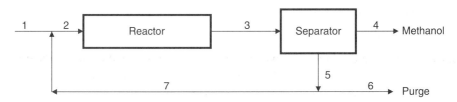

Figure P2.13   Methanol rector with a separator.

conversion over the entire process is 90%, and the methane mole fraction at the reactor inlet (stream 2) is 2.87%. Calculate:

a. The production rate of methanol
b. The flow rate of the purge stream (stream 6)
c. The composition of the purge stream
d. The portion of the recycle stream (stream 5) that is purged
e. The CO conversion per pass in the reactor

2.14$_3$ Solve Problem 2.13 for the following information. The CO conversion per pass is 15%, and the methane mole fraction at the reactor inlet (stream 2) is 2.5%. Calculate:

a. The production rate of methanol
b. The flow rate of the purge stream (stream 6)
c. The composition of the purge stream
d. The portion of the recycle stream (stream 5) that is purged
e. The CO conversion in the process

2.15$_2$ The following reactions are believed to take place during the catalytic oxidation of methane:

$$2CH_4 + 3O_2 \longrightarrow 2CO + 4H_2O$$
$$CH_4 + 2O_2 \longrightarrow CO_2 + 2H_2O$$
$$2CH_4 + O_2 \longrightarrow 2CH_3OH$$
$$CH_4 + O_2 \longrightarrow HCHO + H_2O$$
$$CO + H_2O \longrightarrow CO_2 + H_2$$
$$CO + 2H_2 \longrightarrow CH_3OH$$
$$CO + H_2 \longrightarrow HCHO$$
$$2CH_4 + O_2 \longrightarrow 2CO + 4H_2$$
$$2CO + O_2 \longrightarrow 2CO_2$$

a. Determine the number of independent reactions.
b. Identify a set of independent reactions from the reactions above.
c. Express the dependent reactions in terms of the independent reactions.

2.16$_2$ Selective oxidation of hydrocarbons is a known method to produce alcohols. However, the alcohols react with the oxygen to produce aldehydes, and the latter react with oxygen to produce organic acids. A 50 mol/s stream consisting of 90% ethane and 10% nitrogen is mixed with a

40 mol/s airstream and fed into a catalytic reactor. The following reactions take place in the reactor:

$$2C_2H_6 + O_2 \longrightarrow 2C_2H_5OH$$

$$2C_2H_5OH + O_2 \longrightarrow 2CH_3CHO + 2H_2O$$

$$2CH_3CHO + O_2 \longrightarrow 2CH_3COOH$$

The oxygen conversion is 80%, and the concentration of the ethanol in the product stream is three times that of the aldehyde and four times that of the acetic acid. Calculate:

a. The ethane conversion

b. The production rate of the ethanol

2.17₃ Ammonia is being produced by reacting nitrogen and hydrogen in the system shown in Figure P2.17. The reaction is

$$N_2 + 3H_2 \longrightarrow 2NH_3$$

The ammonia is removed completely in the separator. A feed stream consisting of 2% argon (by mole) and stoichiometric proportion of the reactants is fed into the system at the rate of 100 mol/min. The mole fraction of argon in the purge stream (stream 6) is 5%, and the conversion per pass in the reactor is 10%. Determine:

a. The production rate of ammonia

b. The overall (process) conversions of nitrogen and hydrogen

c. The flow rate of the recycle stream (stream 7)

2.18₃ A gaseous fuel consisting of a mixture of methane (CH_4) and ethane (C_2H_6) is fed into a burner in a proportion of 1 mol of fuel per 18 mol of air. The following reactions are believed to take place in the reactor:

$$CH_4 + 2O_2 \longrightarrow CO_2 + 2H_2O$$

$$2CH_4 + 3O_2 \longrightarrow 2CO + 4H_2O$$

$$2C_2H_6 + 5O_2 \longrightarrow 4CO + 6H_2O$$

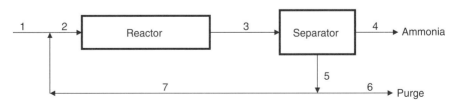

Figure P2.17 Ammonia rector with a separator.

$$2C_2H_6 + 7O_2 \longrightarrow 4CO_2 + 6H_2O$$

$$2CO + O_2 \longrightarrow 2CO_2$$

An analysis of the flue gas (dry basis) indicates that it consists of 7.71% CO_2, 0.51% CO, 7.41% O_2, 84.32% of N_2, and a trace amount of methane and ethane (% mole). Calculate:

a. The composition of the fuel

b. The conversion of oxygen

c. The conversion of methane

d. The conversion of ethane

e. The excess amount of air used

2.19₃ Toluene ($C_6H_5CH_3$) is hydrogenated according to the two simultaneous chemical reactions

$$C_6H_5CH_3 + H_2 \longrightarrow C_6H_6 + CH_4 \tag{a}$$

$$2C_6H_5CH_3 + H_2 \longrightarrow (C_6H_6)_2 + 2CH_4 \tag{b}$$

Initially, the reactor contains 40% toluene and 60% H_2 (% mole). At the end of the operation, the reactor contains 10% toluene and 30% CH_4 (% mole). Find:

a. The mole fraction of the diphenyl, $(C_6H_5)_2$, at the end of the operation

b. The conversion of H_2

c. Solve (a) and (b) when the following final composition is specified: The reactor contains 10% toluene and 40% H_2 (% mole).

BIBLIOGRAPHY

Other treatments of stoichiometric coefficients, extent of chemical reactions, and independent reactions can be found in:

R. Aris, *Introduction to the Analysis of Chemical Reactors*, Prentice-Hall, Englewood Cliffs, NJ, 1965.

S. I. Sandler, *Chemical and Engineering Thermodynamics*, 2nd ed. Wiley, New York, 1989.

G. V. Reklaitis, *Introduction to Material and Energy Balances*, Wiley, New York, 1983.

3

CHEMICAL KINETICS

This chapter covers the second fundamental concept used in chemical reaction engineering—chemical kinetics. The kinetic relationships used in the analysis and design of chemical reactors are derived and discussed. In Section 3.1, we discuss the various definitions of the species formation rates. In Section 3.2, we define the rates of chemical reactions and discuss how they relate to the formation (or depletion) rates of individual species. In Section 3.3, we discuss the rate expression that provides the relationship between the reaction rate, the temperature, and species concentrations. Without going into the theory of chemical kinetics, we review the common forms of the rate expressions for homogeneous and heterogeneous reactions. In the last section, we introduce and define a measure of the reaction rate—the characteristic reaction time. In Chapter 4 we use the characteristic reaction time to reduce the reactor design equations to dimensionless forms.

3.1 SPECIES FORMATION RATES

Consider a batch reactor with volume V where species j is formed (or consumed) by chemical reactions. We denote the total moles of species j in the reactor by N_j; hence, the rate species j being formed in the reactor is dN_j/dt. The magnitude of dN_j/dt depends, of course, on the size of the reactor. To define an intensive quantity for the formation rate of species j, we consider the nature of the chemical reactions that are taking place. For *homogeneous* chemical reactions (reactions that take

Principles of Chemical Reactor Analysis and Design, Second Edition. By Uzi Mann
Copyright © 2009 John Wiley & Sons, Inc.

place throughout a given phase), the *volume-based* species formation rate, (r_j), is defined by

$$(r_j) = (r_j)_V \equiv \frac{1}{V}\frac{dN_j}{dt} \tag{3.1.1a}$$

Thus, (r_j) is the formation rate of species j expressed in terms of moles of species j formed per unit time per unit volume. For *heterogeneous* chemical reactions (reactions that take place on the surface of a solid catalyst or at the interface of the two phases), the *surface-based* formation rate of species j, $(r_j)_S$, is defined by

$$(r_j)_S \equiv \frac{1}{S}\frac{dN_j}{dt} \tag{3.1.1b}$$

where $(r_j)_S$ is expressed in terms of moles of species j formed per unit time per unit surface area. In some cases, it is convenient to define the *mass-based* formation rate of species j (on the basis of the mass of solid catalyst):

$$(r_j)_W \equiv \frac{1}{W}\frac{dN_j}{dt} \tag{3.1.1c}$$

Here, $(r_j)_W$ is the mass-based formation rate of species j expressed in moles of species j formed per unit time per unit mass of catalyst.

From Eqs. 3.1.1a–3.1.1c, it is easy to show that

$$(r_j) = \left(\frac{S}{V}\right)(r_j)_S \tag{3.1.2}$$

$$(r_j) = \left(\frac{W}{V}\right)(r_j)_W \tag{3.1.3}$$

Hence, when any one of these rates is known, the other two can be determined if the properties of the reactor (mass of solid catalyst per unit volume or catalyst surface per unit volume) are provided.

Note that all these definitions do not relate to any specific chemical reaction but rather to the net formation of species j, which may be formed simultaneously by some reactions and consumed by others. Also, when species j is consumed, (r_j) is negative.

3.2 RATES OF CHEMICAL REACTIONS

Consider a specific chemical reaction and focus on the rate that it progresses. As discussed in Section 2.2, we first have to select the chemical formula for this reaction (select a basis for the calculation) and then express the progress of the reaction in terms of the extent of that formula. For a homogeneous chemical reaction, the volume-based rate of a reaction is defined by

$$r \equiv \frac{1}{V}\frac{dX}{dt} \qquad (3.2.1a)$$

where r denotes the reaction rate expressed in terms of moles extent per unit time per unit volume. For heterogeneous reactions, the surface-based rate of the chemical reaction is defined by

$$r_S \equiv \frac{1}{S}\frac{dX}{dt} \qquad (3.2.1b)$$

where r_S is expressed in terms of moles extent per unit time per unit surface area. Similarly, the mass-based rate of the chemical reaction is defined by

$$r_W \equiv \frac{1}{W}\frac{dX}{dt} \qquad (3.2.1c)$$

where r_W is expressed in terms of moles extent per unit time per unit mass. As before, relationships among these three reaction rate definitions can be easily obtained:

$$r = \left(\frac{S}{V}\right) r_S \qquad (3.2.2)$$

$$r = \left(\frac{W}{V}\right) r_W \qquad (3.2.3)$$

Next, we consider the relationship between the formation rates of individual species and the rates of the chemical reactions. When a *single* chemical reaction takes place, we can easily relate the rate of the chemical reaction, r, to the formation rate of a specific species, say species j, (r_j). Differentiating the definition of the reaction extent (Eq. 2.3.1) with time,

$$\frac{dX}{dt} = \frac{1}{s_j}\frac{dn_j}{dt} \qquad (3.2.4)$$

and by substituting this into Eq. 3.2.1a,

$$(r_j) = s_j r \qquad (3.2.5)$$

Equation 3.2.5 provides a relationship between the reaction rate and the rate of formation (or depletion) of any species j by that chemical reaction.

When *multiple* chemical reactions take place simultaneously, species j may participate in several reactions, formed by some and consumed by others. The change in the molar content of species j in the reactor relates to its formation by all chemical reactions. Using the definition of the reaction extent (Eq. 2.3.1), the molar content relates to the reaction extents by

$$N_j(t) - N_j(0) = \sum_{i=1}^{n_R} (s_j)_i X_i(t)$$

where n_R is the number of reactions (reaction pathways). Differentiating this expression with respect to time and substituting into Eq. 3.1.1a, the formation rate of species j is

$$(r_j) = \sum_{i=1}^{n_R} (s_j)_i r_i \tag{3.2.6}$$

where $(s_j)_i$ is the stoichiometric coefficient of species j in the ith chemical reaction, and r_i is the rate of the ith reaction. Equation 3.2.6 relates the formation rate of any species to the rates of the chemical reactions. It is important to note that the summation in Eq. 3.2.6 is over *all* the chemical reactions that *actually* take place in the reactor (both dependent and independent). Also, note that when a single chemical reaction takes place, Eq. 3.2.6 reduces to Eq. 3.2.5.

Example 3.1   The heterogeneous catalytic chemical reaction

$$2C_2H_4 + O_2 \longrightarrow 2C_2H_4O$$

is investigated in a packed-bed reactor. The reactor is filled with catalytic pellets whose surface area is $7\,m^2/g$, and the density of the bed is 1.4 kg/L. The measured consumption rate of ethylene is 0.35 mol/h g catalyst. Determine:
a. The volume-based and surface-based formation rates of ethylene
b. The volume-based rate of the chemical reaction
c. The volume-based formation rate of oxygen
d. The volume-based formation rate of ethylene oxide

Solution   The stoichiometric coefficients of the chemical reaction are

$$s_{C_2H_4} = -2 \quad s_{O_2} = -1 \quad s_{C_2H_4O} = 2$$

The given mass-based formation rate of ethylene is -0.35 mol/h g (the minus sign indicates that ethylene is consumed).

a. Using Eq. 3.1.3, the volume-based formation rate of ethylene is

$$(r_{C_2H_4}) = \left(\frac{W}{V}\right)(r_{C_2H_4})_W = \rho_{bed}(r_{C_2H_4})_W$$

$$= \left(1.4\frac{kg}{L}\right)\left(-0.35\frac{mol\ Et}{h\ g}\right)\left(\frac{1000\ g}{kg}\right) = -490\frac{mol\ Et}{h\ L} \quad \text{(a)}$$

To determine the surface-based formation rate of ethylene, we first calculate the specific surface in the reactor:

$$\left(\frac{S}{V}\right) = \left(\frac{W}{V}\right)\left(\frac{S}{W}\right) = \rho_{bed}\left(\frac{S}{W}\right)$$

$$\left(\frac{S}{V}\right) = (1.4\ kg/L)(7.0\ m^2/kg)(1000\ g/kg) = 9800\ m^2/L \quad \text{(b)}$$

Using Eq. 3.1.2, (a), and (b), the surface-based formation rate of ethylene is

$$(r_{C_2H_4})_S = \frac{-490\ mol\ Et/h\ L}{9.80 \times 10^3\ m^2/L} = -5.00 \times 10^{-2}\ mol\ Et/h\ m^2 \quad \text{(c)}$$

b. The volume-based rate of the reaction is determined by using Eq. 3.2.5 and (a):

$$r = \frac{(r_{C_2H_4})}{s_{C_2H_4}} = \frac{-490\ mol\ Et/h\ L}{-2\ mol\ Et/mol\ extent} = 245\ mol\ extent/h\ L \quad \text{(d)}$$

c. The volume-based formation rate of oxygen is determined by using Eq. 3.2.5 and (d):

$$r_{O_2} = \left(-1\frac{mol\ oxygen}{mol\ extent}\right)\left(245\frac{mol\ extent}{h\ L}\right)$$

$$= -245\ mol\ oxygen/h\ L \quad \text{(e)}$$

The negative sign indicates that oxygen is consumed.

d. The volume-based formation rate of ethylene oxide is determined by using Eq. 3.2.5 and (d):

$$r_{C_2H_4O} = \left(2\frac{mol\ oxide}{mol\ extent}\right)\left(245\frac{mol\ extent}{h\ L}\right) = 490\ mol\ oxide/h\ L \quad \text{(f)}$$

3.3   RATE EXPRESSIONS OF CHEMICAL REACTIONS

The rate of a chemical reaction is a function of the temperature, the composition of the reacting mixture, and, if present, the catalyst. The relationship between the reaction rate and these parameters is commonly called the *rate expression* or, sometimes, the *rate law*. Chemical kinetics is the branch of chemistry that deals with reaction mechanisms and provides a theoretical basis for the rate expression. When such information is available, we use it to obtain the rate expression. In many instances, the reaction rate expression is not available and should be determined experimentally.

For most chemical reactions, the rate expression is a product of two functions, one of temperature, $k(T)$, and the second of species concentrations, $h(C_j\text{'s})$:

$$r = k(T)h(C_j\text{'s}) \tag{3.3.1}$$

The function $k(T)$ is commonly called the *reaction rate constant*. However, note that the reaction rate depends on the temperature. The term *rate constant* comes about because $k(T)$ is independent of the *composition* and is constant at isothermal operations.

For most chemical reactions, $k(T)$ relates to the temperature by the Arrhenius equation:

$$k(T) = k_0 e^{-E_a/RT} \tag{3.3.2}$$

where E_a is a parameter called the activation energy, k_0 is a parameter called the frequency factor or the preexponential coefficient, and R is the universal gas constant. Both parameters are characteristic of the chemical reaction and the presence (or absence) of a catalyst. The value of E_a indicates the sensitivity of the reaction rate to changes in temperature. When the activation energy is relatively large (200–400 kJ/mol), the reaction rate is sensitive to temperature. Such values are typical of combustion and gasification reactions that take place at high temperatures and are very slow at room temperature. On the other hand, when the value of E_a is relatively low (20–40 kJ/mol), the reaction rate is not insensitive to temperature. These values are typical of biological and enzymatic reactions that take place at room temperature. Figure 3.1 shows the relation between k and T for large and small activation energies.

The value of the rate constant at a given temperature is readily calculated when both parameters in the Arrhenius equation, k_0 and E_a, are known. However, it is convenient to calculate the value of the rate constant at one temperature on the basis of its value at a different temperature, using only the activation energy. To obtain a relationship between the values of the reaction rate constant at two temperatures, T_1 and T_2, we take the log of Eq. 3.3.2 for each and combine the two equations to obtain

$$\ln \frac{k(T_2)}{k(T_1)} = -\frac{E_a}{R}\left(\frac{1}{T_2} - \frac{1}{T_1}\right) \tag{3.3.3}$$

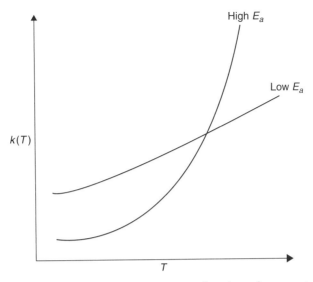

Figure 3.1 Reaction rate constant as a function of temperature.

The values of the parameters in the Arrhenius equation, k_0 and E_a, are determined from experimental measurements of the reaction rate constant, $k(T)$, at different temperatures. Using Eq. 3.3.3, the plot $\ln k(T)$ versus $1/T$ gives a straight line with a slope of $-E_a/R$, as shown schematically in Figure 3.2. The value of k_0 can be determined from the intercept. However, this procedure is not accurate because it involves an extrapolation far beyond the temperature range of the experimental data. Instead, we first determine the slope and then use Figure 3.2 to obtain the value of the rate constant at the average temperature, T_m. We then calculate k_0 from Eq. 3.3.2 using the value of E_a obtained from the slope (see Fig. 3.2).

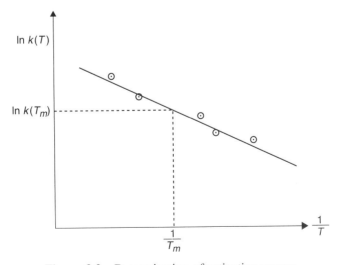

Figure 3.2 Determination of activation energy.

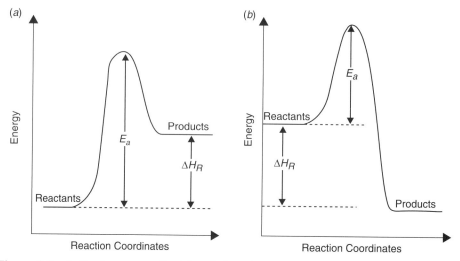

Figure 3.3 Activation energy for (*a*) endothermic reactions and (*b*) exothermic reactions.

Physically, the activation energy represents an energy barrier that should be overcome as the reaction proceeds. This barrier and its relationship to the heat of reaction is shown schematically in Figure 3.3 for exothermic and endothermic reactions. Note that for reversible chemical reactions, the heat of reaction is the difference between the activation energy of the forward reaction and the activation energy of the backward reaction, $\Delta H_R = (E_a)_{\text{for}} - (E_a)_{\text{back}}$.

As will be discussed later, we formulate the design equations of chemical reactors in terms of dimensionless quantities and would like to express the reaction rate constants in terms of them. We define dimensionless temperature

$$\theta \equiv \frac{T}{T_0} \tag{3.3.4}$$

where T_0 is a conveniently selected reference temperature. We also define a dimensionless activation energy, γ, by

$$\gamma \equiv \frac{E_a}{RT_0} \tag{3.3.5}$$

which is a characteristic of the chemical reaction and the reference temperature. Using Eqs. 3.3.4 and 3.3.5, the Arrhenius equation reduces to

$$k(\theta) = k(T_0)e^{\gamma(\theta-1)/\theta} \tag{3.3.6}$$

where $k(T_0)$ is the reaction rate constant at the reference temperature T_0.

Example 3.2   The rate constant of a chemical reaction is determined experimentally at two temperatures. Based on the data below, determine:

a. The activation energy

b. The dimensionless activation energy if the reference temperature is 30°C

c. The preexponential coefficient

Data:
T (°C)	30	50
k (min^{-1})	0.25	1.4

Solution

a. To determine the activation energy, write Eq. 3.3.3 for the two given temperatures, $T_1 = 30 + 273.15 = 303.15$ K and $T_2 = 50 + 273.15 = 323.15$ K,

$$\ln\left(\frac{1.4}{0.25}\right) = -\frac{E_a}{R}\left(\frac{1}{323.15} - \frac{1}{303.15}\right) \tag{a}$$

to obtain $E_a/R = 8438$ K. Hence,

$$E_a = (8438 \text{ K})(1.987 \text{ cal/mol K}) = 16.767 \text{ kcal/mol} \tag{b}$$

b. The reference temperature is $T_0 = 303.15$ K, and, using Eq. 3.3.5, the dimensionless activation energy is

$$\gamma = \frac{16,767}{1.987 \times 303.15} = 27.83$$

c. To determine the preexponential coefficient, we write Eq. 3.3.2 for one of the given temperatures. For $T_1 = 303.15$ K,

$$(0.25 \text{ min}^{-1}) = k_0 e^{-8438/303.15}$$

to obtain $k_0 = 3.06 \times 10^{11} \text{ min}^{-1}$.

Next, we consider the dependence of the rate expression on the species composition–function $h(C_j\text{'s})$ in Eq. 3.3.1. For many homogeneous chemical reactions, $h(C_j\text{'s})$ is expressed as a power relation of the species concentrations:

$$h(C_j\text{'s}) = C_A{}^\alpha C_B{}^\beta \tag{3.3.7}$$

where C_A, C_B, ... are the concentrations of the different species. The powers in Eq. 3.3.7 are called the orders of the reaction: α is the order of the reaction with respect to species A, β is the order of the reaction with respect to species B, and so on, and $\alpha + \beta + \cdots$ is the overall order of the reaction. The orders can be either integers or fractions and should be determined experimentally. Chemical kinetics provides a theoretical basis for the values of the orders for some reactions. For elementary reactions (reactions that take place in a single step without formation of intermediates), the orders of the species are identical to the absolute value of their respective stoichiometric coefficients when the reaction is written in its simplest form. However, in general, we consider Eq. 3.3.7 to be an empirical expression, and the orders are parameters that should be determined experimentally. Combining Eqs. 3.3.1 and 3.3.7 and using Eq. 3.3.6, the rate expression for many homogeneous reactions with two reactants, A and B, is

$$r = k(T)C_A{}^\alpha C_B{}^\beta = k(T_0)e^{\gamma(\theta-1)/\theta}C_A{}^\alpha C_B{}^\beta \tag{3.3.8}$$

Other forms of composition function $h(C_j\text{'s})$ are used for different types of homogenous and heterogeneous reactions. A written chemical reaction is merely a reaction pathway presentation of many elementary reaction steps that usually involve unstable intermediate species (e.g., free radicals). Hence, the rate of a chemical reaction depends on the rates of the individual steps and results in different forms. For example, the rate of many biological and enzymatically catalyzed reactions is described by an expression of the form (known as the Michaelis–Menten expression),

$$r = k(T)\frac{C_A}{K_m + C_A} \tag{3.3.9}$$

where $k(T)$ is the reaction rate constant, and K_m is a parameter to be determined experimentally. Another example is the rate expression of the reaction between hydrogen and bromine to form hydrogen bromide:

$$r = k(T)\frac{C_{H_2}C_{Br_2}^{0.5}}{1 + K\dfrac{C_{HBr}}{C_{Br_2}}} \tag{3.3.10}$$

For most gas–solid heterogeneous catalytic reactions, rate expressions with the following general forms are usually used:

$$r = k(T)\frac{C_A}{1 + K_A C_A} \tag{3.3.11a}$$

$$r = k(T)\frac{C_A C_B}{1 + K_A C_A + K_B C_B} \tag{3.3.11b}$$

$$r = k(T)\frac{C_A C_B}{(1 + K_A C_A + K_B C_B)^2} \tag{3.3.11c}$$

where K_A and K_B are parameters that should be determined experimentally. In many instances, the species concentrations are expressed in terms of the partial pressures.

The dimensions and units of the reaction rate constant, $k(T)$, depend on the form of function $h(C_j\text{'s})$. For chemical reactions whose rate expressions are power functions of the species concentrations, one can determine the units of the rate constant. From Eq. 3.3.8, the rate can be written as

$$r = k(T)C^n$$

where $n = \alpha + \beta + \cdots$ is the overall order of the reaction. In terms of units,

$$\left(\frac{\text{mole}}{\text{volume} \cdot \text{time}}\right) [=] k \left(\frac{\text{mole}}{\text{volume}}\right)^n$$

Thus, the dimensions of the rate constant are

$$k \; [=] \; \left(\frac{\text{mole}}{\text{volume}}\right)^{1-n} \left(\frac{1}{\text{time}}\right) \tag{3.3.12}$$

3.4 EFFECTS OF TRANSPORT PHENOMENA

The rate expressions derived above describe the dependence of the reaction rate expressions on kinetic parameters related to the chemical reactions. These rate expressions are commonly called the *intrinsic* rate expressions of the chemical reactions. However, as discussed in Chapter 1, in many instances, the local species concentrations depend also on the rate that the species are transported in the reaction medium. Hence, the actual reaction rates are affected by the transport rates of reactants and products. This is manifested in two general cases: (i) gas–solid heterogeneous reactions, where species diffusion through the pore plays an important role, and (ii) gas–liquid reactions, where interfacial species mass-transfer rate as well as solubility and diffusion play an important role. Considering the effect of transport phenomena on the global rates of the chemical reactions represents a very difficult task in the design of many chemical reactors. These topics are beyond the scope of this text, but the reader should remember to take them into consideration.

3.5 CHARACTERISTIC REACTION TIME

In the preceding sections, we characterized the rates of chemical reactions in terms of their variation with changes in temperature and their dependency on species compositions. The progress of chemical reactions is a rate process that can be characterized by a time constant. Reactions with the same rate expression (say, first order, second order, etc.) exhibit a similar behavior with respect to concentration variations—the difference between fast and slow reactions is merely the time scale over which the reaction takes place. To describe the generic behavior of chemical reactions and to derive dimensionless design equations for chemical reactors, we define the characteristic reaction time. The characteristic reaction time is a measure of the time scale over which the reaction takes place (second, minute, etc.).

For volume-based reaction rate expressions, the characteristic reaction time, t_{cr}, is defined by

$$t_{cr} \equiv \frac{\text{Reference concentration}}{\text{Reference reaction rate}} = \frac{C_0}{r_0} \tag{3.5.1}$$

where C_0 is a characteristic concentration and r_0 is a characteristic reaction rate, both conveniently selected. We select them in a way so that they can be applied for all forms of rate expressions. For *batch* reactors, the reference concentration, C_0, is defined by

$$C_0 \equiv \frac{(N_{tot})_0}{V_{R_0}} \tag{3.5.2a}$$

where $(N_{tot})_0$ and V_{R_0} are, respectively, the total number of moles and the volume of a conveniently-selected reference state. For *flow* reactors, C_0 is defined by

$$C_0 \equiv \frac{(F_{tot})_0}{v_0} \tag{3.5.2b}$$

where $(F_{tot})_0$ and v_0 are, respectively, the total molar flow rate and the volumetric flow rate of a conveniently selected reference stream. Note that $(N_{tot})_0$ and $(F_{tot})_0$ were used to define the dimensionless extents in Eqs. 2.7.1 and 2.7.2, respectively.

The selection of the reference reaction rate, r_0, should be done in a way that can be used for all forms of rate expressions. Further, since the reaction rate depends on the initial (or inlet) composition, r_0 should be selected such that it is independent of the specific composition of the reference state (or stream). To define r_0, we express the reaction rate at any instance by

$$\begin{pmatrix} \text{Reaction} \\ \text{rate} \end{pmatrix} = \begin{pmatrix} \text{Reference} \\ \text{rate, } r_0 \end{pmatrix} \begin{pmatrix} \text{Dimensionless} \\ \text{correction} \\ \text{due to} \\ \text{temperature} \end{pmatrix} \begin{pmatrix} \text{Dimensionless} \\ \text{correction} \\ \text{due to} \\ \text{composition} \end{pmatrix} \tag{3.5.3}$$

The dimensionless correction term due to temperature variation is given by Eq. 3.3.6. The dimensionless correction term due to changes in composition is readily derived by expressing the species compositions in terms of dimensionless species concentrations. The reference reaction rate, r_0, is defined as the value for which the two dimensionless correction factors are equal to 1.

For convenience, the following procedure is adopted in calculating the characteristic reaction time:

1. Write the reaction rate expression, Eq. 3.3.1.
2. Express the reaction rate constant, $k(T)$, in terms of Eq. 3.3.6.
3. Express the species composition in terms of a dimensionless relation. Either substitute

$$C_j \equiv \left(\frac{C_j}{C_0}\right) C_0$$

   or express C_j in term of the dimensionless extents.
4. Combine all the dimensionless terms and equate them to 1.
5. The remaining term, the dimensional term, is r_0. Determine its value.
6. Determine the characteristic reaction time using Eq. 3.5.1.

For an nth order reaction (overall), the characteristic reaction time is

$$t_{cr} = \frac{1}{k(T_0)C_0^{n-1}} \qquad (3.5.4)$$

The four examples below illustrate how the characteristic reaction time is determined.

Example 3.3 The homogeneous chemical reaction

$$A + B \longrightarrow C$$

is carried out in a batch reactor. The reaction is first order with respect to reactant A, and the order of reactant B is 1.5. Initially, the reactor contains 2 mol of A, 3 mol of B, and 0.5 mol of C, and its initial volume is 2 L. The reaction rate constant at the initial temperature is 0.1 $(L/mol)^{1.5}$ min^{-1}. Determine the characteristic reaction time.

Solution First, calculate the reference concentration. Selecting the initial state as the reference state,

$$(N_{tot})_0 = N_{tot}(0) = N_A(0) + N_B(0) + N_C(0) = 5.50 \text{ mol}$$

and $V_{R_0} = V_R(0) = 2$ L. Hence, using Eq. 3.5.2a,

$$C_0 = \frac{(N_{tot})_0}{V_R(0)} = \frac{5.5}{2.0} = 2.75 \text{ mol/L} \qquad (a)$$

The rate expression of the chemical reaction is

$$r = k(T_0)e^{\gamma(\theta-1)/\theta}C_A C_B^{1.5} \tag{b}$$

Following the procedure, we express the species concentrations in dimensionless form:

$$C_A = C_0\left(\frac{C_A}{C_0}\right) \qquad C_B = C_0\left(\frac{C_B}{C_0}\right)$$

and substitute in (b)

$$r = k(T_0)C_0^{2.5}e^{\gamma(\theta-1)/\theta}\left(\frac{C_A}{C_0}\right)\left(\frac{C_B}{C_0}\right)^{1.5} \tag{c}$$

The dimensional portion of the rate expression is

$$r_0 = k(T_0)/C_0^{2.5}$$

Using Eq. 3.5.1, the characteristic reaction time is

$$t_{cr} = \frac{C_0}{k(T_0)C_0^{2.5}} = \frac{1}{(0.1)(2.75)^{1.5}} = 2.19 \text{ min}$$

Example 3.4   A biological waste, A, is decomposed by an enzymatic reaction

$$A \longrightarrow B + C$$

in aqueous solution. The rate expression of the reaction is

$$r = \frac{kC_A}{K_m + C_A}$$

An aqueous solution with a concentration of 2 mol/L of A is charged into a batch reactor. For the enzyme type and concentration used, $k = 0.1$ mol/L min and $K_m = 4$ mol/L. Determine the characteristic reaction time.

Solution   Selecting the initial state as the reference state, and since A is the only species charged into the reactor, $C_0 = C_A(0)$. To express C_A in dimensionless form,

$$C_A = C_0\left(\frac{C_A}{C_0}\right) \tag{a}$$

and substituting (a) into the rate expression,

$$r = k\left[\frac{C_A/C_0}{(K_m/C_0) + (C_A/C_0)}\right] \tag{b}$$

The dimensional term (the characteristic reaction rate) is

$$r_0 = k \qquad \text{(c)}$$

Using Eq. 3.5.1, the characteristic reaction time is

$$t_{cr} = \frac{C_0}{k} = \frac{2 \text{ mol/L}}{0.1 \text{ mol/L min}} = 20 \text{ min} \qquad \text{(d)}$$

Example 3.5 The heterogeneous, gas-phase catalytic chemical reaction

$$A + B \longrightarrow C$$

is carried out in a flow reactor. The volume-based rate expression of the chemical reaction is

$$r = k(T) \frac{C_A C_B}{(1 + K_A C_A + K_B C_B)^2}$$

A gas stream at 2 atm and 230°C is fed into a reactor. At this temperature, $k = 80 \text{ L/mol min}$. Determine the characteristic reaction time.

Solution Note that the units of K_A and K_B are $(\text{mol/L})^{-1}$. We select the inlet stream as the reference stream. For gas-phase reactions, and assuming ideal gas behavior,

$$C_0 = \frac{P_0}{RT_0} = \frac{2 \text{ atm}}{(0.08206 \text{ L} \cdot \text{atm/mol} \cdot \text{K})(503 \text{ K})} = 4.85 \times 10^{-2} \text{ mol/L} \quad \text{(a)}$$

To express the species concentrations in dimensionless form

$$C_A = C_0 \left(\frac{C_A}{C_0}\right) \qquad C_B = C_0 \left(\frac{C_B}{C_0}\right) \qquad \text{(b)}$$

Substituting (b) into the rate expression,

$$r = k(T_0)C_0^2 \left[e^{\gamma(\theta-1)/\theta} \frac{(C_A/C_0)(C_B/C_0)}{[1 + K_A C_0(C_A/C_0) + K_B C_0(C_B/C_0)]^2} \right] \qquad \text{(c)}$$

The dimensional portion of (c) (the characteristic reaction rate) is

$$r_0 = k(T_0)C_0^2$$

Using Eq. 3.5.1, the characteristic reaction time is

$$t_{cr} = \frac{C_0}{k(T_0)C_0^2} = \frac{1}{(80 \text{ L/mol min})(4.85 \times 10^{-2} \text{ mol/L})} = 0.258 \text{ min} \quad (d)$$

Example 3.6   Allyl chloride is produced at 400°F by a homogeneous gas-phase reaction between chlorine and propylene:

$$Cl_2 + C_3H_6 \longrightarrow C_3H_5Cl + HCl$$

Based on experimental runs, the chlorine depletion rate is

$$(-r_{Cl_2}) = k(T)P_{Cl_2}P_{C_3H_6}$$

where $(-r_{Cl_2})$ is in lbmol/h ft^3, and P is in psia. A gas stream at 30 psia and 400°F, consisting of 50% propylene and 50% chlorine, is fed into the reactor. The reaction rate constant at 400°F is 0.02 lbmol/h ft^3 psia^2. Determine the characteristic reaction time.

Solution   Since the chlorine depletion rate is provided (rather than the reaction rate), first we convert the species rate to a reaction rate. Using Eq. 3.2.5, the reaction rate is

$$r = -\frac{-r_{Cl_2}}{s_{Cl_2}} = -r_{Cl_2} \quad (a)$$

To determine the characteristic reaction time, first the reference concentration, C_0, is calculated. Selecting the inlet stream as the reference stream, for gas-phase reactions, assuming ideal-gas behavior

$$C_0 = \frac{P_0}{RT_0} = \frac{30 \text{ psia}}{(10.73 \text{ psia ft}^3/\text{lbmol°R})(860°R)} = 3.25 \times 10^{-3} \text{ lbmol/ft}^3 \quad (b)$$

To express the species partial pressure in dimensionless form

$$P_{Cl_2} = P_0\left(\frac{P_{Cl_2}}{P_0}\right) \qquad P_{C_2H_6} = P_0\left(\frac{P_{C_2H_6}}{P_0}\right) \quad (c)$$

where P_0 is the total reference pressure. Substituting these into the rate expression,

$$r = k(T_0)P_0^2 e^{\gamma(\theta-1)/\theta}\left(\frac{P_{Cl_2}}{P_0}\right)\left(\frac{P_{C_2H_6}}{P_0}\right) \quad (d)$$

The dimensional portion of the rate expression is

$$r_0 = k(T_0)P_0{}^2 \tag{e}$$

Substituting (b) and (e) into Eq. 3.5.1, the characteristic reaction time is

$$t_{cr} = \frac{1}{k(T_0)RT_0} = 1.81 \times 10^{-4}\ h \tag{f}$$

3.6   SUMMARY

In this chapter, the main concepts and relations of chemical kinetics that are used in reactor design were discussed. We covered the following topics:

1. Definition of species formation rates on the basis of volume, mass, and surface and the relations among them
2. Definition of reaction rates and their relation to the species formation rates
3. The general forms of the reaction rate expressions for most homogeneous reactions (power expression)
4. The reaction activation energy and how to determine it
5. The orders of the reaction
6. Different forms of the rate expressions for biological and heterogeneous catalytic reactions
7. Definition and determination of the characteristic reaction time as a measure of the reaction time scale

For convenience, Table A.2 in the Appendix provides a summary of the kinetic definitions and relations.

PROBLEMS*

3.1$_1$ The rate of a heterogeneous reaction was measured in a rotating basket reactor. The volume of the basket was 100 mL and it was filled with 240 g of catalytic particles whose specific surface area was $9.5\ m^2/g$ of

---

*Subscript 1 indicates simple problems that require application of equations provided in the text. Subscript 2 indicates problems whose solutions require some more in-depth analysis and modification of given equations. Subscript 3 indicates problems whose solutions require more comprehensive analysis and involve application of several concepts. Subscript 4 indicates problems that require the use of a mathematical software or the writing of a computer code to obtain numerical solutions.

catalyst. At $100°C$, the measured volume-based reaction rate constant was 2.5 s^{-1}. Determine:

a. The order of the reaction, based on the units of the reaction rate constant

b. The mass-based rate constant

c. The surface-based rate constant

3.2₁ For a first-order homogeneous reaction, the following experimental data were recorded:

T (°C)	0	20	40	60	80
$k \times 10^3$ (min^{-1})	0.3	6.18	86.4	879	6900

Determine:

a. The activation energy

b. The rate constant at $25°C$

c. The frequency factor

d. The dimensionless activation energy if the reference temperature is $25°C$

e. The characteristic reaction time at $40°C$

3.3₁ A common rule-of-thumb for many chemical reactions states that, at normal conditions, the reaction rate doubles for each increase of $10°C$. Assuming a "normal" temperature of $20°C$, estimate the activation energy of a "typical" chemical reaction. If we select $20°C$ as the reference temperature, what is a "typical" dimensionless activation energy?

3.4₂ Fine and Beall (*Chemistry for Engineers and Scientists*, 1990, p. 854) provide the kinetic data for a particular proton transfer reaction:

T (K)	273	283	293	303
k (L mol^{-1}s^{-1})	35	150	500	1800

Determine:

a. The overall order of the reaction from the units of k

b. The activation energy

c. If we select 300 K as the reference temperature, what is the dimensionless activation energy?

d. If $C_0 = 2$ mol/L, what is the characteristic reaction time at 300 K?

3.5₂ The use of a pressure cooker is a common method to increase the reaction rate and to shorten the cooking time. A typical pressure cooker operates at 15 psig, thus raising the cooking temperature from 212 to $249°F$.

Neglecting the cooking that occurs during the heating-up period and assuming the rate constant is inversely proportional to the cooking time,

a. Estimate the cooking activation energy of the foods listed below.

b. If we select 212°F as the reference temperature, what is the dimensionless activation energy of each food?

c. Assuming first-order reaction, what are the characteristic reaction times at 212°F?

	Normal Cooking Time	Cooking Time in a Pressure Cooker
Green beans	12 min	3 min
Corn beef	4 h	60 min
Chicken stew	2.5 h	20 min

BIBLIOGRAPHY

More detailed treatments of chemical kinetics, reaction mechanisms, catalysis, and the theoretical basis for the rate expression can be found in:

S. W. Benson, *The Foundation of Chemical Kinetics*, McGraw-Hill, New York, 1960.

M. Boudart, *Kinetics of Chemical Processes*, Prentice-Hall, Englewood Cliffs, NJ, 1968.

J. H. Espenson, *Chemical Kinetics and Reaction Mechanisms*, McGraw-Hill, New York, 1981.

W. C. Gardiner, *Rates and Mechanisms of Chemical Reactions*, Benjamin, New York, 1969.

O. A. Hougen and K. M. Watson, *Chemical Process Principles III; Kinetics and Catalysis*, Wiley, New York, 1947.

K. J. Laidler, *Chemical Kinetics*, 3rd, Harper & Row, New York, 1987.

R. I. Masel, *Principles of Adsorption and Reaction on Solid Surfaces*, Wiley Interscience, New York, 1996.

R. I. Masel, *Chemical Kinetics and Catalysis*, Wiley Interscience, Hoboken, NJ, 2001.

Compilations of experimental and theoretical reaction rate data can be found in:

C. H. Bamford, and C. F. H. Tipper, eds., *Comprehensive Chemical Kinetics*, Elsevier, Amsterdam, 1989.

J. A. Kerr, and M. J. Parsonage, *Evaluated Kinetic Data of Gas-Phase Addition Reactions*, Butterworth, London, 1972.

V. N. Kondratiev, *Rate Constants of Gas-Phase Reactions—Reference Book*, Trans. L. J. Holtschlag, R. M. Fristrom, ed., National Bureau of Standards, NITI Service, Springfield, VA, 1972.

NIST, *NIST Chemical Kinetic Database on Diskette*, NIST, Gaithersburg, MD, 1993.

4

SPECIES BALANCES AND DESIGN EQUATIONS

This chapter covers the fundamental concept that leads to the formulation of the design equations of chemical reactors—conservation of mass. Since we deal here with operations involving changes in composition, we carry out mass balances for individual species. Also, since the operations involve chemical reactions, it is convenient to express the amount of a species in terms of moles rather than mass. In general, the species balances are carried out in one of two ways—as *microscopic* balances or *macroscopic* balances. Microscopic species balances, often referred to as the "species continuity equations," are carried out over a differential element and describe what takes place at a given "point" in a reactor. When integrated over the volume of the reactor, they provide the species-based design equations that describe the reactor operation. Macroscopic balances are carried out over the entire reactor, or a large portion of it, and provide the design equations.

The general conservation statement for species j over a stationary system is

$$\begin{Bmatrix} \text{Molar flow} \\ \text{rate of} \\ \text{species } j \\ \text{into system} \end{Bmatrix} + \begin{Bmatrix} \text{Rate moles} \\ \text{of species } j \\ \text{formed} \\ \text{inside system} \end{Bmatrix} = \begin{Bmatrix} \text{Molar flow} \\ \text{rate of} \\ \text{species } j \text{ out} \\ \text{of system} \end{Bmatrix} + \begin{Bmatrix} \text{Rate moles} \\ \text{of species } j \\ \text{accumulate} \\ \text{in system} \end{Bmatrix} \quad (4.0.1)$$

Appendix B provides the derivation of the design equation from the species continuity equation. In Section 4.1, we carry out macroscopic species balances to derive the species-based design equation of any chemical reactor. In Section 4.2,

---

Principles of Chemical Reactor Analysis and Design, Second Edition. By Uzi Mann

we apply the general species-based design equations to reactor configurations commonly used in reactor analysis—ideal batch reactor, continuous stirred-tank reactor (CSTR), and plug-flow reactor. In Section 4.3, the reaction-based design equations are derived for the three ideal reactor configurations. In Section 4.4, the reaction-based design equations are reduced to dimensionless forms.

4.1 MACROSCOPIC SPECIES BALANCES—GENERAL SPECIES-BASED DESIGN EQUATIONS

To perform a macroscopic species balance, we conduct a species balance over an entire reactor. Consider an arbitrary reactor with one inlet and one outlet and volume V_R, shown schematically in Figure 4.1a. Fluid flows through the system, and chemical reactions take place inside the system. We impose no restrictions on the system except the assumption that the chemical reactions are homogeneous (i.e., they take place throughout the system).

The balance equation of species j over the reactor is

$$F_{j_{\text{in}}} + G_j = F_{j_{\text{out}}} + \frac{dN_j}{dt} \tag{4.1.1}$$

where G_j is the rate species j is formed inside the entire reactor by chemical reactions. To obtain useful relations, we should express the generation term, G_j. When different conditions exist at different points in the reactor, the species formation rates, (r_j)'s, vary from point to point in the reactor. Consider a differential volume element, dV, and express the rate species j is generated in that element, dG_j, by

$$dG_j = (r_j)\,dV$$

where (r_j) is defined by Eq. 3.1.1a. Therefore, the rate species j is generated in the entire reactor is

$$G_j = \int_{V_R} dG_j = \int_{V_R} (r_j)\,dV \tag{4.1.2}$$

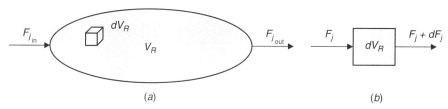

$$(a) \qquad\qquad (b)$$

Figure 4.1 Schematic description of a reacting system: (a) overall system and (b) differential reactor.

where V_R is the volume of the reactor. When the same conditions (temperature and concentrations) exist throughout the system (due to good mixing), the formation rate of species j, (r_j), is the same everywhere in the reactor, and Eq. 4.1.2 becomes

$$G_j = (r_j)V_R$$

Substituting Eq. 4.1.2, Eq. 4.1.1 reduces to

$$F_{j_{in}} - F_{j_{out}} + \int_{V_R} (r_j) \, dV = \frac{dN_j}{dt} \tag{4.1.3}$$

Equation 4.1.3 is the integral form of the general species-based design equation of chemical reactors, written for species j.

The differential form of the general species-based design equation can be derived either by a formal differentiation of Eq. 4.1.3 or by conducting a balance for species j over a differential reactor shown schematically in Figure 4.1b. For a differential reactor, the balance over species j is

$$F_j - (F_j + dF_j) + (r_j) \, dV_R = \frac{dN_j}{dt}$$

which becomes

$$(r_j) \, dV_R = dF_j + \frac{dN_j}{dt} \tag{4.1.4}$$

Equation 4.1.4 is the differential form of the general, species-based design equation for all chemical reactors.

For reactors with single chemical reactions, it has been customary to write the species-based design equation for the limiting reactant A. Hence,

$$F_{A_{in}} = F_{A_{out}} + \int_{V_R} (-r_A) \, dV + \frac{dN_A}{dt} \tag{4.1.5}$$

where $(-r_A)$ is the *depletion* rate of reactant A, defined by Eq. 3.1.1a. Note that since A is a reactant, $(-r_A)$ is a positive quantity. Similarly, from Eq. 4.1.4, we can readily obtain the differential form of the general species-based design equation, written for limiting reactant A:

$$-dF_A = (-r_A) \, dV + \frac{dN_A}{dt} \tag{4.1.6}$$

To obtain useful expressions from the general species-based design equation, we should know the formation rate of species j, (r_j), at any point in the reactor. To express (r_j), the local concentrations of all species as well as the local temperatures should be provided. To obtain these quantities, we should solve the overall continuity equation, the individual species continuity equations, and the energy balance equation. This is a formidable task, and, in most situations, we cannot reduce those equations to useful forms and have to resort to numerical or approximate solution methods that are beyond the scope of this book. Therefore, in most instances, we apply the design equations to simplified reactor configurations (or mathematical models) that approximate the operations of actual reactors. In this text, we discuss the application of the reactor design equations to the following reactor configurations:

- Well-mixed batch reactor (ideal batch reactor)
- Steady, well-mixed continuous reactor (CSTR)
- Steady plug-flow reactor (PFR)
- Certain other special reactor configurations

In Section 4.2, we derive the design equations for the first three ideal reactor models. Other reactor configurations are discussed in Chapter 9.

4.2 SPECIES-BASED DESIGN EQUATIONS OF IDEAL REACTORS

4.2.1 Ideal Batch Reactor

For a batch reactor, $F_{j_{out}} = F_{j_{in}} = 0$, and Eq. 4.1.3 becomes

$$\frac{dN_j}{dt} = \int\limits_{V_R} (r_j)\, dV$$

If the reactor is well mixed, the same conditions (concentration and temperature) exist everywhere; hence, (r_j) is the same throughout the reactor. Thus, Eq. 4.1.3 becomes

$$\frac{dN_j}{dt} = (r_j)V_R(t) \qquad (4.2.1)$$

Equation 4.2.1 is the species-based design equation of an *ideal* batch reactor, written for species j. To obtain the operating time, we separate the variables and integrate Eq. 4.2.1:

$$t = \int\limits_{N_j(0)}^{N_j(t)} \frac{dN_j}{(r_j)V_R} \qquad (4.2.2)$$

Equation 4.2.2 is the integral form of the species-based design equation for an ideal batch reactor, written for species j. It provides a relation between the operating time, t, the amount of the species in the reactor, $N_j(t)$ and $N_j(0)$, the species formation rate, (r_j), and the reactor volume, V_R. Note that when the reactor volume does not change during the operation, Eq. 4.2.2 reduces to

$$t = \int\limits_{C_j(0)}^{C_j(t)} \frac{dC_j}{(r_j)} \qquad (4.2.2a)$$

For reactors with single chemical reactions, it has been customary to write the species-based design equation for the limiting reactant A, and Eq. 4.2.1 reduces to

$$-\frac{dN_A}{dt} = (-r_A)V_R(t) \tag{4.2.3}$$

Using the conversion definition (Eq. 2.6.1a),

$$-\frac{dN_A}{dt} = N_A(0)\frac{df_A}{dt}$$

and Eq. 4.2.1 becomes

$$N_A(0)\frac{df_A}{dt} = (-r_A)V_R(t) \tag{4.2.4}$$

To obtain the integral form of the design, we separate the variables and integrate Eq. 4.2.4:

$$t = N_A(0) \int_0^{f_A(t)} \frac{df_A}{(-r_A)V_R} \tag{4.2.5}$$

When the reactor volume does not change during the operation, $V_R = V_R(0)$, Eq. 4.2.5 reduces to

$$t = C_A(0) \int_0^{f_A(t)} \frac{df_A}{-r_A} \tag{4.2.5a}$$

4.2.2 Continuous Stirred-Tank Reactor (CSTR)

A CSTR is a reactor model based on two assumptions: (i) steady-state operation and (ii) the same conditions exist everywhere inside the reactor (due to good mixing). For steady operations, the accumulation term in the design equation vanishes. Since the same conditions exist everywhere, the rate (r_j) is the same throughout the reactor and is equal to the rate at the reactor effluent, $(r_j)_{out}$. Hence, the general, species-based design equation Eq. (4.1.3) reduces to

$$F_{j_{out}} - F_{j_{in}} = (r_j)_{out}V_R \tag{4.2.6}$$

We can rearrange Eq. 4.2.6 as

$$V_R = \frac{F_{j_{out}} - F_{j_{in}}}{(r_j)_{out}} \tag{4.2.7}$$

Equation 4.2.7 is the species-based design equations for a CSTR, written for species j. It provides a relation between the species flow rate at the inlet and outlet of the reactor, $F_{j_{in}}$ and $F_{j_{out}}$, the species formation rate (r_j), and the reactor volume, V_R.

For a CSTR with single chemical reactions, the species-based design equation is usually written for the limiting reactant A, and Eq. 4.2.7 reduces to

$$V_R = \frac{F_{A_{\text{in}}} - F_{A_{\text{out}}}}{(-r_A)_{\text{out}}} \qquad (4.2.8)$$

Using the conversion definition (Eq. 2.6.1b) and applying it to the reference stream, F_{A_0}:

$$F_{A_{\text{in}}} = F_{A_0}(1 - f_{A_{\text{in}}})$$

$$F_{A_{\text{out}}} = F_{A_0}(1 - f_{A_{\text{out}}})$$

and Eq. 4.2.8 becomes

$$\frac{V_R}{F_{A_0}} = \frac{f_{A_{\text{out}}} - f_{A_{\text{in}}}}{(-r_A)_{\text{out}}} \qquad (4.2.9)$$

Equation 4.2.9 is the species-based design equation of a CSTR, expressed in terms of the conversion of reactant A.

4.2.3   Plug-Flow Reactor (PFR)

A PFR is a reactor model based on two assumptions: (i) steady-state operation and (ii) a flat velocity profile (plug flow) with the same conditions existing at any given cross-section area (no concentration or temperature gradients in the direction perpendicular to the flow). Since the species compositions change along the reactor and the species formation rate, (r_j), varies along the reactor, we consider a differential reactor of volume dV_R. For steady operations, the accumulation term vanishes, and Eq. 4.1.4 reduces to

$$dF_j = (r_j)\, dV_R \qquad (4.2.10)$$

Equation 4.2.10 is the species-based differential design equations for a PFR, written for species j. To obtain the reactor volume of a PFR, we integrate Eq. 4.2.10,

$$V_R = \int_{F_{j_{\text{in}}}}^{F_{j_{\text{out}}}} \frac{dF_j}{(r_j)} \qquad (4.2.11)$$

Equation 4.2.11 provides a relation among the species flow rate at the inlet and outlet of the reactor, $F_{j_{\text{in}}}$ and $F_{j_{\text{out}}}$, the species formation rate, (r_j), and the volume of the reactor, V_R, for a plug-flow reactor.

For PFRs with single chemical reactions, it has been customary to write the species-based design equation for the limiting reactant A, and Eq. 4.2.10 reduces to

$$dV_R = -\frac{dF_A}{(-r_A)} \tag{4.2.12}$$

Differentiating the conversion definition (Eq. 2.6.1b),

$$dF_A = -F_{A_0}\, df_A$$

and Eq. 4.2.12 reduces to

$$dV_R = F_{A_0}\frac{df_A}{(-r_A)} \tag{4.2.13}$$

Equation 4.2.13 is the species-based differential design equation of a plug-flow reactor, expressed in terms of the conversion of reactant A. To obtain the integral form of the design equation, we separate the variables and integrate Eq. 4.2.13:

$$\frac{V_R}{F_{A_0}} = \int_{f_{A_{in}}}^{f_{A_{out}}} \frac{df_A}{-r_A} \tag{4.2.14}$$

Equation 4.2.14 is the species-based integral design equation of a plug-flow reactor, expressed in terms of the conversion of reactant A.

4.3   REACTION-BASED DESIGN EQUATIONS

To analyze and design chemical reactors more effectively and to obtain insight into the operation, we adopt reaction-based design formulation. In this section, we derive the reaction-based design equations for the three ideal reactor models. Reaction-based design equations of other reactor configurations are derived in Chapter 9.

4.3.1   Ideal Batch Reactor

For an ideal batch reactor, the species-based design equation, written for species j, is given by Eq. 4.2.1,

$$\frac{dN_j}{dt} = (r_j)V_R(t) \tag{4.3.1}$$

Using stoichiometric relation Eq. 2.3.3, $N_j(t)$ is expressed in terms of the extents of the independent reactions by

$$N_j(t) = N_j(0) + \sum_m^{n_I} (s_j)_m X_m(t)$$

which, upon differentiation with time, becomes

$$\frac{dN_j}{dt} = \sum_{m}^{n_I} (s_j)_m \frac{dX_m}{dt} \tag{4.3.2}$$

Using Eq. 3.2.6 to express the formation rate of species j in terms of the rates of the chemical reactions, we obtain

$$(r_j) = \sum_{i}^{n_R} (s_j)_i r_i \tag{4.3.3}$$

Substituting Eqs. 4.3.2 and 4.3.3 into Eq. 4.3.1, the design equation becomes

$$\sum_{m}^{n_I} (s_j)_m \frac{dX_m}{dt} = \sum_{i}^{n_R} (s_j)_i r_i V_R(t) \tag{4.3.4}$$

Note that the summation on the left-hand side of Eq. 4.3.4 is over *independent* reactions only, whereas the summation on the right-hand side is over *all* reactions that take place in the reactor. Using Eq. 2.4.8, we write the right-hand side of Eq. 4.3.4 as a sum of dependent reactions and independent reactions,

$$\sum_{m}^{n_I} (s_j)_m \frac{dX_m}{dt} = \left(\sum_{m}^{n_I} (s_j)_m r_m + \sum_{k}^{n_D} (s_j)_k r_k \right) V_R(t) \tag{4.3.5}$$

The stoichiometric coefficient of species j in the kth-dependent reaction, $(s_j)_k$, relates to the stoichiometric coefficients of species j in the independent reactions, $(s_j)_m$, by Eq. 2.4.9,

$$\sum_{m}^{n_I} \alpha_{km}(s_j)_m = (s_j)_k$$

where α_{km} is the multiplier of the mth-independent reaction to obtain the kth-dependent reaction. Substituting this relation into Eq. 4.3.5, we obtain

$$\sum_{m}^{n_I} (s_j)_m \frac{dX_m}{dt} = \left[\sum_{m}^{n_I} (s_j)_m r_m + \sum_{k}^{n_D} \sum_{m}^{n_I} \alpha_{km}(s_j)_m r_k \right] V_R(t) \tag{4.3.6}$$

Since all multipliers α_{km} are constants, we can switch the order of the two summations in the second term in the parentheses and obtain

$$\sum_{m}^{n_I} (s_j)_m \frac{dX_m}{dt} = \left[\sum_{m}^{n_I} (s_j)_m r_m + \sum_{m}^{n_I} (s_j)_m \sum_{k}^{n_D} \alpha_{km} r_k \right] V_R(t) \tag{4.3.7}$$

Since the coefficients and summations of all the terms are identical, Eq. 4.3.7 reduces to

$$\frac{dX_m}{dt} = \left(r_m + \sum_{k}^{n_D} \alpha_{km} r_k \right) V_R(t) \tag{4.3.8}$$

Equation 4.3.8 is the reaction-based, differential design equation of an ideal batch reactor, written for the *m*th-independent reaction. As will be discussed below, to describe the operation of a reactor with multiple chemical reactions, we have to write Eq. 4.3.8 for each of the independent reactions. Note that the reaction-based design equation is invariant of the specific species used in the derivation.

For an ideal batch reactor with a *single* chemical reaction, Eq. 4.3.8 reduces to

$$\frac{dX}{dt} = rV_R(t) \tag{4.3.9}$$

When a single chemical reaction takes place, the extent of the reaction is proportional to the conversion of reactant A, f_A, given by Eq. 2.6.2:

$$f_A(t) = -\frac{s_A}{N_A(0)}X(t)$$

where $N_A(0)$ is the mole content of reactant A initially in the reactor. Hence, we can readily express the design equation in terms of the conversion of the limiting reactant. We differentiate Eq. 2.6.2 and obtain

$$\frac{dX}{dt} = -\frac{N_A(0)}{s_A}\frac{df_A}{dt}$$

Also, using Eq. 3.2.5, the depletion rate of reactant A, $(-r_A)$, relates to the reaction rate by

$$-r_A = -s_A r \tag{4.3.10}$$

Substituting these two relations into Eq. 4.3.9, we obtain

$$N_A(0)\frac{df_A}{dt} = (-r_A)V_R(t)$$

which is identical to Eq. 4.2.4, the differential, species-based design equations of an ideal batch reactor with a single chemical reaction, expressed in terms of the conversion of reactant A.

4.3.2   Plug-Flow Reactor

For a differential steady-flow reactor, the species-based design equation, written for species *j*, is given by Eq. 4.2.10:

$$dF_j = (r_j)\,dV_R \tag{4.3.11}$$

We use Eq. 2.3.11 to express the local molar flow rate of species *j*, F_j, in terms of extents of the independent reactions:

$$F_j = F_{j_0} + \sum_m^{n_I} (s_j)_m \dot{X}_m \tag{4.3.12}$$

Differentiating Eq. 4.3.12,

$$dF_j = \sum_m^{n_I} (s_j)_m \, d\dot{X}_m$$

and substituting this and Eq. 4.3.3 into Eq. 4.3.11, the design equation becomes

$$\sum_m^{n_I} (s_j)_m \frac{d\dot{X}_m}{dV_R} = \sum_i^{n_R} (s_j)_i r_i \qquad (4.3.13)$$

Note that here too, the summation on the left-hand side is over the independent reactions only, whereas the summation on the right-hand side is over all the reactions that take place in the reactor. We follow the same procedure as in the case of the ideal batch reactor. We first write the summation on the right as two sums over dependent reactions and independent reactions. Next, we express the stoichiometric coefficient of species j in the kth-dependent reaction, $(s_j)_k$, in terms of the stoichiometric coefficients of species j in the independent reactions, $(s_j)_m$, using Eq. 2.4.9, and then switch the order of the summations to obtain

$$\frac{d\dot{X}_m}{dV_R} = r_m + \sum_k^{n_D} \alpha_{km} r_k \qquad (4.3.14)$$

Equation 4.3.14 is the reaction-based, differential design equation for steady-flow reactors, written for the mth-independent reaction. As will be discussed below, to describe the operation of the reactor with multiple reactions, we have to write Eq. 4.3.14 for each of the independent reactions.

For steady-flow reactors with a single chemical reaction, Eq. 4.3.14 reduces to

$$\frac{d\dot{X}}{dV_R} = r \qquad (4.3.15)$$

When a single chemical reaction takes place, we can readily express the design equation in terms of the conversion of the limiting reactant using Eq. 2.6.5:

$$f_A = -\frac{s_A}{F_{A_{iu}}} \dot{X}$$

where the molar flow rate of reactant A is in the reactor inlet. We differentiate this relation,

$$d\dot{X} = -\frac{F_{A_{iu}}}{s_A} df_A$$

using Eq. 4.3.10, and substitute them into Eq. 4.3.15 to obtain

$$\frac{df_A}{dV_R} = \frac{(-r_A)}{F_{A_{iu}}} \qquad (4.3.16)$$

which is identical to Eq. 4.2.13, the differential, species-based design equation of a plug-flow reactor with a single chemical reaction, expressed in terms of the conversion of reactant A.

4.3.3   Continuous Stirred-Tank Reactor (CSTR)

For steady, continuous stirred-tank reactors, the species-based design equation is given by Eq. 4.2.7:

$$F_{j_{\text{out}}} - F_{j_{\text{in}}} = (r_j)_{\text{out}} V_R \tag{4.3.17}$$

Using Eq. 4.3.12, the species molar flow rate at the inlet and the outlet are

$$F_{j_{\text{in}}} = F_{j0} + \sum_{m}^{n_I} (s_j)_m \dot{X}_{m_{\text{in}}} \tag{4.3.18}$$

$$F_{j_{\text{out}}} = F_{j0} + \sum_{m}^{n_I} (s_j)_m \dot{X}_{m_{\text{out}}} \tag{4.3.19}$$

and

$$F_{j_{\text{out}}} - F_{j_{\text{in}}} = \sum_{m}^{n_I} (s_j)_m \left(\dot{X}_{m_{\text{out}}} - \dot{X}_{m_{\text{in}}} \right) \tag{4.3.20}$$

where F_{j0} is the molar flow rate of species j in a reference stream, and $\dot{X}_{m_{\text{in}}}$ and $\dot{X}_{m_{\text{out}}}$ are, respectively, the extents per time of the mth-independent reaction between the reference stream and the reactor inlet and outlet. Substituting Eq. 4.3.20 into Eq. 4.3.17, the design equation becomes

$$\sum_{m}^{n_I} (s_j)_m \left(\dot{X}_{m_{\text{out}}} - \dot{X}_{m_{\text{in}}} \right) = \sum_{i}^{n_R} (s_j)_i r_{i_{\text{out}}} V_R \tag{4.3.21}$$

Here too, the summation on the left-hand side is over the independent reactions only, whereas the summation on the right-hand side is over all the reactions that take place in the reactor. We follow the same procedure as in the case of the ideal batch reactor. We first write the summation on the right as two sums over dependent reactions and independent reactions. Next, we express the stoichiometric coefficient of species j in the kth-dependent reaction, $(s_j)_k$, in terms of the stoichiometric coefficients of species j in the independent reactions, $(s_j)_m$, using Eq. 2.4.9, and then switch the order of the summations to obtain

$$\dot{X}_{m_{\text{out}}} - \dot{X}_{m_{\text{in}}} = \left(r_{m_{\text{out}}} + \sum_{k}^{n_D} \alpha_{km} r_{k_{\text{out}}} \right) V_R \tag{4.3.22}$$

Equation 4.3.22 is the reaction-based design equation for CSTRs, written for the mth-independent reaction. To describe the operation of the reactor with multiple reactions, we have to write Eq. 4.3.22 for each independent reaction.

For a CSTR with a single chemical reaction, Eq. 4.3.22 reduces to

$$\dot{X}_{\text{out}} - \dot{X}_{\text{in}} = rV_R \qquad (4.3.23)$$

For a single chemical reaction, the extent of the reaction is proportional to the conversion of reactant A, f_A, given by Eq. 2.6.5. Hence,

$$\dot{X}_{\text{in}} = -\frac{F_{A_0}}{s_A} f_{A_{\text{in}}} \qquad \dot{X}_{\text{out}} = -\frac{F_{A_0}}{s_A} f_{A_{\text{out}}}$$

and, using Eq. 4.3.10, the depletion rate of reactant A is $(-r_A) = -s_A r$. Substituting these into Eq. 4.3.23, we obtain

$$V_R = F_{A_0} \frac{f_{A_{\text{out}}} - f_{A_{\text{in}}}}{(-r_A)_{\text{out}}} \qquad (4.3.24)$$

which is identical to Eq. 4.2.9, the species-based design equation of a CSTR with a single chemical reaction, expressed in terms of the conversion of reactant A.

4.3.4 Formulation Procedure

The reaction-based design equations (Eqs. 4.3.8, 4.4.14, and 4.3.22) are written for the mth-independent reaction. Since all state variables of the reactor (composition, temperature, enthalpy, etc.) depend on the extents of the independent reactions, to design a chemical reactor with multiple reactions, we have to write a design equation for each of the independent chemical reactions. Since there are always more chemical species than independent reactions, by formulating the design in terms of reaction-based design equations, we express the design by the *smallest* number of design equations.

As indicated in Chapter 2, we can select different sets of independent reactions. The question then arises as to what is the most appropriate set of independent reactions for the design formulation. Since the design equations include the rates of all chemical reactions that actually take place in the reactor, by selecting a set of independent reactions among them, we minimize the number of terms in each design equation. Hence, we adopt the following *heuristic rule*:

> **Select a set of independent reactions among the chemical reactions whose rate expressions are provided.**
>
> Or
>
> **Do not select a set of independent reactions that includes a chemical reactions whose rate expressions are not provided.**

By adopting this heuristic rule, the design equations consist of the *least* number of rate terms. Considering that each of them is a function of temperature, and in many instances, a stiff function, we formulate the design in terms of the *most robust* set of algebraic or differential equations for numerical solutions (see Example 4.3).

4.4 DIMENSIONLESS DESIGN EQUATIONS AND OPERATING CURVES

The reaction-based design equations derived in the previous section are expressed in terms of extensive quantities such as reaction extents, reactor volume, molar flow rates, and the like. To describe the generic behavior of chemical reactors, we would like to express the design equations in terms of intensive, dimensionless variables. This is done in two steps:

- Selecting a convenient framework (or "basis") for the calculation
- Selecting a characteristic time constant

Below, we reduce the design equations of the three ideal reactors to dimensionless forms. Dimensionless design equations for other reactor configurations are derived in Chapter 9.

To reduce the design equation of an ideal batch reactor, Eq. 4.3.8, to dimensionless form, we first select a *reference state* of the reactor (usually, the initial state) and use the dimensionless extent, Z_m, of the mth-independent reaction, defined by Eq. 2.7.1:

$$Z_m \equiv \frac{X_m}{(N_{tot})_0} \qquad (4.4.1)$$

where $(N_{tot})_0$ is the total number of moles of the reference state, which is usually the initial state of the reactor. For *liquid-phase* reactions, we usually define a system that consists of all the species that participate in the chemical reactions and leave out inert species (e.g., solvents). Hence, $(N_{tot})_0$ is the sum of the reactants and products that are initially in the reactor. For *gas-phase* reactions, we define a system that consists of all species in the reactor, including inert species. Hence, $(N_{tot})_0$ is the sum of *all* species that are initially in the reactor. We also take the initial reactor volume as the reference volume and define the reference concentration, C_0, by

$$C_0 \equiv \frac{(N_{tot})_0}{V_{R_0}} \qquad (4.4.2)$$

where V_{R_0} is the volume of the reference state. Next, we define the dimensionless operating time by

$$\tau \equiv \frac{\text{Operating time}}{\text{Characteristic reaction time}} = \frac{t}{t_{cr}} \qquad (4.4.3)$$

where t_{cr} is the characteristic reaction time, defined by Eq. 3.5.1. To reduce the reaction-based design equation to dimensionless form, differentiate Eqs. 4.4.1 and 4.4.2:

$$dX_m = (N_{tot})_0 \, dZ_m$$

$$dt = t_{cr} \, d\tau$$

and the volume is expressed

$$V_R = V_{R_0} \left(\frac{V_R}{V_{R_0}} \right)$$

Substituting these into Eq. 4.3.8,

$$\frac{dZ_m}{d\tau} = \left(r_m + \sum_{k}^{n_D} \alpha_{km} r_k \right) \left(\frac{V_R}{V_{R_0}} \right) \left(\frac{t_{cr}}{C_0} \right) \tag{4.4.4}$$

Equation 4.4.4 is the dimensionless, reaction-based design equation of an ideal batch reactor, written for the mth-independent reaction. The factor (t_{cr}/C_0) is a scaling factor that converts the design equation to dimensionless form. Its physical significance is discussed below (Eqs. 4.4.13–4.4.15).

To reduce the design equations of flow reactors to dimensionless forms, we select a convenient *reference stream* as a basis for the calculation. In most cases, it is convenient to select the inlet stream into the reactor as the reference stream, but, in some cases, it is more convenient to select another stream, even an imaginary stream. There is no restriction on the selection of the reference stream, except that we should be able to relate the reactor composition to it in terms of the reaction extents. Once we select the reference stream, we use the dimensionless extent, Z_m, of the mth-independent reaction, defined by Eq. 2.7.2,

$$Z_m = \frac{\dot{X}_m}{(F_{tot})_0} \tag{4.4.5}$$

where $(F_{tot})_0$ is the total molar flow rate of a reference stream, usually, the inlet stream. For liquid-phase reactions, we usually define a reference stream that consists only of species that participate in the chemical reactions and leave out inert species (e.g., solvents). For gas-phase reactions, we define a system that consists of all species in the reactor, including inert species.

The difficulty in analyzing flow reactors is that the transformations of chemical reactions take place over space (volume), whereas the reaction operation is a rate process. To relate the reactor volume to a time domain, we select a reference volumetric flow rate for the reference stream, v_0, and define the reactor *space time*, t_{sp}, by

$$t_{sp} \equiv \frac{\text{Reactor volume}}{\text{Volumetric flow rate of a reference stream}} = \frac{V_R}{v_0} \tag{4.4.6}$$

where v_0 is a conveniently selected volumetric flow rate. Note that the space time is an artificial quantity that depends on the selection of the reference stream and v_0. Also note that t_{sp} is not the residence time of the fluid in the reactor. The residence time is given by

$$t_{res} = \int_{V_R} \frac{dV}{v}$$

where v is the local volumetric flow rate. The residence time is equal to the space time only when the volumetric flow rate through the reactor is constant and equal to v_0. Once v_0 is selected, the reference concentration, C_0, is defined by

$$C_0 \equiv \frac{(F_{\text{tot}})_0}{v_0} \tag{4.4.7}$$

For flow reactors, the dimensionless space time is defined by

$$\tau \equiv \frac{\text{Reactor space time}}{\text{Characteristic reaction time}} = \frac{V_R}{v_0 t_{\text{cr}}} \tag{4.4.8}$$

where t_{cr} is the characteristic reaction time, defined by Eq. 3.5.1.

To reduce the reaction-based design equation of a plug-flow reactor to dimensionless form, we differentiate Eqs. 4.4.5 and 4.4.8,

$$d\dot{X}_m = (F_{\text{tot}})_0 \, dZ_m \tag{4.4.9}$$

$$dV_R = t_{\text{cr}} v_0 \, d\tau \tag{4.4.10}$$

and substitute these into Eq. 4.3.14. Using Eq. 4.4.7,

$$\frac{dZ_m}{d\tau} = \left(r_m + \sum_k^{n_D} \alpha_{km} r_k \right) \left(\frac{t_{\text{cr}}}{C_0} \right) \tag{4.4.11}$$

Equation 4.4.11 is the dimensionless, reaction-based design equation of a plug-flow reactor, written for the mth-independent reaction. To reduce the reaction-based design equation of a CSTR to dimensionless form, we substitute Eqs. 4.4.5 and 4.4.8 into Eq. 4.3.21 and, using Eq. 4.4.7,

$$Z_{m_{\text{out}}} - Z_{m_{\text{in}}} = \left(r_{m_{\text{out}}} + \sum_k^{n_D} \alpha_{km} r_{k_{\text{out}}} \right) \tau \left(\frac{t_{\text{cr}}}{C_0} \right) \tag{4.4.12}$$

Equation 4.4.12 is the dimensionless, reaction-based design equation of a CSTR, written for the mth-independent reaction. The factor (t_{cr}/C_0) in Eqs. 4.4.11 and 4.4.12 is a dimensional scaling factor that converts the design equations to dimensionless forms.

To gain insight into the structure of the dimensionless reaction-based design equations, recall the definition of the characteristic reaction time, Eq. 3.5.1, $t_{\text{cr}} = C_0/r_0$. It follows that the scaling factor is $(t_{\text{cr}}/C_0) = 1/r_0$, where r_0 is the reference rate of a selected chemical reaction. Substituting this relation into Eqs. 4.4.4, 4.4.11, and 4.4.12, the design equations reduce, respectively, to

$$\frac{dZ_m}{d\tau} = \left[\left(\frac{r_m}{r_0} \right) + \sum_k^{n_D} \alpha_{km} \left(\frac{r_k}{r_0} \right) \right] \left(\frac{V_R}{V_{R_0}} \right) \tag{4.4.13}$$

$$\frac{dZ_m}{d\tau} = \left(\frac{r_m}{r_0}\right) + \sum_{k}^{n_D} \alpha_{km}\left(\frac{r_k}{r_0}\right) \tag{4.4.14}$$

$$Z_{m_{\text{out}}} - Z_{m_{\text{in}}} = \left[\left(\frac{r_{m_{\text{out}}}}{r_0}\right) + \sum_{k}^{n_D} \alpha_{km}\left(\frac{r_{k_{\text{out}}}}{r_0}\right)\right]\tau \tag{4.4.15}$$

Hence, the dimensionless design equations are expressed in terms of *relative* reaction rates, (r_m/r_0)'s and (r_k/r_0)'s, with respect to a conveniently selected reaction rate.

To solve the dimensionless, reaction-based design equations, we have to express the rates of the chemical reactions, r_m's and r_k's, in terms of the dimensionless extents of the independent reactions. Since the reaction rates depend on species concentrations, we have to express the species concentrations in terms of the extents of the independent reactions. These relationships depend on the reactor configuration and the way the reactants are fed, and can be readily derived by using the stoichiometric relations derived in Chapter 2. This procedure will be discussed in detail in Chapters 5–9, where we cover the design of different chemical reactor configurations. The reaction rates also depend on the temperature, θ, whose variation is obtained by solving simultaneously the energy balance equation. By substituting the concentration relations into the rate expressions and the latter into the design equations, we obtain equations in τ, Z_m's, and θ.

For ideal batch and plug-flow reactors, we obtain a set of nonlinear, first-order, differential equations of the form

$$\frac{dZ_{m_1}}{d\tau} = G_1(\tau, Z_{m_1}, Z_{m_2}, \ldots, Z_{n_I}, \theta)$$

$$\frac{dZ_{m_2}}{d\tau} = G_2(\tau, Z_{m_1}, Z_{m_2}, \ldots, Z_{n_I}, \theta) \tag{4.4.16}$$

$$\frac{dZ_{m_I}}{d\tau} = G_{m_I}(\tau, Z_{m_1}, Z_{m_2}, \ldots, Z_{n_I}, \theta)$$

where τ, the dimensionless operating time (or space time), is the independent variable. We have here a set of n_I equations with n_{I+1} dependent variables. The additional equation is the energy balance equation (derived in Chapter 5) that reduces to the general form

$$\frac{d\theta}{d\tau} = G_{n_{il+1}}(\tau, Z_{m_1}, Z_{m_2}, \ldots, Z_{n_I}, \theta) \tag{4.4.17}$$

We solve these equations for Z_m's and θ as functions of τ, subject to the initial condition that at $\tau = 0$, the dimensionless extents and the dimensionless temperature are specified. For isothermal operations, Eq. 4.4.17 reduces to $d\theta/d\tau = 0$; thus, only the design equations should be solved.

For ideal stirred-tank reactors, the design formulation is a set of nonlinear, homogeneous, algebraic equations of the form

$$G_1(\tau, Z_{m_1}, Z_{m_2}, \ldots, Z_{n_I}, \theta) = 0$$

$$G_2(\tau, Z_{m_1}, Z_{m_2}, \ldots, Z_{n_I}, \theta) = 0 \qquad (4.4.18)$$

$$G_{n_I}(\tau, Z_{m_1}, Z_{m_2}, \ldots, Z_{n_I}, \theta) = 0$$

and the energy balance equation reduces to the form

$$G_{n_{i+1}}(\tau, Z_{m_1}, Z_{m_2}, \ldots, Z_{n_I}, \theta) = 0 \qquad (4.4.19)$$

We solve Eqs. 4.4.18 and 4.4.19 simultaneously for $Z_{m_1}, \ldots, Z_{n_I}$ and θ for different values of dimensionless reactor space times, τ.

The solutions of the design equations (Z_m's versus τ) and the energy balance equation (θ versus τ) provide, respectively, the dimensionless *reaction operating curves* and the dimensionless *temperature curve* of the reactor operation. The dimensionless reaction operating curves describe the progress of the independent chemical reactions without regard to a specific operating time or reactor volume. The dimensionless temperature curve describes the temperature variation during the reactor operation. Using stoichiometric relations derived in Chapter 2, we can readily use the reaction operating curves to obtain the dimensionless operating curves for each species in the reactor, $N_j(\tau)/(N_{tot})_0$ or $F_j/(F_{tot})_0$, versus τ.

The formulation of the reaction-based design equations is illustrated in the three examples below. Example 4.1 shows how a case that is conventionally formulated in terms of 4 species-based design equations with 16 terms is formulated here in terms of 3 reaction-based design equations with 6 terms. Example 4.3 illustrates how, by adopting the heuristic rule on selecting a set of independent reactions, we formulate the design by the most robust set of equations.

Example 4.1 The following three reversible chemical reactions represent gas-phase cracking of hydrocarbons:

Reactions 1 & 2: $A \; \underset{\longrightarrow}{\overset{\longleftarrow}{}} \; 2B$

Reactions 3 & 4: $A + B \; \underset{\longrightarrow}{\overset{\longleftarrow}{}} \; C$

Reactions 5 & 6: $A + C \; \underset{\longrightarrow}{\overset{\longleftarrow}{}} \; D$

Species B is the desired product. Formulate the reaction-based design equations, expressed in terms of the reaction rates, for:

a. Ideal batch reactor

b. Plug-flow reactor

c. CSTR

Solution First, determine the number of independent reactions and select a set of independent reactions. The given chemical reactions are represented by the following six reactions:

$$
\begin{array}{lll}
\text{Reaction 1:} & \quad A & \longrightarrow \quad 2B \\
\text{Reaction 2:} & \quad 2B & \longrightarrow \quad A \\
\text{Reaction 3:} & \quad A + B & \longrightarrow \quad C \\
\text{Reaction 4:} & \quad C & \longrightarrow \quad A + B \\
\text{Reaction 5:} & \quad A + C & \longrightarrow \quad D \\
\text{Reaction 6:} & \quad D & \longrightarrow \quad A + C
\end{array}
$$

We can construct the matrix of stoichiometric coefficients and reduce it to a diagonal form to determine the number of independent reactions. However, in this case, we have three reversible reactions, and, since each of the three forward reactions has a species that does not appear in the other two, we have three independent reactions and three dependent reactions. We select the three forward reactions as the set of independent reactions. Hence, the indices of the independent reactions are $m = 1, 3, 5$, and we describe the reactor operation in terms of their dimensionless extents, Z_1, Z_3, and Z_5. The indices of the dependent reactions are $k = 2, 4, 6$. Since this set of independent reactions consists of chemical reactions whose rate expressions are known, the heuristic rule on selecting independent reactions is satisfied. The stoichiometric coefficients of the selected three independent reactions are

$$
\begin{array}{lllll}
s_{A_1} = -1 & s_{B_1} = 2 & s_{C_1} = 0 & s_{D_1} = 0 & \Delta_1 = 1 \\
s_{A_3} = -1 & s_{B_3} = -1 & s_{C_3} = 1 & s_{D_3} = 0 & \Delta_3 = -1 \\
s_{A_5} = -1 & s_{B_5} = 0 & s_{C_5} = -1 & s_{D_5} = 1 & \Delta_5 = -1
\end{array}
$$

To determine the α_{km} multipliers for each of the dependent reactions, we use stoichiometric relation Eq. 2.4.9. The values are:

$$
\begin{array}{lll}
\alpha_{21} = -1 & \alpha_{23} = 0 & \alpha_{25} = 0 \\
\alpha_{41} = 0 & \alpha_{43} = -1 & \alpha_{45} = 0 \\
\alpha_{61} = 0 & \alpha_{63} = 0 & \alpha_{65} = -1
\end{array}
\tag{a}
$$

a. For an ideal batch reactor, we write Eq. 4.4.4 for each of the three independent reactions. Hence, for this case, the design equations are

$$
\frac{dZ_1}{d\tau} = (r_1 - r_2) \left(\frac{V_R}{V_{R_0}} \right) \left(\frac{t_{cr}}{C_0} \right)
\tag{b}
$$

$$\frac{dZ_3}{d\tau} = (r_3 - r_4)\left(\frac{V_R}{V_{R_0}}\right)\left(\frac{t_{cr}}{C_0}\right) \tag{c}$$

$$\frac{dZ_5}{d\tau} = (r_5 - r_6)\left(\frac{V_R}{V_{R_0}}\right)\left(\frac{t_{cr}}{C_0}\right) \tag{d}$$

Note that $(r_1 - r_2)$, $(r_3 - r_4)$, and $(r_5 - r_6)$ are the *net* forward rates of the three independent reactions. To solve these equations, we have to select a reference state, define reference concentration, C_0, and characteristic reaction time t_{cr}, and then express the individual reaction rates and $V_R/V_R(0)$ in terms of τ, Z_1, Z_3, and Z_5.

b. For a steady plug-flow reactor, we write Eq. 4.4.11 for each of the three independent reactions. Hence, for this case, the design equations are

$$\frac{dZ_1}{d\tau} = (r_1 - r_2)\left(\frac{t_{cr}}{C_0}\right) \tag{e}$$

$$\frac{dZ_3}{d\tau} = (r_3 - r_4)\left(\frac{t_{cr}}{C_0}\right) \tag{f}$$

$$\frac{dZ_5}{d\tau} = (r_5 - r_6)\left(\frac{t_{cr}}{C_0}\right) \tag{g}$$

To solve these equations, we have to select a reference stream, define reference concentration, C_0, and characteristic reaction time t_{cr}, and then express the individual reaction rates in terms of τ, Z_1, Z_3, and Z_5.

c. For a CSTR, we write Eq. 4.4.12 for each of the three independent reactions. Hence, for this case, the design equations are

$$Z_{1_{out}} - Z_{1_{in}} - (r_{1_{out}} - r_{2_{out}})\tau\left(\frac{t_{cr}}{C_0}\right) = 0 \tag{h}$$

$$Z_{3_{out}} - Z_{3_{in}} - (r_{3_{out}} - r_{4_{out}})\tau\left(\frac{t_{cr}}{C_0}\right) = 0 \tag{i}$$

$$Z_{5_{out}} - Z_{5_{in}} - (r_{5_{out}} - r_{6_{out}})\tau\left(\frac{t_{cr}}{C_0}\right) = 0 \tag{j}$$

To solve these design equations, we have to express the individual reaction rates in terms of $Z_{1_{out}}$, $Z_{3_{out}}$, $Z_{5_{out}}$, and then, for different values of dimensionless space time τ, we can solve (h), (i), and (j) simultaneously for $Z_{1_{out}}$, $Z_{3_{out}}$, $Z_{5_{out}}$.

Example 4.2   Acrolein is being produced by catalytic oxidation of propylene on a bismuth molybdate catalyst. The following reactions are taking place in the reactor:

$$
\begin{aligned}
\text{Reaction 1:} \quad & C_3H_6 + O_2 \longrightarrow C_3H_4O + H_2O \\
\text{Reaction 2:} \quad & C_3H_4O + 3.5O_2 \longrightarrow 3CO_2 + 2H_2O \\
\text{Reaction 3:} \quad & C_3H_6 + 4.5O_2 \longrightarrow 3CO_2 + 3H_2O
\end{aligned}
$$

Adams et al. (*J. Catalysis* **3**, 379, 1964) investigated these reactions and expressed the rate of each as second order (first order with respect to each reactant). Formulate the dimensionless, reaction-based design equations for an ideal batch reactor, plug-flow reactor, and a CSTR.

Solution   The purpose of this example is to illustrate the design formulations for chemical reactions where the dependency between dependent reactions and independent reactions is not due to reversible reactions. First, we determine the number of independent reactions and select a set of independent reactions. We construct a matrix of stoichiometric coefficients for the given reactions:

$$
\begin{array}{ccccc}
C_3H_6 & O_2 & C_3H_4O & H_2O & CO_2
\end{array}
$$

$$
\begin{bmatrix}
-1 & -1 & 1 & 1 & 0 \\
0 & -3.5 & -1 & 2 & 3 \\
-1 & -4.5 & 0 & 3 & 3
\end{bmatrix} \tag{a}
$$

Applying Gaussian elimination procedure, Matrix (a) reduces to

$$
\begin{bmatrix}
-1 & -1 & 1 & 1 & 0 \\
0 & -3.5 & -1 & 2 & 3 \\
0 & 0 & 0 & 0 & 0
\end{bmatrix} \tag{b}
$$

Matrix (b) is a reduced matrix since all the elements below the diagonal are zeros, and, because it has two nonzero rows, we have here two independent chemical reactions. We select the first two reactions as a set of independent reactions; hence, for this case, $m = 1, 2$, and $k = 3$. To determine the multipliers α_{km}'s of the dependent reaction (Reaction 3) and the two independent reactions (Reactions 1 and 2), we write Eq. 2.4.9 for propylene and CO_2:

$$
\alpha_{31}(-1) + \alpha_{32}(0) = -1 \tag{c}
$$

$$
\alpha_{31}(0) + \alpha_{32}(3) = 3 \tag{d}
$$

Solving (c) and (d), we obtain $\alpha_{31} = 1$ and $\alpha_{32} = 1$. (Indeed, Reaction 3 is the sum of Reactions 1 and 2.) Now that the α_{km} factors are known, we can formulate the design equations. For an ideal batch reactor, we write Eq. 4.4.4

for each of the two independent reactions. Hence, for this case, the design equations are

$$\frac{dZ_1}{d\tau} = (r_1 + r_3)\left(\frac{V_R}{V_{R_0}}\right)\left(\frac{t_{cr}}{C_0}\right) \tag{e}$$

$$\frac{dZ_2}{d\tau} = (r_2 + r_3)\left(\frac{V_R}{V_{R_0}}\right)\left(\frac{t_{cr}}{C_0}\right) \tag{f}$$

To solve these design equations, we have to express r_1, r_2, r_3, and $V_R/V_R(0)$ in terms of Z_1 and Z_2. For a plug-flow reactor, we write Eq. 4.4.11 for each of the two independent reactions. Hence, for this case, the design equations are

$$\frac{dZ_1}{d\tau} = (r_1 + r_3)\left(\frac{t_{cr}}{C_0}\right) \tag{g}$$

$$\frac{dZ_2}{d\tau} = (r_2 + r_3)\left(\frac{t_{cr}}{C_0}\right) \tag{h}$$

To solve these design equations, we have to express r_1, r_2, and r_3 in terms of Z_1 and Z_2. For a CSTR, we write Eq. 4.4.12 for each of the two independent reactions. Hence, for this case, the design equations are

$$Z_{1_{out}} - Z_{1_{in}} - (r_{1_{out}} + r_{3_{out}})\tau\left(\frac{t_{cr}}{C_0}\right) = 0 \tag{i}$$

$$Z_{2_{out}} - Z_{2_{in}} - (r_{2_{out}} + r_{3_{out}})\tau\left(\frac{t_{cr}}{C_0}\right) = 0 \tag{j}$$

To solve these design equations, we have to express $r_{1_{out}}$, $r_{2_{out}}$, and $r_{3_{out}}$ in terms of $Z_{1_{out}}$ and $Z_{2_{out}}$ and, then, for different values of τ, we solve for $Z_{1_{out}}$ and $Z_{2_{out}}$.

Example 4.3

a. The following gaseous chemical reactions take place in the reactor:

Reaction 1:	$4NH_3 + 5O_2 \longrightarrow 4NO + 6H_2O$
Reaction 2:	$4NH_3 + 3O_2 \longrightarrow 2N_2 + 6H_2O$
Reaction 3:	$2NO + O_2 \longrightarrow 2NO_2$
Reaction 4:	$4NH_3 + 6NO \longrightarrow 5N_2 + 6H_2O$

Formulate the dimensionless, reaction-based design equations for ideal batch reactor, plug-flow reactor, and CSTR using the heuristic rule.

b. Formulate the dimensionless, reaction-based design equations for an ideal batch reactor using a set of independent reactions obtained from the reduced matrix.

Solution   First, we determine the number of independent reactions and select a set of independent reactions. The matrix of stoichiometric coefficients for these reactions was constructed and reduced to a reduced matrix in Example 2.12. Since there are three nonzero rows in the reduced matrix, there are three independent reactions. The nonzero rows in the reduced matrix represent the following set of independent reactions:

$$\text{Reaction 1:} \quad 4NH_3 + 5O_2 \longrightarrow 4NO + 6H_2O$$
$$\text{Reaction 5:} \quad 2NO \longrightarrow O_2 + N_2$$
$$\text{Reaction 6:} \quad 4NO \longrightarrow N_2 + 2NO_2$$

Note that this set of independent reactions includes two reactions (Reactions 5 and 6) that are not among the given reactions.

a. Applying the heuristic rule of selecting a set of independent reactions whose rate expressions are known, we should select a set of three independent reactions from the given chemical reactions. Reactions 1, 2, and 3, represent such a set; hence, $m = 1, 2, 3$ and $k = 4$, and we express the design equations in terms of Z_1, Z_2, and Z_3. The stoichiometric coefficients of the selected independent reactions are

$$(s_{NH_3})_1 = -4 \quad (s_{O_2})_1 = -5 \quad (s_{NO})_1 = 4 \quad (s_{H_2O})_1 = 6 \quad (s_{N_2})_1 = 0 \quad (s_{NO_2})_1 = 0$$
$$(s_{NH_3})_2 = -4 \quad (s_{O_2})_2 = -3 \quad (s_{NO})_2 = 0 \quad (s_{H_2O})_2 = 6 \quad (s_{N_2})_2 = 2 \quad (s_{NO_2})_2 = 0$$
$$(s_{NH_3})_3 = 0 \quad (s_{O_2})_3 = -1 \quad (s_{NO})_3 = -2 \quad (s_{H_2O})_3 = 0 \quad (s_{N_2})_3 = 0 \quad (s_{NO_2})_3 = 2$$

$$\Delta_1 = 1 \qquad \Delta_2 = 1 \qquad \Delta_3 = -1$$

To determine the multipliers α_{km}'s of the dependent reaction (Reaction 4) and the three independent reactions, we write stoichiometric relation Eq. 2.4.9 for NO_2,

$$\alpha_{41}(0) + \alpha_{42}(0) + \alpha_{43}(2) = 0 \tag{a}$$

and obtain $\alpha_{43} = 0$. We write Eq. 2.4.9 for N_2,

$$\alpha_{41}(0) + \alpha_{42}(2) + \alpha_{43}(0) = 5 \tag{b}$$

and obtain $\alpha_{42} = 2.5$. We write Eq. 2.4.9 for NH_3:

$$\alpha_{41}(-4) + \alpha_{42}(-4) + \alpha_{43}(0) = -4 \qquad \text{(c)}$$

and obtain $\alpha_{41} = -1.5$. Now that the α_{km} factors are known, we can formulate the design equations. For an ideal batch reactor, we write Eq. 4.4.4 for each of the three independent reactions. Hence, the design equations are

$$\frac{dZ_1}{d\tau} = (r_1 - 1.5r_4)\left(\frac{V_R}{V_{R_0}}\right)\left(\frac{t_{cr}}{C_0}\right) \qquad \text{(d)}$$

$$\frac{dZ_2}{d\tau} = (r_2 + 2.5r_4)\left(\frac{V_R}{V_{R_0}}\right)\left(\frac{t_{cr}}{C_0}\right) \qquad \text{(e)}$$

$$\frac{dZ_3}{d\tau} = r_3\left(\frac{V_R}{V_{R_0}}\right)\left(\frac{t_{cr}}{C_0}\right) \qquad \text{(f)}$$

For a plug-flow reactor, we write Eq. 4.4.11 for each of the three independent reactions. Hence, the design equations are

$$\frac{dZ_1}{d\tau} = (r_1 - 1.5r_4)\left(\frac{t_{cr}}{C_0}\right) \qquad \text{(g)}$$

$$\frac{dZ_2}{d\tau} = (r_2 + 2.5r_4)\left(\frac{t_{cr}}{C_0}\right) \qquad \text{(h)}$$

$$\frac{dZ_3}{d\tau} = r_3\left(\frac{t_{cr}}{C_0}\right) \qquad \text{(i)}$$

For a CSTR, we write Eq. 4.4.12 for each of the three independent reactions. Hence, for this case, the design equations are

$$Z_{1_{out}} - Z_{1_{in}} - (r_{1_{out}} - 1.5r_{4_{out}})\tau\left(\frac{t_{cr}}{C_0}\right) = 0 \qquad \text{(j)}$$

$$Z_{2_{out}} - Z_{2_{in}} - (r_{2_{out}} + 2.5r_{4_{out}})\tau\left(\frac{t_{cr}}{C_0}\right) = 0 \qquad \text{(k)}$$

$$Z_{3_{out}} - Z_{3_{in}} - r_{3_{out}}\tau\left(\frac{t_{cr}}{C_0}\right) = 0 \qquad \text{(l)}$$

To solve the design equations, we have to express the rates of the individual reactions in terms of the dimensionless extents of the independent reactions, Z_1, Z_2, and Z_3.

b. Taking the reactions of the reduced matrix as a set of independent reactions, $m = 1, 5, 6$, and $k = 2, 3, 4$. Hence, we express the design equations in terms of Z_1, Z_5, and Z_6. The stoichiometric coefficients of the three independent reactions are

$$(s_{NH_3})_1 = -4 \quad (s_{O_2})_1 = -5 \quad (s_{NO})_1 = 4 \quad (s_{H_2O})_1 = 6 \quad (s_{N_2})_1 = 0 \quad (s_{NO_2})_1 = 0$$
$$(s_{NH_3})_5 = 0 \quad (s_{O_2})_5 = 1 \quad (s_{NO})_5 = -2 \quad (s_{H_2O})_5 = 0 \quad (s_{N_2})_5 = 1 \quad (s_{NO_2})_5 = 0$$
$$(s_{NH_3})_6 = 0 \quad (s_{O_2})_6 = 0 \quad (s_{NO})_6 = -4 \quad (s_{H_2O})_6 = 0 \quad (s_{N_2})_6 = 1 \quad (s_{NO_2})_6 = 2$$

$$\Delta_1 = 1 \qquad \Delta_5 = 1 \qquad \Delta_6 = -1$$

To determine the multipliers α_{km}'s of the three dependent reactions (Reactions 2, 3, and 4) and the three independent reactions, we write stoichiometric relation Eq. 2.4.9. For $k = 2$, we write Eq. 2.4.9 for NO_2,

$$\alpha_{21}(0) + \alpha_{25}(0) + \alpha_{26}(2) = 0 \tag{m}$$

and obtain $\alpha_{26} = 0$. We write Eq. 2.4.9 for N_2,

$$\alpha_{21}(0) + \alpha_{25}(1) + \alpha_{26}(1) = 2 \tag{n}$$

and obtain $\alpha_{25} = 2$. We write Eq. 2.4.9 for NH_3,

$$\alpha_{21}(-4) + \alpha_{25}(0) + \alpha_{26}(0) = -4 \tag{o}$$

and obtain $\alpha_{21} = 1$. For $k = 3$, we write Eq. 2.4.9 for NO_2,

$$\alpha_{31}(0) + \alpha_{35}(0) + \alpha_{36}(2) = 2 \tag{p}$$

and obtain $\alpha_{36} = 1$. We write Eq. 2.4.9 for N_2,

$$\alpha_{31}(0) + \alpha_{35}(1) + \alpha_{36}(1) = 0 \tag{q}$$

and obtain $\alpha_{35} = -1$. We write Eq. 2.4.9 for NH_3,

$$\alpha_{31}(-4) + \alpha_{35}(0) + \alpha_{36}(0) = 0 \tag{r}$$

and obtain $\alpha_{31} = 0$. For $k = 4$, we write Eq. 2.4.9 for NO_2,

$$\alpha_{41}(0) + \alpha_{45}(0) + \alpha_{46}(2) = 0 \tag{s}$$

and obtain $\alpha_{46} = 0$. We write Eq. 2.4.9 for N_2,

$$\alpha_{41}(0) + \alpha_{45}(1) + \alpha_{46}(1) = 5 \tag{t}$$

and obtain $\alpha_{45} = 5$. We write Eq. 2.4.9 for NH_3,

$$\alpha_{41}(-4) + \alpha_{45}(0) + \alpha_{46}(0) = -4 \qquad \text{(u)}$$

and obtain $\alpha_{41} = 1$. Hence, the α_{kn} factors for the dependent reactions are

$$\alpha_{21} = 1 \qquad \alpha_{25} = 2 \qquad \alpha_{26} = 0$$

$$\alpha_{31} = 0 \qquad \alpha_{35} = -1 \qquad \alpha_{36} = 1$$

$$\alpha_{41} = 1 \qquad \alpha_{45} = 5 \qquad \alpha_{46} = 0$$

Now that the α_{kn} factors are known, we can formulate the design equations. For an ideal batch reactor, we write Eq. 4.4.4 for each of the three independent reactions. Noting that independent Reactions 5 and 6 do not actually take place $r_5 = r_6 = 0$, and the design equations are

$$\frac{dZ_1}{d\tau} = (r_1 + r_2 + r_4)\left(\frac{V_R}{V_{R_0}}\right)\left(\frac{t_{cr}}{C_0}\right) \qquad \text{(v)}$$

$$\frac{dZ_5}{d\tau} = (2r_2 - r_3 + 5r_4)\left(\frac{V_R}{V_{R_0}}\right)\left(\frac{t_{cr}}{C_0}\right) \qquad \text{(w)}$$

$$\frac{dZ_6}{d\tau} = r_3\left(\frac{V_R}{V_{R_0}}\right)\left(\frac{t_{cr}}{C_0}\right) \qquad \text{(x)}$$

Note that, in this case, the design equations have seven terms (seven rate expressions), whereas in the formulation in part (a), design equations (d), (e), and (f), have only five terms. This illustrates that, by adopting the heuristic rule, we minimize the number of terms in the design equations. To solve the design equations, we have to express the rates of the individual reactions in terms of the dimensionless extents of the independent reactions, Z_1, Z_5, and Z_6.

4.5 SUMMARY

In this chapter, we discussed the application of the conservation of mass principle to derive design equations for chemical reactors. The following topics were covered:

- We derived (in Appendix B) the species continuity equations, describing the variations in species concentrations at any point in the reactor.
- We integrated the continuity equation over the reactor volume and obtained the general, species-based design equation.

- We noticed the difficulty of solving the general form of the design equation.
- We derived the species-based design equations for three ideal reactor models: ideal batch reactor, plug-flow reactor, and CSTR.
- We derived the reaction-based design equations for three ideal reactors.
- We converted the reaction-based design equations to dimensionless forms that, upon solution, provide the dimensionless operating curves.

PROBLEMS*

4.1$_2$ A $200 \, L/min$ stream of freshwater is fed to and withdrawn from a well-mixed tank with a volume of $1000 \, L$. Initially, the tank contains pure water. At time $t = 0$, the feed is switched to a brine stream with a concentration of $180 \, g/L$ that is also fed at a rate of $200 \, L/min$.

    a. Derive a differential equation for the salt concentration in the tank.

    b. Separate the variables and integrate to obtain an expression for the salt concentration as a function of time.

    c. At what time will the salt concentration in the tank reach a level of $100 \, g/L$?

4.2$_2$ Many specialty chemicals are produced in semibatch reactors where a reactant is added gradually into a batch reactor. This problem concerns the governing equations of such operations without considering chemical reactions. A well-mixed batch reactor initially contains $200 \, L$ of pure water. At time $t = 0$, we start feeding a brine stream with a salt concentration of $180 \, g/L$ into the tank at a constant rate of $50 \, L/min$. Calculate:

    a. The time the salt concentration in the tank is $60 \, g/L$

    b. The volume of the tank at that time

    c. The time and salt concentration in the tank when the volume of the tank is $600 \, L$

Assume the density of the brine is the same as the density of the water.

4.3$_3$ Solve Problem 4.2 when the feed rate of the brine is not constant. Consider the case where the feed rate is a function of time given by $v_{in}(t) = 100 - 10t$ (t is in minute and v is in liter/minute), and the exit flow rate is constant at $20 \, L/min$.

    a. Plot a graph of the salt concentration as a function of time.

    b. What is the salt concentration when the amount of solution in the tank is the maximum?

    c. What is the salt concentration of the last drop in the tank?

*Subscript 1 indicates simple problems that require application of equations provided in the text. Subscript 2 indicates problems whose solutions require some more in-depth analysis and modifications of given equations. Subscript 3 indicates problems whose solutions require more comprehensive analysis and involve application of several concepts. Subscript 4 indicates problems that require the use of a mathematical software or the writing of a computer code to obtain numerical solutions.

Hint: You may resort to numerical solution, but you have to show the equations you solve and the initial conditions.

4.4₁ Ethylene oxide is produced by a catalytic reaction of ethylene and oxygen at controlled conditions. However, side reactions cannot be completely prevented, and it is believed that the following reactions take place:

$$2C_2H_4 + O_2 \longrightarrow 2C_2H_4O$$

$$C_2H_4 + 3O_2 \longrightarrow 2CO_2 + 2H_2O$$

$$C_2H_4 + 2O_2 \longrightarrow 2CO + 2H_2O$$

$$2CO + O_2 \longrightarrow 2CO_2$$

$$CO + H_2O \longrightarrow CO_2 + H_2$$

Formulate the dimensionless, reaction-based design equations for an ideal batch reactor, plug-flow reactor, and a CSTR.

4.5₁ Cracking of naphtha to produce olefins is a common process in the petrochemical industry. The cracking reactions are represented by the simplified elementary gas-phase reactions:

$$C_{10}H_{22} \longrightarrow C_4H_{10} + C_6H_{12}$$

$$C_4H_{10} \longrightarrow C_3H_6 + CH_4$$

$$C_6H_{12} \longrightarrow C_2H_4 + C_4H_8$$

$$C_4H_8 \longrightarrow 2C_2H_4$$

Formulate the dimensionless, reaction-based design equations for a plug-flow reactor.

4.6₁ The cracking of propane to produce ethylene is represented by the simplified kinetic model:

$$C_3H_8 \longrightarrow C_2H_4 + CH_4$$

$$C_3H_8 \longrightarrow C_3H_6 + H_2$$

$$C_3H_8 + C_2H_4 \longrightarrow C_2H_6 + C_3H_6$$

$$2C_3H_6 \longrightarrow 3C_2H_4$$

$$C_3H_6 \longrightarrow C_2H_2 + CH_4$$

$$C_2H_4 + C_2H_2 \longrightarrow C_4H_6$$

Formulate the dimensionless, reaction-based design equations for a plug-flow reactor.

4.7₁ The following reactions are believed to take place during the direct oxidation of methane:

$$2CH_4 + 3O_2 \longrightarrow 2CO + 4H_2O$$

$$CH_4 + 2O_2 \longrightarrow CO_2 + 2H_2O$$

$$2CH_4 + O_2 \longrightarrow 2CH_3OH$$

$$CH_4 + O_2 \longrightarrow HCHO + H_2O$$

$$CO + H_2O \longrightarrow CO_2 + H_2$$

$$CO + 2H_2 \longrightarrow CH_3OH$$

$$CO + H_2 \longrightarrow HCHO$$

$$2CH_4 + O_2 \longrightarrow 2CO + 4H_2$$

$$2CO + O_2 \longrightarrow 2CO_2$$

Formulate the dimensionless, reaction-based design equations for a plug-flow reactor.

4.8₁ In the reforming of methane, the following chemical reactions occur:

$$CH_4 + 2H_2O \longrightarrow CO_2 + 4H_2$$

$$CH_4 + 2O_2 \longrightarrow CO_2 + 2H_2O$$

$$2CH_4 + O_2 \longrightarrow 2CH_3OH$$

$$2CH_4 + 3O_2 \longrightarrow 2CO + 4H_2O$$

$$CO_2 + 3H_2 \longrightarrow CH_3OH + H_2O$$

$$2CH_3OH + 3O_2 \longrightarrow 2CO_2 + 4H_2O$$

$$2CO + O_2 \longrightarrow 2CO_2$$

$$CO + H_2O \longrightarrow CO_2 + H_2$$

Formulate the dimensionless, reaction-based design equations for a plug-flow reactor.

BIBLIOGRAPHY

More comprehensive treatments of the species continuity equations and the molar fluxes can be found in:

R. B. Bird, The Basic Concepts in Transport Phenomena, *Chem. Eng. Ed.*, Winter, 102, 1994.

R. B. Bird, W. E. Stewart, and E. N. Lightfoot, *Transport Phenomena*, 2nd ed., Wiley, Hoboken, NJ, 2002.

J. R. Welty, C. E. Wicks, and R. E. Wilson, *Fundamentals of Momentum Heat and Mass Transfer*, 2nd ed., Wiley, New York, 1984.

Applications of the continuity equations for turbulent flows can be found in:

J. O. Hinze, *Turbulence*, 2nd ed., McGraw-Hill, New York, 1975.

H. Tennekes, and J. L. Lumley, *A First Course in Turbulence*, MIT Press, Cambridge, MA, 1972.

Applications of the species continuity equation in reacting systems is found in:

L. A. Belfiore, *Transport Phenomena for Chemical Reactor Design*, Wiley-Intersciences, Hoboken, NJ, 2003.

D. E. Rosner, *Transport Processes in Chemically Reacting Flow Systems*, Butterworth-Heineman, Boston, 1986.

5

ENERGY BALANCES

This chapter covers the fourth fundamental concept used in the analysis and design of chemical reactors—conservation of energy (the first law of thermodynamics). We use energy balances to express the temperature variations during reactor operations. Section 5.1 provides a brief overview of basic thermodynamic quantities and relations used in reactor design. We define the *heat of reaction* and equilibrium constant of a chemical reaction and discuss how they vary with temperature and pressure. In Section 5.2, we apply the first law of thermodynamics to derive the energy balance equation for closed and open systems. We reduce these equations to dimensionless forms and derive the energy balance equations for the three ideal reactor configurations: ideal batch reactor, plug-flow reactor, and continuous stirred-tank reactor (CSTR).

5.1 REVIEW OF THERMODYNAMIC RELATIONS

5.1.1 Heat of Reaction

The heat of reaction, or more accurately, the enthalpy change during a chemical reaction, $\Delta \hat{H}_R^{\circ}$, indicates the amount of energy being absorbed or released when a chemical transformation takes place at given operating atmosphere and temperature of 298 K. The standard molar heat of reaction of a chemical reaction (expressed in energy per mole extent) is denoted by $\Delta \hat{H}_{R_{298}}^{\circ}$. It is calculated by

$$\Delta \hat{H}_{R_{298}}^{\circ} = \sum_{j}^{J} s_j H_{f_j}^{\circ} \tag{5.1.1}$$

Principles of Chemical Reactor Analysis and Design, Second Edition. By Uzi Mann
Copyright © 2009 John Wiley & Sons, Inc.

where $\hat{H}^{\circ}_{f_j}$ is the standard molar heat of formation of species j at 298 K, and s_j is the stoichiometric coefficient of species j. The superscript "o" indicates a standard pressure of 1 atm. The values of $\hat{H}^{\circ}_{f_j}$'s of many species are tabulated in thermodynamic textbooks and handbooks. We adopt the common convention in applying the first law of thermodynamics—heat added to a system is positive and heat removed from a system is negative. Therefore, for exothermic reactions, $\Delta\hat{H}^{\circ}_R$ is negative, and, for endothermic reactions, $\Delta\hat{H}^{\circ}_R$ is positive.

The heat of reaction is a function of temperature and pressure. Since enthalpy is a state quantity, we relate the heat of reaction at any temperature and the standard pressure of 1 atm, to the standard heat of reaction at 298 K by

$$\Delta\hat{H}^{\circ}_R(T) = \Delta\hat{H}^{\circ}_{R_{298}} + \int_{298}^{T} \sum_{j}^{J} s_j \hat{c}_{p_j}\, dT \tag{5.1.2}$$

where $\hat{c}_{p_j}$ is the specific molar heat capacity of species j. Note that Eq. 5.1.2 represents the calculation of $\Delta\hat{H}^{\circ}_R$ by cooling the reactants from T to 298 K, carrying out the reaction at 298 K, and then heating up the products from 298 K to T, as shown schematically in Figure 5.1. Once the heat of reaction at temperature T and at standard pressure (1 atm) is known, we can calculate the heat of reaction at T and any pressure P by

$$\Delta\hat{H}_R(T, P) = \Delta\hat{H}^{\circ}_R(T) + \sum_{j}^{J} s_j \int_{1}^{P} \left[\hat{V}_j - T\left(\frac{\partial\hat{V}_j}{\partial T}\right)_P \right] dP \tag{5.1.3}$$

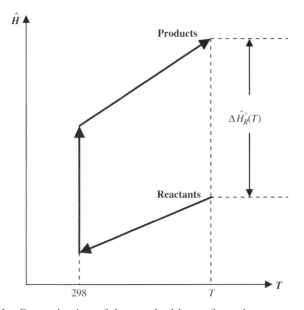

Figure 5.1   Determination of the standard heat of reaction at temperature T.

where $\hat{V}_j$ is the specific molar volume of species j. The term inside the integral indicates the change in the enthalpy of species j due to the change in pressure at temperature T. In most applications, this term is small, and, unless the reactor is operated at a very high pressure (several hundred bars or higher), we can approximate the heat of reaction by $\Delta\hat{H}_R^\circ(T)$, the heat of reaction at 1 atm.

Example 5.1   Determine the heat of combustion of n-octane at 620°C and 1 atm. The heats of formations of the various species and their heat capacities are given below. *Data*: The standard heats of formation of the species (at 298 K) are

$$C_8H_{18}(g) = -208.70 \text{ J/mol}$$

$$CO_2(g) = -393.509 \text{ J/mol}$$

$$H_2O(g) = -241.818 \text{ J/mol}$$

The specific molar heat capacities $\hat{c}_{p_j}$ (assumed constant) are

$$C_8H_{18}(g) = -254 \text{ J/mol K}$$

$$O_2(g) = -30.2 \text{ J/mol K}$$

$$CO_2(g) = -40.3 \text{ J/mol K}$$

$$H_2O(g) = -34.0 \text{ J/mol K}$$

Solution   The chemical reaction is

$$C_8H_{18}(g) + 12.5O_2(g) \longrightarrow 8CO_2(g) + 9H_2O(g)$$

and the stoichiometric coefficients are

$$s_{C_8H_{18}} = -1 \qquad s_{O_2} = -12.5 \qquad s_{CO_2} = 8 \qquad s_{H_2O} = 9 \qquad \Delta = 3.5$$

We determine the standard heat of reaction by Eq. 5.1.1, noting that the heat of formation of an element is, by definition, zero:

$$\Delta\hat{H}_{R_{298}}^\circ = \sum_j^J s_j\hat{H}_{f_j}^\circ = (-1)\hat{H}_{f_{C_8H_{18}}}^\circ + (-12.5)\hat{H}_{f_{O_2}}^\circ + (8)\hat{H}_{f_{CO_2}}^\circ + (9)\hat{H}_{f_{H_2O}}^\circ$$

$$= (-1)(-208{,}750) + (-12.5)(0) + (8)(-393{,}509) + (9)(-241.818)$$

$$= -5115.68 \text{ kJ/mol} \tag{a}$$

Since $\Delta \hat{H}_R^\circ$ is negative, the reaction is exothermic. Next, we determine the heat of reaction at 620°C by using (a) and Eq. 5.1.2. For this case, the individual species heat capacities are constant; therefore, the term inside the integral is independent of temperature:

$$\sum_j^J s_j \hat{c}_{p_j} = (-1)(254) + (-12.5)(30.2) + (8)(40.3) + (9)(34.0)$$

$$= -3.10 \text{ J/mol K} \qquad (b)$$

Substituting the values from (a) and (c) into Eq. 5.1.2,

$$\Delta \hat{H}_R^\circ (893 \text{ K}) = -5115.68 \times 10^3 + \int_{298}^{893} (-3.10) \, dT = -5117.52 \text{ kJ/mol} \qquad (c)$$

Note that, in this case, because the difference between the heat capacity of the reactants and the heat capacity of the products is small, the heat of reaction changes very slightly with temperature.

5.1.2 Effect of Temperature on Reaction Equilibrium Constant

The equilibrium constant, K, of a chemical reaction relates to the heat of reaction by

$$\frac{d(\ln K)}{dT} = -\frac{\Delta \hat{H}_R^\circ(T)}{RT^2} \qquad (5.1.4)$$

To express how the reaction equilibrium constant varies with temperature, we separate the variables and integrate:

$$\ln\left(\frac{K(T)}{K(T_0)}\right) = -\int_{T_0}^{T} \frac{\Delta \hat{H}_R^\circ(T)}{RT^2} \, dT \qquad (5.1.5)$$

In many instances, the heat of reaction is essentially constant, and Eq. 5.1.5 reduces to

$$\ln\left(\frac{K(T)}{K(T_0)}\right) = -\frac{\Delta \hat{H}_R^\circ(T_0)}{R}\left(\frac{1}{T} - \frac{1}{T_0}\right) \qquad (5.1.6)$$

The equilibrium constant, K, relates to the composition of the individual species at equilibrium. For a general chemical reaction of the form $bB + cC \rightarrow rR + sS$,

$$K = \frac{a_R^r a_S^s}{a_B^b a_C^c} \qquad (5.1.7)$$

where a_j is the activity coefficient of species j.

5.2 ENERGY BALANCES

The first law of thermodynamics is concerned with the conservation of energy. In its most general form, it is written as the following statement:

$$
\left\{ \begin{array}{c} \text{Rate} \\ \text{energy} \\ \text{enters} \\ \text{system by} \\ \text{streams} \end{array} \right\} + \left\{ \begin{array}{c} \text{Rate heat} \\ \text{energy} \\ \text{enters} \\ \text{system} \end{array} \right\} = \left\{ \begin{array}{c} \text{Rate} \\ \text{energy} \\ \text{leaves} \\ \text{system by} \\ \text{streams} \end{array} \right\} + \left\{ \begin{array}{c} \text{Rate} \\ \text{work} \\ \text{done by} \\ \text{system} \end{array} \right\}
$$

$$
+ \left\{ \begin{array}{c} \text{Rate} \\ \text{energy} \\ \text{accumulates} \\ \text{in system} \end{array} \right\} \tag{5.2.1}
$$

When applying Eq. 5.2.1, we have to consider *all* forms of energy and account for *all* types of work. We usually derive simplified and useful relations by imposing certain assumptions and incorporating other thermodynamic relations (equation of state, Maxwell's equations, etc.) into the energy balance equation. Therefore, it is essential to identify all the assumptions made and examine whether they are valid.

Let E denote the total amount of energy, in all its forms, contained in a system, and let e denote the specific energy (energy per unit mass). In most chemical processes, electric, magnetic, and nuclear energies are negligible. Therefore, we restrict the treatment of the first law of thermodynamics to systems where energy is present in three forms: internal energy (U), kinetic energy (KE), and potential (gravitational) energy (PE); hence,

$$
E = U + \text{KE} + \text{PE} \tag{5.2.2}
$$

The specific energy (energy per unit mass) is

$$
e = u + \tfrac{1}{2}v^2 + gz \tag{5.2.3}
$$

where u is the specific internal energy, v is the linear velocity, and z is the vertical elevation.

Below, we apply the energy balances for macroscopic systems. First, we derive the energy balance equation for closed systems (batch reactors) and then for open systems (flow reactors). Microscopic energy balances, used to describe point-to-point temperature variations inside a chemical reactor, are outside the scope of this book.

5.2.1   Batch Reactors

For batch reactors (closed systems), the energy balance is written in the common form of the first law of thermodynamics:

$$\Delta E(t) = Q(t) - W(t) \qquad (5.2.4)$$

where $\Delta E(t) = E(t) - E(0)$ is the change in the energy of the system, $Q(t)$ is the heat added to the system, and $W(t)$ is the work done by the system on the surroundings during operating time t. For a *stationary system*, the only energy changed is internal energy; hence, $\Delta E(t) = \Delta U(t)$, and Eq. 5.2.4 reduces to

$$\Delta U(t) = Q(t) - W(t) \qquad (5.2.5)$$

For batch reactors, the work term consists of two components: shaft work (work done by a mechanical device such as a stirrer) and work done by expanding the boundaries of the system against the surroundings:

$$W = W_{\mathrm{sh}} + \int P\,dV \qquad (5.2.6)$$

Expressing the internal energy in terms of the enthalpy (recall, $H \equiv U + PV$ and $dH = dU + p\,dV + V\,dp$)

$$\Delta U = \Delta H - \int P\,dV - \int V\,dP$$

and, substituting into Eq. 5.2.6, the energy balance becomes

$$\Delta H(t) = Q(t) - W_{\mathrm{sh}}(t) + \int V\,dP \qquad (5.2.7)$$

Most chemical reactors operate at isobaric or near isobaric conditions. Also, in general, the enthalpy is a weak function of the pressure. Hence, the last term is relatively small, and the energy balance equation reduces to

$$\Delta H(t) = Q(t) - W_{\mathrm{sh}}(t) \qquad (5.2.8)$$

In many instances, the reacting fluid is not viscous, and the shaft work is small in comparison with the heat added to the system; hence

$$\Delta H(t) \approx Q(t) \qquad (5.2.9)$$

It is important to note that Eq. 5.2.9 is applicable only when all the assumptions made in its derivation are valid: *negligible kinetic, potential and electric energies, negligible effect of pressure, and negligible viscous work.*

For closed systems with chemical reactions, the enthalpy varies due to changes in both composition and temperature. We usually select a reference state at some convenient composition and temperature T_0. Assuming no phase change, the change in the enthalpy of the reactor in operating time t is

$$\Delta H(t) = \sum_{m}^{n_I} \Delta H_{R_m}(T_0) X_m(t) + \int_{T_0}^{T(t)} \sum_{j}^{J} (N_j \hat{c}_{p_j})_t \, dT - \int_{T_0}^{T(0)} \sum_{j}^{J} (N_j \hat{c}_{p_j})_0 \, dT \quad (5.2.10)$$

where $\sum_{j}^{J} (N_j \hat{c}_{p_j})_t$ and $\sum_{j}^{J} (N_j \hat{c}_{p_j})_0$ indicate the heat capacity of the reacting fluid initially and at time t, respectively. The first term on the right-hand side of Eq. 5.2.10 represents the change in enthalpy due to composition changes (by chemical reactions) at the reference temperature, T_0, whereas the other two terms indicate the change in the "sensible heat" (enthalpy change due to variation in temperature). Note that since the enthalpy is a state quantity, the calculation of the enthalpy difference is shown schematically in Figure 5.2. Also note that the summation in the first term on the right is over the *independent reactions*.

Substituting Eq. 5.2.10 into Eq. 5.2.8, the general energy balance equation of closed systems becomes

$$Q(t) = \sum_{m}^{n_I} \Delta H_{R_m}(T_0) X_m(t) + \int_{T_0}^{T(t)} \sum_{j}^{J} (N_j \hat{c}_{p_j})_t \, dT - \int_{T_0}^{T(0)} \sum_{j}^{J} (N_j \hat{c}_{p_j})_0 \, dT$$

$$+ \, W_{\text{sh}}(t) \quad\quad (5.2.11)$$

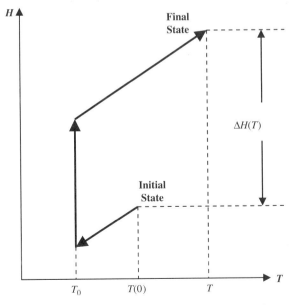

Figure 5.2 Determination of enthalpy change for reacting systems.

It is convenient to select the initial temperature as the reference temperature, $T_0 = T(0)$, and Eq. 5.2.11 reduces to

$$Q(t) = \sum_{m}^{n_I} \Delta H_{R_m}(T_0) X_m(t) + \int_{T_0}^{T(t)} \sum_{j}^{J} (N_j \hat{c}_{p_j})\, dT + W_{sh}(t) \tag{5.2.12}$$

Equation 5.2.12 is the integral form of the energy balance equation for batch reactors, relating the reactor temperature, $T(t)$, to the extents of the independent reactions, $X_m(t)$, the heat added to the reactor, $Q(t)$, and the mechanical work done by the system, $W_{sh}(t)$, during operating time t.

The differential form of the energy balance equation is obtained by differentiating Eq. 5.2.12 with respect to time:

$$\frac{dQ}{dt} = \dot{Q}(t) = \sum_{m}^{n_I} \Delta H_{R_m}(T_0)\frac{dX_m}{dt} + \sum_{j}^{J} (N_j \hat{c}_{p_j})\frac{dT}{dt} + \dot{W}_{sh}(t) \tag{5.2.13}$$

where $\dot{Q}(t)$ is the rate heat is added to the reactor, and $\dot{W}_{sh}(t)$ is the rate of mechanical work done by the reactor at time t. For an ideal batch reactor, the temperature is uniform throughout the reactor, and the rate of heat transfer is expressed by

$$\dot{Q}(t) = US(t)[T_F - T(t)] \tag{5.2.14}$$

where U is the overall heat transfer coefficient, $S(t)$ is the heat transfer area at time t, and T_F is the temperature of the external heating (or cooling) medium. The heat transfer area $S(t)$ relates to the reactor volume by

$$S(t) = \left(\frac{S}{V}\right) V_R(t) \tag{5.2.15}$$

where (S/V) is the heat transfer area per unit volume, a system characteristic that depends on the geometry of the reactor and the heat exchange area (wall, coil, etc.). Substituting Eqs. 5.2.14 and 5.2.15 into Eq. 5.2.13 and rearranging,

$$\frac{dT}{dt} = \frac{UV_R(t)}{\sum_{j}^{J}(N_j\hat{c}_{p_j})}\left(\frac{S}{V}\right)[T_F - T(t)] - \sum_{m}^{n_I}\frac{\Delta H_{R_m}(T_0)}{\sum_{j}^{J}(N_j\hat{c}_{p_j})}\frac{dX_m}{dt} - \frac{\dot{W}_{sh}(t)}{\sum_{j}^{J}(N_j\hat{c}_{p_j})} \tag{5.2.16}$$

Equation 5.2.16 expresses the changes in the reactor temperature during the reactor operation. The first term on the right-hand side represents the rate heat transferred to the reactor at time t divided by the heat capacity of the reacting fluid:

$$\frac{\dot{Q}(t)}{\sum_{j}^{J}(N_j\hat{c}_{p_j})} = \frac{UV_R(t)}{\sum_{j}^{J}(N_j\hat{c}_{p_j})}\left(\frac{S}{V}\right)[T_F - T(t)] \tag{5.2.17}$$

To reduce Eq. 5.2.16 to a dimensionless form, we use the dimensionless temperature, defined by Eq. 3.3.4, $\theta = T/T_0$, the dimensionless extent, defined by Eq. 2.7.1, $Z_m = X_m/(N_{tot})_0$, and the dimensionless operating time, defined by Eq. 4.4.3, $\tau = t/t_{cr}$. Dividing both sides of Eq. 5.2.16 by T_0 and $(N_{tot})_0$ and multiplying both sides by the characteristic reaction time, t_{cr}, we obtain

$$\frac{d\theta}{d\tau} = \frac{UV_R t_{cr}}{\sum_j^J (N_j \hat{c}_{p_j})} \left(\frac{S}{V}\right)(\theta_F - \theta) - \frac{(N_{tot})_0}{T_0 \sum_j^J (N_j \hat{c}_{p_j})} \times \sum_m^{n_I} \Delta H_{R_m}(T_0) \frac{dZ_m}{d\tau}$$

$$- \frac{1}{T_0 \sum_j^J (N_j \hat{c}_{p_j})} \frac{dW_{sh}}{d\tau} \tag{5.2.18}$$

Equation 5.2.18 is the *dimensionless*, differential energy balance equation of ideal batch reactors, relating the reactor dimensionless temperature, $\theta(\tau)$, to the dimensionless extents of the independent reactions, $Z_m(\tau)$, at dimensionless operating time τ. Note that individual $dZ_m/d\tau$'s are expressed by the reaction-based design equations derived in Chapter 4.

Equation 5.2.18 is not conveniently used because the heat capacity of the reacting fluid, $\sum_j^J (N_j \hat{c}_{p_j})$, is a function of the temperature and reaction extents, and consequently, it varies during the operation. To simplify the equation and obtain dimensionless quantities for heat transfer, we define a heat capacity of the reference state and relate the heat capacity of the reacting fluid at any instant to it by

$$\begin{pmatrix} \text{Heat capacity} \\ \text{of the reacting} \\ \text{fluid} \end{pmatrix} \equiv \begin{pmatrix} \text{Heat capacity} \\ \text{of the} \\ \text{reference state} \end{pmatrix} \begin{pmatrix} \text{Correction} \\ \text{factor} \end{pmatrix}$$

Mathematically, it is written as

$$\sum_j^J (N_j \hat{c}_{p_j}) \equiv (N_{tot})_0 \hat{c}_{p_0} \text{CF}(Z_m, \theta) \tag{5.2.19}$$

where $\hat{c}_{p_0}$ is the specific molar heat capacity of the reference state and $\text{CF}(Z_m, \theta)$ is a correction factor that adjusts the value of the heat capacity of the reacting fluid as Z_m and θ vary. We discuss the determination of $\hat{c}_{p_0}$ in detail below. Substituting Eq. 5.2.19 into Eq. 5.2.18 and noting that $(N_{tot})_0 = C_0 V_{R_0}$,

$$\frac{d\theta}{d\tau} = \frac{1}{\text{CF}(Z_m, \theta)} \left[\frac{U t_{cr}}{C_0 \hat{c}_{p_0}} \left(\frac{S}{V}\right) \left(\frac{V_R}{V_{R_0}}\right)(\theta_F - \theta) \right]$$

$$- \frac{1}{\text{CF}(Z_m, \theta)} \left\{ \sum_m^{n_I} \frac{\Delta H_{R_m}(T_0)}{T_0 \hat{c}_{p_0}} \frac{dZ_m}{d\tau} + \frac{d}{d\tau} \left[\frac{W_{sh}}{T_0 (N_{tot})_0 \hat{c}_{p_0}} \right] \right\} \tag{5.2.20}$$

Equation 5.2.20 is a dimensionless differential energy balance equation of batch reactors and can be simplified further by defining two dimensionless groups:

$$\frac{d\theta}{d\tau} = \frac{1}{\mathrm{CF}(Z_m, \theta)} \left\{ \mathrm{HTN}\left(\frac{V_R}{V_{R_0}}\right)(\theta_F - \theta) - \sum_m^{n_I} \mathrm{DHR}_m \frac{dZ_m}{d\tau} - \frac{d}{d\tau}\left[\frac{W_{\mathrm{sh}}}{(N_{\mathrm{tot}})_0 \hat{c}_{p_0} T_0}\right] \right\}$$

(5.2.21)

where HTN is the dimensionless heat-transfer number of the reactor, defined by

$$\mathrm{HTN} \equiv \frac{U t_{\mathrm{cr}}}{C_0 \hat{c}_{p_0}}\left(\frac{S}{V}\right)$$

(5.2.22)

and DHR_m is the dimensionless heat of reaction of the mth-independent reaction, defined by

$$\mathrm{DHR}_m \equiv \frac{\Delta H_{R_m}(T_0)}{T_0 \hat{c}_{p_0}}$$

(5.2.23)

The dimensionless heat-transfer number lumps together all the effects related to the heat transfer in relation to the time scale of the chemical reaction. Note that the HTN is proportional to the heat transfer coefficient, U, which depends on the flow conditions, the properties of the fluid, and to the heat-transfer area per unit volume (S/V). The DHR_m is a characteristic of the chemical reaction, the reference temperature, and the composition of the reference state.

The first term in the bracket of Eq. 5.2.21 represents the dimensionless heat-transfer rate with a dimensionless driving force, $(\theta_F - \theta)$,

$$\frac{d}{d\tau}\left(\frac{Q}{(N_{\mathrm{tot}})_0 \hat{c}_{p_0} T_0}\right) = \mathrm{HTN}\left(\frac{V_R(\tau)}{V_{R_0}}\right)(\theta_F - \theta)$$

(5.2.24)

where

$$\frac{Q}{(N_{\mathrm{tot}})_0 \hat{c}_{p_0} T_0}$$

(5.2.25)

is the dimensionless heat transferred to the reactor. The second term in the bracket of Eq. 5.2.21 represents the dimensionless heat of reactions. The third term in the bracket of Eq. 5.2.21 represents the dimensionless shaft work,

$$\frac{W_{\mathrm{sh}}}{(N_{\mathrm{tot}})_0 \hat{c}_{p_0} T_0}$$

(5.2.26)

The definition of the specific molar heat capacity of the reference state, $\hat{c}_{p_0}$, and the determination of the correction factor, $CF(Z_m, \theta)$, are discussed next. The specific molar heat capacity of the reference state, $\hat{c}_{p_0}$, is defined by

$$\begin{pmatrix} \text{Specific molar} \\ \text{heat capacity of} \\ \text{the reference state} \end{pmatrix} \equiv \frac{\text{Heat capacity of reference state}}{\text{Total number of moles of the reference state}} \quad (5.2.27)$$

and mathematically it is written as

$$\hat{c}_{p_0} \equiv \frac{\sum_j^J (N_j c_{p_j})_0}{(N_{\text{tot}})_0} \quad (5.2.28)$$

As discussed in Section 2.7, for gas-phase reactions, $(N_{\text{tot}})_0$ is taken as the total number of moles, including inert species, and

$$\sum_j^J (N_j \hat{c}_{p_j})_0 = (N_{\text{tot}})_0 \sum_j^J y_{j_0} \hat{c}_{p_j}(T_0)$$

Hence, from Eq. 5.2.28

$$\hat{c}_{p_0} = \sum_j^J y_{j_0} \hat{c}_{p_j}(T_0) \quad (5.2.29)$$

where y_{j_0} is the molar fraction of species j in the reference state.

For liquid-phase reactions, the heat capacity is expressed on a mass basis. Hence,

$$\sum_j^J (N_j \hat{c}_{p_j})_0 \equiv M_0 \bar{c}_{p_0} \quad (5.2.30)$$

where M_0 is the total mass of the reference state, and $\bar{c}_{p_0}$ is its specific mass-based heat capacity. The specific molar-based heat capacity of the reference state is

$$\hat{c}_{p_0} = \frac{M_0 \bar{c}_p}{(N_{\text{tot}})_0} \quad (5.2.31)$$

Note that Eq. 5.2.31 converts the specific mass-based heat capacity to molar-based specific heat capacity. Also recall from Section 2.7 that for liquid-phase reactions, $(N_{\text{tot}})_0$ includes all the reactants and products, but not a solvent (inert species). Hence, the numerical value of $\hat{c}_{p_0}$ is usually large since it also accounts for the heat capacity of the solvent.

Next, we derive an expression for the correction factor of the heat capacity, $CF(Z_m, \theta)$. First consider gas-phase reactions. From Eq. 2.7.3 the content of species j in the reactor is

$$N_j = (N_{\text{tot}})_0 \left\{ \left[\frac{N_{\text{tot}}(0)}{(N_{\text{tot}})_0} \right] y_j(0) + \sum_m^{n_I} (s_j)_m Z_m \right\}$$

and the heat capacity of the reacting fluid is

$$\sum_j^J (N_j \hat{c}_{p_j}) = (N_{\text{tot}})_0 \left\{ \left[\frac{N_{\text{tot}}(0)}{(N_{\text{tot}})_0} \right] \sum_j^J y_j(0) \hat{c}_{p_j}(\theta) + \sum_j^J \hat{c}_{p_j}(\theta) \sum_m^{n_I} (s_j)_m Z_m \right\}$$

Using Eq. 5.2.27 and Eq. 5.2.28,

$$CF(Z_m, \theta) = \frac{1}{\hat{c}_{p0}} \left\{ \left[\frac{N_{\text{tot}}(0)}{(N_{\text{tot}})_0} \right] \sum_j^J y_j(0) \hat{c}_{p_j}(\theta) + \sum_j^J \hat{c}_{p_j}(\theta) \sum_m^{n_I} (s_j)_m Z_m \right\} \quad (5.2.32)$$

For liquid-phase reactions, the heat capacity of the reacting fluid usually does not change. Hence, for batch reactors

$$\sum_j^J (N_j \hat{c}_{p_j}) = M \bar{c}_p$$

where M is the mass of the reacting fluid and Eq. 5.2.28 reduces to

$$CF(Z_m, \theta) = \frac{M \bar{c}_p}{M_0 \bar{c}_{p0}} \quad (5.2.33)$$

Note that when the initial state is taken as the reference state and the heat capacity of the liquid is constant, $CF(Z_m, \theta) = 1$.

In practice, we encounter three levels of complexities in applying the energy balance equation:

1. The heat capacity of the reacting fluid is *independent* of *both* the composition and the temperature. In this case, $CF(Z_m, \theta) = 1$. This situation commonly occurs in liquid-phase reactions as well as in gas-phase reactions where the heat capacities of the products are close to that of the reactants and do not vary with temperature.

2. The heat capacity of the reacting fluid *depends* on the composition but is *independent* of temperature. In this case, the species heat capacities do not

vary with temperature. Therefore, $\hat{c}_{p_j}(\theta) = \hat{c}_{p_j}(T_0)$ for all species, and Eq. 5.2.32 reduces to

$$CF(Z_m, \theta) = 1 + \frac{1}{\hat{c}_{p0}} \sum_{j}^{J} \hat{c}_{p_j}(T_0) \sum_{m}^{n_I} (s_j)_m Z_m \qquad (5.2.34)$$

This situation occurs in gas-phase reactions when the specific heat capacities of the individual species do not vary with temperature.

3. The heat capacity of the reacting fluid depends on *both* the composition and the temperature. This situation occurs in gas-phase reactions, where the species heat capacities vary with temperature. In this case, the specific heat capacities of the individual species are usually expressed in the form

$$\frac{\hat{c}_{p_j}}{R} = A_j + B_j T + C_j T^2 + D_j T^{-2} \qquad (5.2.35)$$

where T is the temperature and A_j, B_j, C_j, and D_j are tabulated constants, characteristic to each species, and R is the universal gas constant. The specific molar heat capacity of species j at dimensionless temperature θ is

$$\hat{c}_{p_j}(\theta) = R\left[A_j + (B_j T_0)\theta + (C_j T_0^2)\theta^2 + \left(\frac{D_j}{T_0^2}\right)\theta^{-2}\right] \qquad (5.2.36)$$

and these functions should be substituted in Eq. 5.2.32 to determine the values of $CF(Z_m, \theta)$. This is readily done numerically. Once $CF(Z_m, \theta)$ is known, the energy balance equation, Eq. 5.2.21, can be solved simultaneously with the dimensionless design equations.

For *isothermal operations*, $d\theta/d\tau = 0$, and for batch reactors with negligible shaft work, Eq. 5.2.26 reduces to

$$\frac{d}{d\tau}\left[\frac{Q}{(N_{tot})_0 \hat{c}_{p0} T_0}\right] = \sum_{m}^{n_I} DHR_m \frac{dZ_m}{d\tau} \qquad (5.2.37)$$

which, using Eq. 5.2.25, can be further simplified to

$$dQ = (N_{tot})_0 \sum_{m}^{n_I} \Delta H_{R_m}(T_0)\, dZ_m \qquad (5.2.38)$$

Equation 5.2.38 provides the heating (or cooling) load needed to maintain isothermal conditions.

For *adiabatic operations* HTN = 0 (no heat-transfer area), and assuming negligible shaft work Eq. 5.2.21 reduces to

$$\frac{d\theta}{d\tau} = \frac{-1}{CF(Z_m, \theta)} \sum_{m}^{n_I} DHR_m \frac{dZ_m}{d\tau} \qquad (5.2.39)$$

which can be further simplified to

$$d\theta = \frac{-1}{CF(Z_m, \theta)} \sum_{m}^{n_I} DHR_m \, dZ_m \qquad (5.2.40)$$

Equation 5.2.40 relates changes in the temperature to the extents of the independent reactions in adiabatic operations.

Example 5.2 The following gas-phase chemical reactions take place in a batch reactor:

$$\begin{aligned} \text{Reaction 1:} &\quad A + B \longrightarrow C \\ \text{Reaction 2:} &\quad C + B \longrightarrow D \end{aligned}$$

Initially, the reactor is at 420 K and 2 atm and it contains a mixture of 45% A, 45% B, and 10% I (by mole). Based on the data below, determine:
a. The average specific molar heat capacity of the reference state.
b. The dimensionless heat of reaction of each chemical reaction.
c. The correction factor of heat capacity.
d. Derive the energy balance equation for adiabatic operation.

Data: At 420 K, $\Delta H_{R_1} = -11,000$ cal/mol extent
$\quad\quad \Delta H_{R_2} = -8000$ cal/mol extent
$\quad\quad \hat{c}_{pA} = 20$ cal/mol K $\quad\quad \hat{c}_{pB} = 7$ cal/mol K $\quad\quad \hat{c}_{pC} = 22$ cal/mol K
$\quad\quad \hat{c}_{pD} = 25$ cal/mol K $\quad\quad \hat{c}_{pI} = 14$ cal/mol K

Solution The two chemical reactions are independent, and their stoichiometric coefficients are

$$\begin{aligned} s_{A_1} = -1 \quad & s_{B_1} = -1 \quad s_{C_1} = 1 \quad s_{D_1} = 0 \quad \Delta_1 = -1 \\ s_{A_2} = 0 \quad & s_{B_2} = -1 \quad s_{C_2} = -1 \quad s_{D_2} = 1 \quad \Delta_2 = -1 \end{aligned}$$

We select the initial temperature as the reference temperature; hence, $T_0 = T(0) = 420$ K.

a. For gas-phase reactions, the specific molar heat capacity of the reference state is determined using Eq. 5.2.29:

$$\hat{c}_{p_0} = y_{A_0}\hat{c}_{p_A}(1) + y_{B_0}\hat{c}_{p_B}(1) + y_{C_0}\hat{c}_{p_C}(1) + y_{D_0}\hat{c}_{p_D}(1) + y_{I_0}\hat{c}_{p_I}(1)$$

$$= (0.45)20 + (0.45)7 + (0.0)22 + (0.1)14 = 13.55 \text{ cal/mol K}$$

b. Using Eq. 5.2.23, the dimensionless heat of reactions of the two reactions are

$$\text{DHR}_1 = \frac{\Delta H_{R_1}(T_0)}{T_0\hat{c}_{p_0}} = \frac{(-11 \times 10^3)}{(420)(13.55)} = -1.933$$

$$\text{DHR}_2 = \frac{\Delta H_{R_2}(T_0)}{T_0\hat{c}_{p_0}} = \frac{(-8 \times 10^3)}{(420)(13.55)} = -1.406$$

c. For gas-phase reactions, the correction factor of heat capacity, $\text{CF}(Z_m, \theta)$, is determined by Eq. 5.2.32. In this case, the species heat capacities are independent of the temperature; hence

$$\text{CF}(Z_m, \theta) = 1 + \frac{1}{\hat{c}_{p_0}}\left(\sum_{j}^{J}\hat{c}_{p_j}\sum_{m}^{n_I}(s_j)_m Z_m\right)$$

$$= 1 + \frac{1}{13.55}\left\{\begin{array}{l}\hat{c}_{p_A}s_{A_1}Z_1 + \hat{c}_{p_B}(s_{B_1}Z_1 + s_{B_2}Z_2)\\ +\hat{c}_{p_C}(s_{C_1}Z_1 + s_{C_2}Z_2) + \hat{c}_{p_D}s_{D_2}Z_2\end{array}\right\}$$

$$= 1 + \frac{1}{13.55}\left\{\begin{array}{l}20(-1)Z_1 + 7[(-1)Z_1 + (-1)Z_2]\\ + 22[(1)Z_1 + (-1)Z_2] + 25(1)Z_2\end{array}\right\}$$

$$= \frac{13.55 - 5Z_1 - 4Z_2}{13.55}$$

d. For adiabatic operations, $\text{HTN} = 0$. Substituting the values calculated above in Eq. 5.2.26, when the shaft work is negligible, the energy balance equation is

$$\frac{d\theta}{d\tau} = \frac{13.55}{13.555Z_1 - 4Z_2}\left[(1.933)\frac{dZ_1}{d\tau} + (1.406)\frac{dZ_2}{d\tau}\right]$$

where $dZ_1/d\tau$ and $dZ_2/d\tau$ are the design equations.

Example 5.3   The following simultaneous chemical reactions take place in an aqueous solution in a batch reactor:

$$\text{Reaction 1:}\quad 2A \longrightarrow B$$
$$\text{Reaction 2:}\quad 2B \longrightarrow C$$

Two hundreds liters of a 4-mol/L solution of reactant A are charged into a batch reactor. The density of the solution is 1.05 kg/L, and its initial temperature is 310 K. Based on the data below, determine:

a. The specific molar heat capacity of the reference state.

b. The dimensionless heat of reaction of each chemical reaction.

c. The correction factor of heat capacity.

d. Derive the energy balance equation for adiabatic operation.

Data: At 310 K, $\Delta H_{R_1} = -9000$ cal/mol A, $\Delta H_{R_2} = -8000$ cal/mol B

The heat capacity of the solution is that of water (1 kcal/kg K) and it does not vary with the solution composition or the temperature.

Solution   The two chemical reactions are independent, and their stoichiometric coefficients are

$$s_{A_1} = -2 \qquad s_{B_1} = 1 \qquad s_{C_1} = 0$$
$$s_{A_2} = 0 \qquad s_{B_2} = -2 \qquad s_{C_2} = 1$$

We select the initial reactor content as the reference state; hence $T_0 = T(0) = 310$ K, $V_{R_0} = V_R(0) = 200$ L, and the mass of the reacting fluid is $M = M_0 = V_{R_0}\rho = 210$ kg. Since only reactant A is initially present in the reactor, the reference concentration is $C_0 = C_A(0) = 4$ mol/L. Therefore, $(N_{\text{tot}})_0 = C_0 V_{R_0} = 800$ mol.

a. For liquid-phase reactions, the specific molar heat capacity of the reference state is defined by Eq. 5.2.31:

$$\hat{c}_{p_0} = \frac{M_0}{(N_{\text{tot}})_0}\,\bar{c}_p = \frac{210\,\text{kg}}{(800\,\text{mol})}(1000\,\text{cal/kg K}) = 262.5\,\text{cal/mol K} \qquad \text{(a)}$$

b. The heats of reactions are given per mole of reactants A and B, respectively, but the selected reactions contain 2 mol of A and B, respectively. Hence, for the selected reactions, $\Delta \hat{H}_{R_1} = -18,000$ cal/mol extent and $\Delta \hat{H}_{R_2} = -16,000$ cal/mol extent. Using Eq. 5.2.24, the dimensionless heats of reactions of the two chemical reactions are

$$DHR_1 = \frac{\Delta H_{R_1}(T_0)}{T_0 \hat{c}_{p_0}} = \frac{(-18 \times 10^3)}{(310)(262.5)} = -0.221$$

$$DHR_2 = \frac{\Delta H_{R_2}(T_0)}{T_0 \hat{c}_{p_0}} = \frac{(-16 \times 10^3)}{(310)(262.5)} = -0.197$$

c. For liquid-phase reactions with the initial state selected as the reference state and a fluid whose heat capacity is independent of composition and temperature, $CF(Z_m, \theta) = 1$.

d. For adiabatic operations, $HTN = 0$. Substituting the values calculated above in Eq. 5.2.26, the energy balance equation for negligible shaft work is

$$\frac{d\theta}{d\tau} = (0.221)\frac{dZ_1}{d\tau} + (0.197)\frac{dZ_2}{d\tau}$$

where $dZ_1/d\tau$ and $dZ_2/d\tau$ are formulated by the design equations.

5.2.2 Flow Reactors

Consider a general flow system with one inlet and one outlet as shown schematically in Figure 5.3. We write Eq. 5.2.1 for this system:

$$\frac{dE_{sys}}{dt} = \dot{Q} - \dot{W} + \left(u + \tfrac{1}{2}v^2 + gz\right)_{in}\dot{m}_{in} - \left(u + \tfrac{1}{2}v^2 + gz\right)_{out}\dot{m}_{out} \qquad (5.2.41)$$

where E_{sys} is the total energy of the system, and $\dot{m}_{in}$ and $\dot{m}_{out}$ are, respectively, the mass flow rates in and out of the system. The work rate term, $\dot{W}$, in Eq. 5.2.41, consists of three components: rate of work the system is doing on the surroundings by a

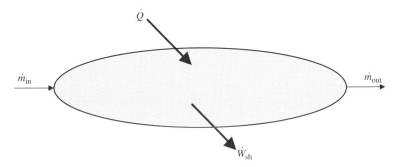

Figure 5.3 General flow system.

mechanical device (shaft work), $\dot{W}_{sh}$, rate of viscous work, $\dot{W}_{vis}$, and work done by pushing the streams in and out of the system:

$$\dot{W} = \dot{W}_{sh} + \dot{W}_{vis} + \left(\frac{P}{\rho}\right)_{in} \dot{m}_{in} - \left(\frac{P}{\rho}\right)_{out} \dot{m}_{out} \qquad (5.2.42)$$

where ρ is the density of the fluid. Using the definition of the specific enthalpy, h, (in energy/mass)

$$h = u + \frac{P}{\rho} \qquad (5.2.43)$$

and substituting Eqs. 5.2.42 and 5.2.43 into Eq. 5.2.41, the energy balance equation becomes

$$\frac{dE_{sys}}{dt} = \dot{Q} - \dot{W}_{sh} - \dot{W}_{vis} + \left(h + \tfrac{1}{2}v^2 + gz\right)_{in} \dot{m}_{in}$$
$$- \left(h + \tfrac{1}{2}v^2 + gz\right)_{out} \dot{m}_{out} \qquad (5.2.44)$$

Equation 5.2.44 is the general energy balance equation for flow systems. For steady operations, $dE_{sys}/dt = 0$, $\dot{m}_{in} = \dot{m}_{out} = \dot{m}$, and Eq. 5.2.44 reduces to

$$\dot{Q} - \dot{W}_{sh} - \dot{W}_{vis} = \left[\left(h + \tfrac{1}{2}v^2 + gz\right)_{out} - \left(h + \tfrac{1}{2}v^2 + gz\right)_{in}\right]\dot{m} \qquad (5.2.45)$$

For most chemical processes, the kinetic and potential energy of the streams are negligible in comparison to the enthalpy, and the viscous work is usually small; hence, Eq. 5.2.45 reduces to

$$\dot{Q} - \dot{W}_{sh} = (h_{out} - h_{in})\dot{m} = \Delta\dot{H} \qquad (5.2.46)$$

For flow systems with chemical reactions, the enthalpy varies due to changes in composition and temperature. We select the reference state as one atmosphere and temperature T_0. Assuming no phase change, the enthalpy difference between the outlet and the inlet is

$$\Delta\dot{H} = \sum_{m}^{n_I} \Delta H_{R_m}(T_0)(\dot{X}_{m_{out}} - \dot{X}_{m_{in}}) + \int_{T_0}^{T_{out}} \sum_{j}^{J} (F_j \hat{c}_{p_j})_{out}\, dT$$
$$- \int_{T_0}^{T_{in}} \sum_{j}^{J} (F_j \hat{c}_{p_j})_{in}\, dT \qquad (5.2.47)$$

where $(\dot{X}_{m_{\text{out}}} - \dot{X}_{m_{\text{in}}})$ is the extent per unit time of the mth-independent reaction in the reactor. Hence, the general energy balance equation for steady flow reactors is

$$\dot{Q} - \dot{W}_{\text{sh}} = \sum_{m}^{n_I} \Delta H_{R_m}(T_0)(\dot{X}_{m_{\text{out}}} - \dot{X}_{m_{\text{in}}}) + \int_{T_0}^{T_{\text{out}}} \sum_{j}^{J} (F_j \hat{c}_{p_j})_{\text{out}} \, dT$$

$$- \int_{T_0}^{T_{\text{in}}} \sum_{j}^{J} (F_j \hat{c}_{p_j})_{\text{in}} \, dT \tag{5.2.48}$$

Equation 5.2.48 is the integral form of the energy balance equation for steady-flow reactors. In most cases, it is convenient to select the inlet temperature as the reference temperature, and in this case the last term vanishes.

The differential energy balance equation for steady-flow reactors with no mechanical work is obtained by differentiating Eq. 5.2.48 over a reactor volume element dV_R:

$$\frac{d\dot{Q}}{dV_R} = \sum_{m}^{n_I} \Delta H_{R_m}(T_0) \frac{d\dot{X}_m}{dV_R} + \sum_{j}^{J} (F_j \hat{c}_{p_j}) \frac{dT}{dV_R} \tag{5.2.49}$$

The rate heat added to the differential reactor element, $d\dot{Q}$, is

$$d\dot{Q} = U \, dS(T_F - T) \tag{5.2.50}$$

where U is the overall heat-transfer coefficient, dS, is the heat-transfer area, and T_F is the temperature of the external heating (or cooling) fluid. The heat-transfer area, dS, relates to the reactor volume by

$$dS = \left(\frac{S}{V}\right) dV_R \tag{5.2.51}$$

where (S/V) is the heat-transfer area per unit volume, a characteristic that depends on the geometry of the reactor. For cylindrical reactors of diameter D, $(S/V) = 4/D$. Substituting Eqs. 5.2.50 and 5.2.51 into Eq. 5.2.49 and rearranging, we obtain

$$\frac{dT}{dV_R} = \frac{U}{\sum_{j}^{J}(F_j \hat{c}_{p_j})} \left(\frac{S}{V}\right)(T_F - T) - \sum_{m}^{n_I} \frac{\Delta H_{R_m}(T_0)}{\sum_{j}^{J}(F_j \hat{c}_{p_j})} \frac{d\dot{X}_m}{dV_R} \tag{5.2.52}$$

Equation 5.2.52 relates the changes in temperature to the reactor volume.

To reduce Eq. 5.2.52 to dimensionless form, we use dimensionless temperature, defined by Eq. 3.3.4, $\theta = T/T_0$, the dimensionless extent, defined by Eq. 2.7.2, $Z_m = \dot{X}_m/(F_{tot})_0$, and the dimensionless space time, defined by Eq. 4.4.8, $\tau = V_R/v_0 \, t_{cr}$. Dividing both sides of Eq. 5.2.52 by T_0 and $(F_{tot})_0$ and multiplying both sides by the characteristic reaction time, t_{cr}, we obtain

$$\frac{d\theta}{d\tau} = \frac{U v_0 t_{cr}}{\sum_j^J (F_j \hat{c}_{p_j})} \left(\frac{S}{V}\right) (\theta_F - \theta) - \frac{(F_{tot})_0}{T_0} \sum_m^{n_I} \frac{\Delta H_{R_m}(T_0)}{\sum_j^J (F_j \hat{c}_{p_j})} \frac{dZ_m}{d\tau} \qquad (5.2.53)$$

Equation 5.2.53 is the *dimensionless*, differential energy balance equation for steady-flow reactors, relating the temperature, θ, to the extents of the independent reactions, Z_m, as functions of dimensionless space time τ. Note that individual $dZ_m/d\tau$'s are expressed in terms of $\theta(t)$ and $Z_m(\tau)$ by the reaction-based design equations, as described in Chapter 4.

As in the case of batch reactors, dimensionless energy balance Eq. 5.2.53 is not conveniently used because the heat capacity of the reacting fluid, $\sum_j^J (F_j \hat{c}_{p_j})$, is a function of the temperature and reaction extents and, consequently, varies along the reactor. To simplify the equation and obtain dimensionless quantities for heat transfer, we define the heat capacity of the reference stream and relate the heat capacity at any point in the reactor to it by

$$\begin{pmatrix} \text{Heat capacity} \\ \text{of the} \\ \text{reacting fluid} \end{pmatrix} \equiv \begin{pmatrix} \text{Heat capacity} \\ \text{of the} \\ \text{reference stream} \end{pmatrix} \begin{pmatrix} \text{Correction} \\ \text{factor} \end{pmatrix}$$

or, in mathematical symbols

$$\sum_j^J (F_j \hat{c}_{p_j}) \equiv (F_{tot})_0 \hat{c}_{p_0} \mathrm{CF}(Z_m, \theta) \qquad (5.2.54)$$

where $\hat{c}_{p_0}$ is the specific molar heat capacity of the reference stream, and $\mathrm{CF}(Z_m, \theta)$ is a correction factor that adjusts the value of the heat capacity as Z_m and θ vary. Substituting Eq. 5.2.54 into Eq. 5.2.53 and noting that $(F_{tot})_0 = C_0 v_0$, we obtain

$$\frac{d\theta}{d\tau} = \frac{1}{\mathrm{CF}(Z_m, \theta)} \left[\mathrm{HTN}(\theta_F - \theta) - \sum_m^{n_I} \mathrm{DHR}_m \frac{dZ_m}{d\tau} \right] \qquad (5.2.55)$$

where HTN is the heat-transfer number, defined by Eq. 5.2.22, and DHR_m is the dimensionless heat of reaction of the mth independent reaction, defined by Eq. 5.2.23. Here, C_0, T_0, and $\hat{c}_{p_0}$ represent the reference concentration, reference temperature, and the specific heat capacity of the reference stream.

Equation 5.2.55 is a simplified dimensionless differential energy balance equation of steady-flow reactors, where each term is divided by $[(F_{tot})_0\hat{c}_{p_0}T_0]$, the *reference thermal energy rate*. The first term in the bracket of Eq. 5.2.55 represents the dimensionless heat-transfer rate and its relationship to the dimensionless driving force $(\theta_F - \theta)$, where

$$\frac{d}{d\tau}\left[\frac{\dot{Q}}{(F_{tot})_0\hat{c}_{p_0}T_0}\right] = \mathrm{HTN}(\theta_F - \theta) \tag{5.2.56}$$

The definition of the specific molar heat capacity of the reference state, $\hat{c}_{p_0}$, and the determination of the correction factor, $\mathrm{CF}(Z_m, \theta)$, deserve a closer examination.

The specific molar-based heat capacity of the reference stream, $\hat{c}_{p_0}$, is defined by

$$\begin{pmatrix} \text{Specific molar} \\ \text{heat capacity of the} \\ \text{reference stream} \end{pmatrix} \equiv \frac{\text{Heat capacity of the reference stream}}{\text{Total molar flow rate of the reference stream}}$$

and it is expressed mathematically by

$$\hat{c}_{p_0} \equiv \frac{\sum_j^J (F_j\hat{c}_{p_j})_0}{(F_{tot})_0} \tag{5.2.57}$$

As discussed in Section 2.7, for gas-phase reactions, $(F_{tot})_0$, is taken as the total molar flow rate of the reference stream, including inerts, and

$$\sum_j^J (F_j\hat{c}_{p_j})_0 = (F_{tot})_0 \sum_j^J y_{j_0}\hat{c}_{p_j}(T_0)$$

where y_{j_0} is the molar fraction of species j in the reference stream. Hence, Eq. 4.2.57 reduces to

$$\hat{c}_{p_0} = \sum_j^J y_{j_0}\hat{c}_{p_j}(T_0) \tag{5.2.58}$$

For liquid-phase reactions, the heat capacity of the reacting stream is usually defined on a mass basis. Hence,

$$\sum_j^J (F_j\hat{c}_{p_j})_0 = \dot{m}_0\bar{c}_p \tag{5.2.59}$$

where $\dot{m}_0$ is the mass flow rate of the reference stream, $\bar{c}_p$ is its mass-based specific heat capacity. Substituting in Eq. 5.2.58,

$$\hat{c}_{p_0} \equiv \frac{\dot{m}_0 \bar{c}_p}{(F_{tot})_0} \tag{5.2.60}$$

Note that Eq. 5.2.60 converts the mass-based specific heat capacity to a molar-based specific heat capacity.

To determine the correction factor of the heat capacity, $CF(Z_m, \theta)$, we use Eq. 5.2.54 and the appropriate definition of the specific molar heat capacity of the reference stream, $\hat{c}_{p_0}$. For gas-phase reactions,

$$F_j = (F_{tot})_0 \left[\left(\frac{(F_{tot})_{in}}{(F_{tot})_0} \right) y_{j_{in}} + \sum_m^{n_I} (s_j)_m Z_m \right]$$

and the heat capacity of the reacting fluid is

$$\sum_j^J (F_j \hat{c}_p) = (F_{tot})_0 \left[\left(\frac{(F_{tot})_{in}}{(F_{tot})_0} \right) \sum_j^J y_{j_{in}} \hat{c}_{p_j}(\theta) + \sum_j^J \hat{c}_{p_j}(\theta) \sum_m^{n_I} (s_j)_m Z_m \right]$$

Using Eq. 5.2.54, the correction factor for *gas-phase* reactions is

$$CF(Z_m, \theta) = \frac{1}{\hat{c}_{p_0}} \left[\left(\frac{(F_{tot})_{in}}{(F_{tot})_0} \right) \sum_j^J y_{j_{in}} \hat{c}_{p_j}(\theta) + \sum_j^J \hat{c}_{p_j}(\theta) \sum_m^{n_I} (s_j)_m Z_m \right] \tag{5.2.61}$$

For *liquid-phase* reactions, the heat capacity of the reacting fluid is

$$\sum_j^J (F_j \hat{c}_{p_j}) = \dot{m} \bar{c}_p$$

where $\dot{m}$ is the mass flow rate of the reacting fluid. Using Eq. 5.2.59, the correction factor is

$$CF(Z_m, \theta) = \frac{\dot{m} \bar{c}_p}{\dot{m}_0 \bar{c}_{p_0}} \tag{5.2.62}$$

Note that when the inlet stream is taken as the reference stream, and assuming constant mass-based specific heat capacity, $CF(Z_m, \theta) = 1$.

For isothermal operations, $d\theta/dt = 0$, for plug-flow reactors, Eq. 5.2.55 reduces to

$$\frac{d}{d\tau} \left[\frac{\dot{Q}}{(F_{tot})_0 \hat{c}_{p_0} T_0} \right] = \sum_m^{n_I} DHR_m \frac{dZ_m}{d\tau} \tag{5.2.63}$$

which can be further simplified to

$$dQ̇ = (F_{tot})_0 \sum_{m}^{n_I} \Delta H_{R_m}(T_0) \, dZ_m \tag{5.2.64}$$

Equation 5.2.64 provides the local heating (or cooling) rate along the reactor needed to maintain isothermal conditions.

For adiabatic operations, HTN $= 0$ (no heat-transfer area), and for plug-flow reactors Eq. 5.2.55 reduces to

$$\frac{d\theta}{d\tau} = \frac{-1}{CF(Z_m, \theta)} \sum_{m}^{n_I} DHR_m \frac{dZ_m}{d\tau} \tag{5.2.65}$$

which can be further simplified to

$$d\theta = \frac{-1}{CF(Z_m, \theta)} \sum_{m}^{n_I} DHR_m \, dZ_m \tag{5.2.66}$$

Equation 5.2.66 relates changes in the temperature to changes in the extents of the independent reactions along the reactor. The application of these equations in the design formulation of plug-flow reactors are discussed in Chapter 7.

To derive the dimensionless energy balance equation for a CSTR, the integral form of the energy balance equation, Eq. 5.2.48 is used. Since the temperature in the reactor is uniform, $T = T_{out}$, the rate heat is transferred to the reactor is

$$Q̇ = U\left(\frac{S}{V}\right) V_R(T_F - T_{out}) \tag{5.2.67}$$

where (S/V) is the heat-transfer area per unit volume of reactor. Substitute Eq. 5.2.67 into Eq. 5.2.48, divide both sides of the equation by T_0 and $(F_{tot})_0$, and multiply both sides by t_{cr}. Using Eq. 5.2.54, the energy balance equation reduces to

$$HTN\,\tau(\theta_F - \theta_{out}) - \frac{\dot{W}_{sh}}{(F_{tot})_0 \hat{c}_{p_0} T_0} = -\sum_{m}^{n_I} DHR_m(Z_{m_{out}} - Z_{m_{in}}) + CF(Z_m, \theta)_{out}(\theta_{out} - 1)$$
$$- CF(Z_m, \theta)_{in}(\theta_{in} - 1) \tag{5.2.68}$$

This is the dimensionless energy balance equation for a CSTR, relating the outlet temperature, θ_{out}, to the inlet temperature, θ_{in}, and the extents of the independent reactions in the reactor, $(Z_{m_{out}} - Z_{m_{in}})$'s.

For isothermal CSTRs with negligible shaft work, and selecting the inlet stream as the reference stream, $\theta_{out} = \theta_{in} = 1$, and, using Eq. 5.2.56, Eq. 5.2.68 reduces to

$$\frac{Q̇}{(F_{tot})_0 \hat{c}_{p_0} T_0} = \sum_{m}^{n_I} DHR_m(Z_{m_{out}} - Z_{m_{in}}) \tag{5.2.69}$$

which can be further simplified to

$$\dot{Q} = (F_{\text{tot}})_0 \sum_m^{n_I} \Delta H_{R_m}(T_0)(Z_{m_{\text{out}}} - Z_{m_{\text{in}}}) \tag{5.2.70}$$

Equation 5.2.70 provides the heating (or cooling) rate needed to maintain isothermal conditions. For adiabatic CSTRs, HTN = 0 (no heat-transfer area), and assuming negligible shaft work, Eq. 5.2.68 reduces to

$$-\sum_m^{n_I} \text{DHR}_m(Z_{m_{\text{out}}} - Z_{m_{\text{in}}}) = \text{CF}(Z_m, \theta)_{\text{out}}(\theta_{\text{out}} - 1)$$

$$- \text{CF}(Z_m, \theta)_{\text{in}}(\theta_{\text{in}} - 1) \tag{5.2.71}$$

Equation 5.2.71 relates the temperature at the reactor outlet to the extents of the independent chemical reactions. The applications of these equations are discussed in Chapter 8.

Example 5.4 The following gas-phase chemical reactions take place in a catalytic reactor producing ethylene oxide:

$$\begin{array}{lll} \text{Reaction 1:} & C_2H_4 + 0.5O_2 & \longrightarrow & C_2H_4O \\ \text{Reaction 2:} & C_2H_4 + 3O_2 & \longrightarrow & 2CO_2 + 2H_2O \end{array}$$

A feed stream at 600 K and 2 atm, consisting of 67.7% C_2H_4 and 33.3% O_2 (by mole), is introduced into the reactor. Based on the data below, determine $\text{CF}(Z_m, \theta)$.

Data: The specific molar heat capacities of the species (in cal/mol K) are

$$\hat{c}_{p_{C_2H_4}} = 2.83 + 28.6 \times 10^{-3}T + 8.73 \times 10^{-6}T^2$$

$$\hat{c}_{p_{O_2}} = 7.83 + 1.005 \times 10^{-3}T - 0.451 \times 10^5 T^{-2}$$

$$\hat{c}_{p_{C_2H_4O}} = -0.765 + 46.62 \times 10^{-3}T + 18.47 \times 10^{-6}T^2$$

$$\hat{c}_{p_{CO_2}} = 10.84 + 2.076 \times 10^{-3}T - 2.30 \times 10^5 T^{-2}$$

$$\hat{c}_{p_{H_2O}} = 6.895 + 2.881 \times 10^{-3}T + 0.240 \times 10^5 T^{-2}$$

Solution This example illustrates how to determine the correction factor $\text{CF}(Z_m, \theta)$ that depends on both Z_m and θ. The two chemical reactions are independent, and their stoichiometric coefficients are

$$\begin{array}{lllll} (s_{C_2H_4})_1 = -1 & (s_{O_2})_1 = -0.5 & (s_{C_2H_4O})_1 = 1 & (s_{CO_2})_1 = 0 & (s_{H_2O})_1 = 0 \\ (s_{C_2H_4})_2 = -1 & (s_{O_2})_2 = -3 & (s_{C_2H_4O})_2 = 0 & (s_{CO_2})_2 = 2 & (s_{H_2O})_2 = 2 \end{array}$$

We select the inlet stream as the reference stream; hence $T_0 = 600\,\mathrm{K}$, $(N_{tot})_0 = (N_{tot})_{in}$ and $y_{j_{in}} = y_{j_0}$. In this case, $(y_{C_2H_4})_0 = 0.667$, $(y_{O_2})_0 = 0.333$, $(y_{CO_2})_0 = 0$, $(y_{CO_2})_0 = 0$, and $(y_{H_2O})_0 = 0$. Using $T = T_0\theta$, we express the specific species molar heat capacities of the species in terms of θ:

$$\hat{c}_{p_{C_2H_4O}}(\theta) = 2.83 + 17.16\theta + 3.143\theta^2$$

$$\hat{c}_{p_{O_2}}(\theta) = 7.23 + 0.603\theta - 0.125\theta^{-2}$$

$$\hat{c}_{p_{C_2H_4O}}(\theta) = -0.765 + 27.97\theta + 6.65\theta^2$$

$$\hat{c}_{p_{CO_2}}(\theta) = 10.84 + 1.246 \times \theta - 0.639 \times \theta^{-2}$$

$$\hat{c}_{p_{H_2O}}(\theta) = 6.895 + 1.729 \times \theta + 0.0667 \times \theta^{-2}$$

For gas-phase reactions, the reference specific molar heat capacity is determined using Eq. 5.2.58:

$$\hat{c}_{p_0} = \sum_j^J y_{j_0}\hat{c}_{p_j}(T_0) = (0.667)23.13 + (0.333)7.71 = 18.0\,\mathrm{cal/mol\,K} \qquad \text{(a)}$$

For this case Eq. 5.2.61 reduces to

$$\mathrm{CF}(Z_m, \theta) = \frac{1}{\hat{c}_{p_0}}\left[\sum_j^J y_{j_0}\hat{c}_{p_j}(\theta) + \sum_j^J \hat{c}_{p_j}(\theta)\sum_m^J (s_j)_m Z_m\right] \qquad \text{(b)}$$

Substituting the $\hat{c}_{p_j}(\theta)$'s, the first term in the bracket on the right reduces to

$$\frac{1}{18}\left[(0.667)\hat{c}_{p_{C_2H_4}}(\theta)+(0.333)\hat{c}_{p_{O_2}}(\theta)\right] = \frac{1}{18}\left\{\begin{array}{l}(0.667)(2.83+17.16\theta+3.243\theta^2)\\+(0.333)(7.23+0.608\theta-0.125\theta^{-2})\end{array}\right\}$$

$$= \frac{4.295+11.65\theta+2.096\theta^2-0.0416\theta^{-2}}{18}$$

Expanding the second term in the bracket on the right of (b),

$$= \frac{1}{\hat{c}_{p_0}}\left\{\begin{array}{l}(-Z_1-Z_2)\hat{c}_{p_{C_2H_4}}(\theta)+Z_1\hat{c}_{p_{C_2H_4O}}(\theta)+(-0.5Z_1-3Z_2)\hat{c}_{p_{O_2}}(\theta)\\+2Z_2\hat{c}_{p_{O_2}}(\theta)+2Z_2\hat{c}_{p_{H_2O}}(\theta)\end{array}\right\}$$

$$= \frac{1}{\hat{c}_{po}} \begin{Bmatrix} (-Z_1 - Z_2)(2.83 + 17.16\theta + 3.143\theta^2) \\ +Z_1(-0.765 + 27.97\theta + 6.65\theta^2) \\ +(-0.5Z_1 - 3Z_2)(7.233 + 0.603\theta - 0.125\theta^2) \\ +2Z_2(10.84 + 1.246\theta - 0.639\theta^{-2}) \\ +2Z_2(6.895 + 1.729\theta - 0.067\theta^{-2}) \end{Bmatrix}$$

$$- \frac{1}{18} \begin{Bmatrix} (-7.21Z_1 + 10.95Z_2) - (10.51Z_1 + 13.02Z_2)\theta \\ +(3.08Z_1 - 3.143Z_2)\theta^2 + (0.0625Z_1 - 0.770Z_2)\theta^{-2} \end{Bmatrix}$$

Combining the two terms, the correction factor is

$$CF(Z_m, \theta) = \frac{1}{18} \begin{Bmatrix} (4295 - 7.21Z_1 + 10.95Z_2) \\ +(11.65 + 10.51Z_1 + 13.02Z_2)\theta \\ +(2.096 + 3.507Z_1 - 3.143Z_2)\theta^2 \\ +(0.0416 - 0.0625Z_1 - 0.770Z_2)\theta^{-2} \end{Bmatrix}$$

5.3 SUMMARY

In this chapter, we reviewed the thermodynamic relations related to reactor operations and derived the energy balance equations for batch and flow reactors. Main topics covered are:

1. Heat of reaction of a chemical reaction and its dependence on temperature and pressure
2. Equilibrium constant of a chemical reaction and its dependence on temperature
3. Definition of the specific molar heat capacity of the reference state, $\hat{c}_{po}$, for gas-phase and liquid-phase reactions
4. Definition of the correction factor of heat capacity, $CF(Z_m, \theta)$, for gas-phase and liquid-phase reactions
5. Definition of the dimensionless heat of reaction, DHR_m
6. Definition of the heat-transfer number, HTN
7. Derivation of the macroscopic energy balance equation of batch reactors and reducing it to a dimensionless form
8. Derivation of the integral and differential energy balance equations for flow reactors and reducing these equations to dimensionless forms

For convenience, Table A.4 in Appendix A provides the energy balance equations and auxiliary relations for the three ideal reactors.

PROBLEMS*

5.1₁ The following liquid-phase chemical reactions take place in a CSTR:

$$A \longrightarrow 2B$$

$$B \longrightarrow C + D$$

Both reactions are first order and the desired product is B. An aqueous solution whose concentration is $C_{A_0} = 3 \, \text{mol/L}$ is fed at a rate of $200 \, \text{L/min}$ into a 500-L CSTR with a heat-transfer area of $20 \, \text{m}^2/\text{m}^3$. The inlet temperature is 60°C, and the temperature of the jacket is 100°C. The density and heat capacity of the solution are essentially the same as those of water ($\rho = 1 \, \text{g/cm}^3$; $\bar{c}_p = 1 \, \text{cal/gC}$). The heat-transfer coefficient is estimated as 5 cal/s cm² °C. Based on the data below, determine:

a. The specific molar heat capacity of the reference stream.

b. The dimensionless heat of reaction of each chemical reaction.

c. The dimensionless heat-transfer number.

d. The correction factor of heat capacity.

e. Formulate the dimensionless energy balance equation for the operation.

Data: At 60°C,

$$\Delta H_{R_1} = 20{,}000 \, \text{cal/mol extent} \qquad \Delta H_{R_2} = 12{,}000 \, \text{cal/mol extent}$$

$$k_1 = 0.8 \, \text{min}^{-1} \qquad k_2 = 0.2 \, \text{min}^{-1}$$

5.2₂ A stream consisting of 40% O_2 and 60% CH_4 is fed into a plug-flow reactor where the following reactions take place:

$$2CH_4 + O_2 \longrightarrow 2CH_3OH$$

$$CH_4 + 2O_2 \longrightarrow CO_2 + 2H_2O$$

The feed temperature is 600 K. Based on the data below, determine:

a. The dimensionless heat of reaction of each chemical reaction at 600 K

b. The specific molar heat capacity of the feed (reference stream)

c. The correction factor of heat capacity, $CF(Z_m, \theta)$

---

*Subscript 1 indicates simple problems that require application of equations provided in the text. Subscript 2 indicates problems whose solutions require some more in-depth analysis and modifications of given equations. Subscript 3 indicates problems whose solutions require more comprehensive analysis and involve application of several concepts. Subscript 4 indicates problems that require the use of a mathematical software or the writing of a computer code to obtain numerical solutions.

Data: The standard heat of formations of the species at 298 K (in joule/mole):

Methane: $-74,520$

Methanol (gas): $-200,600$

CO_2: $-393,509$

Water (gas): $-241,818$

The species heat capacities in Joule/mol K (T in K)

Methane $-1.702 + 9.081 \times 10^{-3}\, T - 2.164 \times 10^{-6}\, T^2$

Methanol $= 2.211 + 12.216 \times 10^{-3}\, T - 3.450 \times 10^{-6}\, T^2$

Oxygen $= 3.639 + 0.506 \times 10^{-3}\, T - 0.227 \times 10^5\, T^{-2}$

$CO_2 = 5.457 + 1.045 \times 10^{-3}\, T - 1.157 \times 10^5\, T^{-2}$

Water $= 3.470 + 1.450 \times 10^{-3}\, T - 0.121 \times 10^5\, T^{-2}$

BIBLIOGRAPHY

More detailed treatment of the energy balance equations and deriving them for microscopic systems can be found in:

R. B. Bird, W. E. Stewart, and E. N. Lightfoot, *Transport Phenomena*, 2nd ed., Wiley, New York, 2002.

J. R. Welty, C. E. Wicks, and R. E. Wilson, *Fundamentals of Momentum Heat and Mass Transfer*, 4th ed., Wiley, New York, 2001.

More detailed treatment of chemical equilibrium and thermodynamic properties can be found in:

J. M. Smith and H. C. Van Ness, *Introduction to Chemical Engineering Thermodynamics*, 7th ed., McGraw-Hill, New York, 2005.

6

IDEAL BATCH REACTOR

A batch reactor is a vessel where chemical reactions take place, and no material is added or withdrawn during the operation. Reactants are charged into the reactor before the commencement of the operation, and the content of the reactor is discharged at the end of the operation. An *ideal* batch reactor, schematically described in Figure 6.1, is a *mathematical model* of "idealized" batch reactors. The model is based on the assumption that, due to good agitation, the same species compositions and temperature exist *everywhere* in the reacting fluid. Hence, at any instant, the same reaction rates prevail throughout the entire reactor volume.

In practice, batch reactors of various sizes are used, ranging from a few cubic centimeters in a test tube, through a beaker agitated by a magnetic stirrer in a chemical laboratory, to a few-liters pilot-scale reactor, to several thousand-gallon production reactors. Achieving good mixing in small reactors is relatively simple, but it is a difficult task in large reactors. Mixing in very large reactors is an important practical issue, but it is beyond the scope of this text. In very large vessels and in reactors with viscous liquids, temperature and concentration gradients may exist. In those cases, the ideal batch reactor model serves only as an approximation of the actual reactor operation. In general, the ideal batch reactor model provides a good representation of the actual operations when the mixing time is short in comparison to the reaction time.

In this chapter, we analyze the operation of ideal batch reactors. In Section 6.1, we review how the design equations are utilized and discuss the auxiliary relations that should be incorporated in order to solve the design equations. In the rest of the

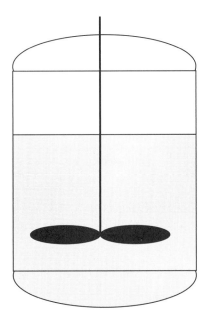

Figure 6.1   Schematic description of batch reactor.

chapter, we cover numerous applications of batch reactors of increased level of complexity. Sections 6.2 and 6.3 are concerned with isothermal operations, where only the design equations should be solved. Section 6.2 covers operations with single reactions, where one-design equations should be solved. Section 6.3 covers isothermal operations with multiple chemical reactions, where we solve multiple design equations without the need to solve the energy balance equation. Section 6.4 covers the general case (nonisothermal with multiple reactions), where we solve multiple design equations simultaneously with the energy balance equation. Since Sections 6.2 and 6.3 cover special, simplified cases of the general case, readers familiar with the operations of ideal batch reactors can proceed directly from Section 6.1 to Section 6.4. Readers who wish to develop an insight of key concepts and methods of the design formulation are encouraged to cover Sections 6.2 and 6.3 first.

6.1   DESIGN EQUATIONS AND AUXILIARY RELATIONS

The differential design equation of an ideal batch reactor, written for the mth-independent reaction, was derived in Section 4.4:

$$\frac{dZ_m}{d\tau} = \left(r_m + \sum_{k}^{n_D} \alpha_{km} r_k \right) \left[\frac{V_R(\tau)}{V_{R_0}} \right] \left(\frac{t_{cr}}{C_0} \right) \tag{6.1.1}$$

Recall that Z_m is the dimensionless extent of the mth-independent chemical reaction defined by

$$Z_m \equiv \frac{X_m}{(N_{\text{tot}})_0} \tag{6.1.2}$$

and τ is the dimensionless operating time defined by

$$\tau \equiv \frac{t}{t_{\text{cr}}} \tag{6.1.3}$$

where t_{cr} is a conveniently selected characteristic reaction time, defined by Eq. 3.5.1 and discussed in Section 3.5. Also, C_0 is the reference concentration defined by

$$C_0 \equiv \frac{(N_{\text{tot}})_0}{V_{R_0}} \tag{6.1.4}$$

where $(N_{\text{tot}})_0$ is the total molar content of the reference state, and V_{R_0} is the volume of a conveniently selected reference state (usually, the initial state).

As discussed in Chapter 4, in order to describe the operation of a reactor with multiple chemical reactions, we have to write the design equation (Eq. 6.1.1) for *each* independent chemical reaction. Also, to solve the design equations (to obtain relationships between Z_m's and τ), we have to express $V_R(\tau)$ and the rates of the chemical reactions, r_m's and r_k's, in terms of Z_m's and τ. The auxiliary relations needed to express the design equations explicitly in terms of Z_m's and τ, are derived next.

The volume-based rate of the ith chemical reaction, r_i, is generally expressed by (see Section 3.3)

$$r_i = k_i(T_0)e^{\gamma_i(\theta-1)/\theta}h_i(C_j\text{'s}) \tag{6.1.5}$$

where $k_i(T_0)$ is the reaction rate constant at the reference temperature T_0, γ_i is the dimensionless activation energy, defined by

$$\gamma_i \equiv \frac{E_{a_i}}{RT_0}$$

where θ is the dimensionless temperature, and $h_i(C_j\text{'s})$ is a function of the species concentrations, given by the rate expression. To express $h_i(C_j\text{'s})$ in terms of Z_m's and τ, we have to relate the species concentrations to Z_m's and τ. For ideal batch reactors, the concentration of species j at operating time τ is

$$C_j(\tau) \equiv \frac{N_j(\tau)}{V_R(\tau)} \tag{6.1.6}$$

Using Eq. 2.7.3, the molar content of species j in the reactor at operating time τ is

$$N_j(\tau) = (N_{tot})_0 \left[\left(\frac{N_{tot}(0)}{(N_{tot})_0} \right) y_j(0) + \sum_{m}^{n_I} (s_j)_m Z_m(\tau) \right] \qquad (6.1.7)$$

where $y_j(0)$ is the molar fraction of species j at the beginning of the operation:

$$y_j(0) \equiv \frac{N_j(0)}{N_{tot}(0)} \qquad (6.1.8)$$

Substituting Eq. 6.1.7 into Eq. 6.1.6, the concentration of species j at time τ is

$$N_j(\tau) = \frac{(N_{tot})_0}{V_R(\tau)} \left[\left(\frac{N_{tot}(0)}{(N_{tot})_0} \right) y_j(0) + \sum_{m}^{n_I} (s_j)_m Z_m(\tau) \right] \qquad (6.1.9)$$

Let us consider now the denominator in Eq. 6.1.9. $V_R(\tau)$ can be expressed by

$$V_R(\tau) = \left[\frac{V_R(\tau)}{V_R(0)} \right] \left[\frac{V_R(0)}{V_{R_0}} \right] V_{R_0} \qquad (6.1.10)$$

where $V_R(0)$ and V_{R_0} are, respectively, the initial volume and the volume of the reference state. If the reactor volume does not vary during the operation, $V_R(\tau) = V_R(0)$, Eq. 6.1.9 becomes

$$C_j(\tau) = C_0 \left[\frac{V_{R_0}}{V_R(0)} \right] \left[\left(\frac{N_{tot}(0)}{(N_{tot})_0} \right) y_j(0) + \sum_{m}^{n_I} (s_j)_m Z_m(\tau) \right] \qquad (6.1.11)$$

where C_0 is the reference concentration defined by Eq. 6.1.4. In most batch reactor applications, it is convenient to select the initial state as the reference state, $V_{R_0} = V_R(0)$, and $(N_{tot})_0 = N_{tot}(0)$, and Eq. 6.1.11 reduces to

$$C_j(\tau) = C_0 \left[y_j(0) + \sum_{m}^{n_I} (s_j)_m Z_m(\tau) \right] \qquad (6.1.12)$$

Equation 6.1.12 provides the species concentrations in terms of the extents of the independent reactions for *constant-volume* batch reactors.

For *gas-phase* variable-volume batch reactors, like the one shown schematically in Figure 6.2, $V_R(\tau)$ varies during the operation. Assuming ideal gas behavior, the

Figure 6.2   Variable-volume gas-phase batch reactor.

reactor volume at time τ is

$$\frac{V_R(\tau)}{V_{R_0}} = \left[\frac{N_{\text{tot}}(\tau)}{(N_{\text{tot}})_0}\right]\left[\frac{T(\tau)}{T_0}\right]\left[\frac{P_0}{P(\tau)}\right] \tag{6.1.13}$$

where T_0 and P_0 are, respectively, the temperature and pressure of the reference state. Using stoichiometric relation (Eq. 2.7.5) to express the total number of moles in terms of the extents of the independent chemical reactions,

$$N_{\text{tot}}(\tau) = (N_{\text{tot}})_0\left[\left(\frac{N_{\text{tot}}(0)}{(N_{\text{tot}})_0}\right) + \sum_m^{n_I}\Delta_m Z_m(\tau)\right]$$

Equation 6.1.13 becomes

$$\frac{V_R(\tau)}{V_{R_0}} = \left[\left(\frac{N_{\text{tot}}(0)}{(N_{\text{tot}})_0}\right) + \sum_m^{n_I}\Delta_m Z_m(\tau)\right]\theta(\tau)\left(\frac{P_0}{P(\tau)}\right) \tag{6.1.14}$$

where $\theta(\tau)$ is the dimensionless temperature at time τ. Substituting Eqs. 6.1.7 and 6.1.14 into Eq. 6.1.6, for gaseous, variable-volume, ideal batch reactors,

$$C_j(\tau) = C_0\frac{[N_{\text{tot}}(0)/(N_{\text{tot}})_0]y_j(0) + \sum_m^{n_I}(s_j)_m Z_m(\tau)}{\{[N_{\text{tot}}(0)/(N_{\text{tot}})_0] + \sum_m^{n_I}\Delta_m Z_m(\tau)\}\theta(\tau)}\left(\frac{P(\tau)}{P_0}\right) \tag{6.1.15}$$

In most applications, we select the initial state as the reference state, $(N_{\text{tot}})_0 = N_{\text{tot}}(0)$, and Eq. 6.1.15 reduces to

$$C_j(\tau) = C_0\frac{y_j(0) + \sum_m^{n_I}(s_j)_m Z_m(\tau)}{[1 + \sum_m^{n_I}\Delta_m Z_m(\tau)]\theta(\tau)}\left[\frac{P(\tau)}{P_0}\right] \tag{6.1.16}$$

Equation 6.1.16 provides the species concentrations in terms of the extents of the independent reactions for variable-volume, gaseous batch reactors.

The design equations and the species concentration relations contain another dependent variable, θ, the dimensionless temperature, whose variation during the reactor operation is expressed by the energy balance equation. For ideal batch reactors with negligible mechanical shaft work, the energy balance equation, derived in Section 5.2, is

$$\frac{d\theta}{d\tau} = \frac{1}{\text{CF}(Z_m, \theta)} \left[\text{HTN}\left(\frac{V_R(\tau)}{V_{R_0}}\right)(\theta_F - \theta) - \sum_m^{n_I} \text{DHR}_m \frac{dZ_m}{d\tau} \right] \qquad (6.1.17)$$

where HTN is the dimensionless heat-transfer number of the reactor, defined by

$$\text{HTN} \equiv \frac{U t_{\text{cr}}}{C_0 \hat{c}_{p_0}}\left(\frac{S}{V}\right) \qquad (6.1.18)$$

DHR_m is of the dimensionless heat of reaction of the mth-independent chemical reaction, defined by

$$\text{DHR}_m \equiv \frac{\Delta H_{R_m}(T_0)}{\hat{c}_{p_0} T_0} \qquad (6.1.19)$$

and $\text{CF}(Z_m, \theta)$ is the correction factor of the heat capacity, defined by Eq. 5.2.19. Recall that the first term inside the bracket of Eq. 6.1.17 represents the rate of heat transfer to the reactor:

$$\frac{d}{d\tau}\left[\frac{Q}{(N_{\text{tot}})_0 \hat{c}_{p_0} T_0}\right] = \text{HTN}\left[\frac{V_R(\tau)}{V_{R_0}}\right](\theta_F - \theta) \qquad (6.1.20)$$

where

$$\frac{Q}{(N_{\text{tot}})_0 \hat{c}_{p_0} T_0} \qquad (6.1.21)$$

is the dimensionless heat transferred to the reactor. The second term inside the bracket of Eq. 6.1.17 represents the heat generated (or consumed) by the chemical reactions. The specific molar heat capacity of the reference state, $\hat{c}_{p_0}$, is defined differently for gas-phase and liquid-phase reactions. For gas phase it is defined by

$$\hat{c}_{p_0} \equiv \sum_j^J y_{j_0} \hat{c}_{p_j}(T_0) \qquad (6.1.22)$$

and for liquid phase it is defined by

$$\hat{c}_{p_0} \equiv \frac{M_0 \bar{c}_p}{(N_{\text{tot}})_0} \tag{6.1.23}$$

To solve the energy balance equation (Eq. 6.1.17), we have to specify the value of HTN. However, its value depends on the selection of the reference state, C_0, and the selection of the characteristic reaction time, t_{cr}. Also note that HTN is proportional to the heat-transfer coefficient, U, which depends on the flow conditions, the properties of the fluid, and the heat-transfer area per unit volume (S/V). These quantities are not known a priori. Therefore, we develop a procedure to estimate the range of HTN. For isothermal operation $(d\theta/d\tau = 0)$, and we can determine the HTN at any instance from Eq. 6.1.17 (taking the operating temperature as the reference temperature, $\theta = 1$):

$$\text{HTN}_{\text{iso}}(\tau) = \left[\frac{V_{R_0}}{V_R(\tau)}\right] \frac{1}{\theta_F - 1} \sum_{m}^{n_I} \text{DHR}_m \frac{dZ_m}{d\tau} \tag{6.1.24}$$

Since $dZ_m/d\tau$ varies with time, the value of HTN varies during the operation. We define an *average* HTN for isothermal operation by

$$\text{HTN}_{\text{ave}} \equiv \frac{1}{\tau_{\text{op}}} \int_0^{\tau_{\text{op}}} \text{HTN}(\tau)\, d\tau \tag{6.1.25}$$

where τ_{op} is the operating time. Recall that for adiabatic operation HTN = 0. Hence, in practice, the heat-transfer number would be

$$0 < \text{HTN} \leq \text{HTN}_{\text{ave}} \tag{6.1.26}$$

Note that Eq. 6.1.25 provides only an estimate on the range of the value of HTN. We select a specific value after examining the reactor performance for different values of HTN. It is important to examine the reactor design for different values of HTN, since, when multiple reactions occur, it is difficult to predict the effect of the heat transfer on the relative rates of the individual reactions. Once the physical reactor vessel has been designed, it is necessary to verify that its configuration (S/V) and the agitation conditions actually provide the desired value of HTN.

For convenience, Tables A.3a and A.3b in Appendix A provide the design equation and the auxiliary relations for ideal batch reactors. Table A.4 provides the energy balance equation.

6.2 ISOTHERMAL OPERATIONS WITH SINGLE REACTIONS

We start the analysis of ideal batch reactors by considering *isothermal* operations with *single* reactions. Note that isothermal operation is a mathematical condition imposed on the design equation and the energy balance equation, $(d\theta/d\tau = 0)$. In practice, isothermal operations rarely occur because they require that, at any instance, the rate of heat generated (or consumed) by the chemical reactions be identical to the rate of heat removal (or supplied). However, examining isothermal operations provides an insight on the application of the design equation and the auxiliary relations.

When a single chemical reaction takes place in the reactor, the operation is described by a single design equation, and Eq. 6.1.1 reduces to

$$\frac{dZ}{d\tau} = r \left[\frac{V_R(\tau)}{V_{R_0}} \right] \left(\frac{t_{cr}}{C_0} \right) \tag{6.2.1}$$

where Z is the dimensionless extent of the reaction and r is its rate. For *isothermal operations*, since the temperature is constant, we have to solve only the design equation. (The energy balance equation provides the heating, or cooling, load necessary to maintain the reactor isothermal.) Furthermore, for isothermal operations, the reaction rates depend only on the species concentrations, and Eq. 6.1.5 reduces to

$$r_i = k_i(T_0)h_i(C_j\text{'s}) \tag{6.2.2}$$

The solution of the design equation, $Z(\tau)$ versus τ, provides the dimensionless reaction operating curve of the reactor. It describes the progress of the chemical reaction with time. Furthermore, once $Z(\tau)$ is known, we can apply stoichiometric relation (Eq. 6.1.7) to obtain the composition of each species at time τ. Also, if one prefers to express the design equation in terms of the actual operating time t, rather than the dimensionless time τ, using Eq. 6.1.3, the design equation becomes

$$\frac{dZ}{dt} = r \left[\frac{V_R(t)}{V_{R_0}} \right] \left[\frac{1}{C_0} \right] \tag{6.2.3}$$

Note that Eq. 6.2.1 has three variables: the operating time τ, the reaction extent Z, and the reaction rate r. The design equation is applied to determine any one of these variables when the other two are known. A typical design problem involves the determination of the operating time necessary to obtain a specified extent for a given reaction rate. The second application involves the determination of the extent obtained in a specified operating time τ for a given reaction rate. The third application involves the determination of the reaction rate when the extent is provided as a function of time. Below, we will consider each of these applications.

6.2.1 Constant-Volume Reactors

First, consider constant-volume batch reactors (reactors whose volumes do not change during the operation), $V_R(\tau) = V_R(0)$. In practice, this condition is satisfied either for gas-phase reactions when the walls of the reactor are stationary or when the reaction takes place in a liquid phase. In the latter case, the assumption is that the density of the liquid does not vary during the operation. For most liquid-phase reactions, the density variations are indeed quite small.

For constant-volume batch reactors with single reactions, and selecting the initial state as the reference state, the design equation, Eq. 6.2.1, reduces to

$$\frac{dZ}{d\tau} = r\left(\frac{t_{cr}}{C_0}\right) \tag{6.2.4}$$

We can solve Eq. 6.2.4 when the reaction rate r is expressed in terms of τ and Z. This is an initial value problem to be solved for the initial value, $Z(0) = 0$, and $Z(\tau)$ indicates the reaction extent during the operation. When the rate expression depends only on Z, the design equation can be solved by separating the variables and integrating

$$\tau = \left(\frac{C_0}{t_{cr}}\right) \int_0^{Z(\tau)} \frac{dZ}{r} \tag{6.2.5}$$

Equation 6.2.5 is the integral form of the design equation for an ideal, constant-volume batch reactor. Figure 6.3 shows the graphical presentation of this design equation. To solve the design equations, we have to express the reaction rate r in

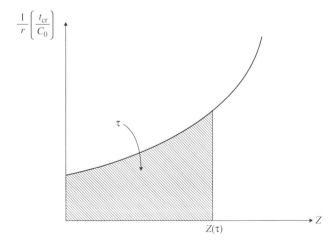

Figure 6.3 Graphical presentation of the design equation for constant-volume batch reactor.

terms of the dimensionless extent Z. Below, we analyze the operation of constant-volume, isothermal reactors with single reactions for chemical reactions with different forms of rate expressions.

First, we consider the application of the design equation for chemical reactions of the general form

$$A \longrightarrow Products$$

whose rate expression is of the form $r = kC_A^\alpha$. For these reactions, $s_A = -1$. If only reactant A is charged into the reactor, and selecting the initial state as the reference state, $C_0 = C_A(0)$ and $y_A(0) = 1$. Using Eq. 6.1.12, the concentration of reactant A is

$$C_A(\tau) = C_0[1 - Z(\tau)] \tag{6.2.6}$$

Hence, the reaction rate is

$$r = kC_0^\alpha(1 - Z)^\alpha \tag{6.2.7}$$

Substituting Eq. 6.2.7 in Eq. 6.2.4, the design equation becomes

$$\frac{dZ}{d\tau} = kC_0^\alpha(1 - Z)^\alpha \left(\frac{t_{cr}}{C_0}\right) \tag{6.2.8}$$

Next, we select the characteristic reaction time, t_{cr}, to be

$$t_{cr} = \frac{1}{kC_0^{\alpha-1}} \tag{6.2.9}$$

(Note that this t_{cr} is identical to the one obtained by the procedure described in Section 3.5.) Using Eq. 6.2.9, the design equation reduces to

$$\frac{dZ}{d\tau} = (1 - Z)^\alpha \tag{6.2.10}$$

where the dimensionless operating time is

$$\tau = \frac{t}{t_{cr}} = kC_0^{\alpha-1}t \tag{6.2.11}$$

We have to solve Eq. 6.2.10, subject to the initial condition that at $\tau = 0$, $Z(0) = 0$. An analytical solution is obtained by separating the variables and integrating,

$$\tau = \int_0^{Z(\tau)} \frac{dZ}{(1 - Z)^\alpha}$$

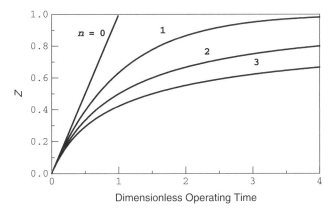

Figure 6.4   Reaction operating curves for nth-order reaction of the form A $\rightarrow$ products.

The solution is

$$Z(\tau) = 1 - e^{-\tau} \qquad \text{for } \alpha = 1 \qquad (6.2.12a)$$

and

$$Z(\tau) = 1 - \left\{ \frac{1}{1 + (\alpha - 1)\tau} \right\}^{1/\alpha - 1} \qquad \text{for } \alpha \neq 1 \qquad (6.2.12b)$$

Figure 6.4 shows the reaction operating curve for different values of α. Note that each reaction order is represented by a curve that is independent of the specific value of the rate constant and the initial reactant concentration. We determine the reaction time needed to achieve a certain extent by reading from the chart the value of τ for the respective order and then calculating the actual operating time t, using Eq. 6.2.11. Once $Z(\tau)$ is known, we can use Eq. 2.7.4 to determine the content of reactant A in the reactor at any time τ:

$$\frac{N_A(\tau)}{(N_{\text{tot}})_0} = 1 - Z(\tau)$$

Example 6.1   Consider the liquid-phase dimerization reaction

$$2A \longrightarrow B$$

The reaction is second order and the rate constant is $k = 0.5 \text{ L/mol h}$. The initial concentration of reactant A is 4 mol/L.
a. Derive and solve the design equation.
b. Derive expressions for the species contents as a function of operating time.

c. Determine the operating times for 70 and 80% conversion, respectfully.
d. Determine the operating times for 70 and 80% conversion, when the initial concentration is 3.5 mol/L.

Solution The stoichiometric coefficients of the chemical reaction are

$$s_A = -2 \qquad s_B = 1 \qquad \Delta = -1$$

We select the initial state as the reference state, and, since only reactant A is charged into the reactor, $C_0 = C_A(0)$, $y_A(0) = 1$, $y_B(0) = 0$.

a. For a constant-volume reactor, using Eq. 6.1.12,

$$C_A = C_0(1 - 2Z)$$

and the reaction rate expression is

$$r = kC_0^2(1 - 2Z)^2 \tag{a}$$

Substituting (a) into Eq. 6.2.1, the design equation becomes

$$\frac{dZ}{d\tau} = kC_0^2(1 - 2Z)^2 \left(\frac{t_{cr}}{C_0}\right) \tag{b}$$

We select the characteristic reaction time to be

$$t_{cr} = \frac{1}{kC_0} \tag{c}$$

and $\tau = kC_0 t$. Substituting (c) into (b), the design equation reduces to

$$\frac{dZ}{d\tau} = (1 - 2Z)^2 \tag{d}$$

We solve (d) subject to the initial condition $Z(0) = 0$, by separating the variables and obtain

$$\tau = \frac{Z(\tau)}{1 - 2Z(\tau)} \tag{e}$$

Hence, the solution is

$$Z(\tau) = \frac{\tau}{1 + 2\tau} \tag{f}$$

Figure E6.1.1 shows the reaction operating curve.

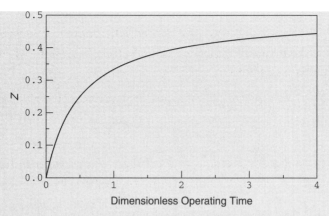

Figure E6.1.1 Reaction curve.

b. Once the reaction curve is known, we use Eq. 2.7.4 to obtain the species curves:

$$\frac{N_A(\tau)}{(N_{\text{tot}})_0} = 1 - 2Z(\tau) \qquad \text{(g)}$$

$$\frac{N_B(\tau)}{(N_{\text{tot}})_0} = Z(\tau) \qquad \text{(h)}$$

Substituting (f) into (g) and (h), the species contents are

$$N_A(\tau) = (N_{\text{tot}})_0 \left(\frac{1}{1 + 2\tau} \right) \qquad \text{(i)}$$

$$N_B(\tau) = (N_{\text{tot}})_0 \left(\frac{\tau}{1 + 2\tau} \right) \qquad \text{(j)}$$

Figure E6.1.2 shows the species operating curves.

c. Using Eq. 2.6.2 to relate the conversion to extent,

$$Z = -\frac{y_A(0)}{s_A} f_A = 0.50 f_A$$

The extents for 70 and 80% conversions are, respectively, $Z = 0.35$ and $Z = 0.4$. Substituting these values in (e), the respective dimensionless operating times are $\tau = 1.167$ and $\tau = 2.0$. Using (c), for $C_0 = 4\,\text{mol/L}$, $t_{\text{cr}} = 0.5\,\text{h}$. The required operating times for 70% conversions is

$$t = \tau t_{\text{cr}} = (1.167)(0.5\,\text{h}) = 0.583\,\text{h}$$

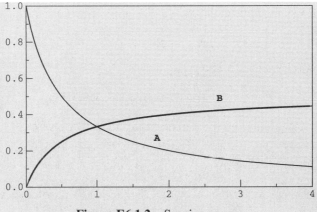

Figure E6.1.2 Species curves.

The required operating times for 80% conversions is

$$t = \tau t_{cr} = (2)(0.5\,\text{h}) = 1\,\text{h}$$

d. Using (c), for $C_0 = 3.5\,\text{mol/L}$, $t_{cr} = 0.571\,\text{h}$, and the required operating times for 70% conversions is

$$t = \tau t_{cr} = (1.167)(0.571\,\text{h}) = 0.666\,\text{h}$$

The required operating times for 80% conversions is

$$t = \tau t_{cr} = (2)(0.571\,\text{h}) = 1.332\,\text{h}$$

Example 6.2 At 800 K, dimethyl ether (CH_3OCH_3) decomposes according to the first-order, irreversible, gas-phase reaction

$$CH_3OCH_3 \longrightarrow CH_4 + CO + H_2$$

Dimethyl ether is introduced into an evacuated, constant-volume batch reactor, and the initial pressure is 310 mm Hg. After 780 min, the reactor pressure is 490 mm Hg.

a. Derive the reaction and species operating curves.
b. Derive an expression for the reactor pressure as a function of operating time.
c. Determine the value of the reaction rate constant.
d. Determine the operating time for 80% conversion.

Solution   For convenience, the chemical reaction is represented symbolically by

$$A \longrightarrow B + C + D$$

and its stoichiometric coefficients are

$$s_A = -1 \qquad s_B = 1 \qquad s_C = 1 \qquad s_D = 1 \qquad \Delta = 2$$

We select the initial state as the reference state, and, since only ether (reactant A) is charged into the reactor, $C_0 = C_A(0)$, $y_A(0) = 1$, $y_B(0) = 0$, $y_C(0) = 0$, and $y_D(0) = 0$.

a. The reaction is first-order, $r = kC_A$. For a constant-volume reactor, using Eq. 6.1.12,

$$C_A = C_0(1 - Z)$$

and

$$r = kC_0(1 - Z) \tag{a}$$

Substituting (a) into Eq. 6.2.4, the design equation reduces to

$$\frac{dZ}{d\tau} = 1 - Z \tag{b}$$

where the characteristic reaction time is

$$t_{cr} = \frac{1}{k} \tag{c}$$

and $\tau = kt$. We solve (b) subject to the initial condition $Z(0) = 0$, and the solution is

$$Z(\tau) = 1 - e^{-\tau} \tag{d}$$

or

$$\tau = -\ln[1 - Z(\tau)] \tag{e}$$

Equation (d) is the expression of the reaction operating curve. Using Eq. 2.7.4, the species curves are

$$\frac{N_A(\tau)}{(N_{tot})_0} = 1 - Z(\tau) = e^{-\tau}$$

$$\frac{N_B(\tau)}{(N_{tot})_0} = \frac{N_C(\tau)}{(N_{tot})_0} = \frac{N_D(\tau)}{(N_{tot})_0} = Z(\tau) = 1 - e^{-\tau}$$

b. Next we derive the relation between the reactor pressure and the extent. At high temperature and low pressure, ideal gas behavior can be assumed,

$$PV_R = N_{tot}RT$$

and, for a constant-volume isothermal reactor

$$\frac{P(\tau)}{P(0)} = \frac{N_{tot}(\tau)}{N_{tot}(0)} \tag{f}$$

Using the relation between the total number of moles and the extent, Eq. 2.7.6,

$$N_{tot}(\tau) = (N_{tot})_0[1 + \Delta Z(\tau)] \tag{g}$$

and substituting in (f),

$$\frac{P(\tau)}{P(0)} = 1 + 2Z(\tau) \tag{h}$$

Substituting (d) into (h), we obtain

$$\frac{P(\tau)}{P(0)} = 3 - 2e^{-\tau} = 3 - 2e^{-k \cdot t} \tag{i}$$

c. At $t = 780$ min,

$$\frac{P(\tau)}{P(0)} = \frac{490}{310} = 1.58$$

Hence, from (h), $Z(\tau) = 0.29$, and from (e), $\tau = 0.343$, and using (c),

$$k = \frac{\tau}{t} = \frac{0.343}{780 \text{ min}} = 4.4 \times 10^{-4} \text{ min}^{-1}$$

d. Using the stoichiometric relation between the conversion and extent (Eq. 2.6.2), the extent for 90% conversion is

$$Z = -\frac{y_A(0)}{s_A}f_A = -\frac{1}{-1}0.90 = 0.90$$

Substituting $Z = 0.90$ in (e), $\tau = 2.303$. To determine the actual operating time t, use the definition of the dimensionless time, (e),

$$t = \tau t_{cr} = (2.303)(20 \text{ min}) = 46.06 \text{ min}$$

Example 6.3 A biological waste, A, is decomposed by an enzymatic reaction

$$A \longrightarrow B + C$$

in aqueous solution. The rate expression of the reaction (Michaelis–Menten equation) is

$$r = \frac{kC_A}{K_m + C_A}$$

A solution with a concentration of 2 mol A/L is charged into a batch reactor. For the enzyme type and concentration used, $k = 0.1$ mol/L min and $K_m = 4$ mol/L.

a. Derive and plot the reaction operating curve.

b. Derive and plot the operating curves for species A and B.

c. Determine how long should we operate the reactor to achieve 80% conversion.

Solution   This example illustrates how to apply the design equation for ideal batch reactors with reactions whose rate expressions are not power functions of the species concentrations.

a. For the chemical reaction, the stoichiometric coefficients are $s_A = -1$, $s_B = 1$, and $s_C = 1$. We select the initial state as the reference state, and, since A is the only species charged into the reactor, $C_0 = C_A(0)$ and $y_A(0) = 1$, $y_B(0) = 0$, $y_C(0) = 0$. The design equation is

$$\frac{dZ}{d\tau} = r\left(\frac{t_{cr}}{C_0}\right) \tag{a}$$

Using Eq. 6.1.12, the concentration of reactant A is

$$C_A = C_0(1 - Z) \tag{b}$$

and, substituting (b), the reaction rate expression is

$$r = k\frac{1 - Z}{K_m/C_0 + (1 - Z)} \tag{c}$$

Substituting (c) into (a), the design equation becomes

$$\frac{dZ}{d\tau} = k\frac{1 - Z}{K_m/C_0 + (1 - Z)}\left(\frac{t_{cr}}{C_0}\right) \tag{d}$$

We select the characteristic reaction time to be

$$t_{cr} = \frac{C_0}{k} = 20 \text{ min} \tag{e}$$

(the same as the one obtained by the procedure described in Section 3.5). Substituting (e) into (d), the design equation reduces to

$$\frac{dZ}{d\tau} = \frac{1-Z}{K_m/C_0 + (1-Z)} \tag{f}$$

To obtain the reaction operating curve, $Z(\tau)$ versus τ, (f) is solved, subject to the initial condition that $Z(0) = 0$. In this case an analytical solution is obtained by separating the variables and integrating:

$$\tau = Z(\tau) - \left(\frac{K_m}{C_0}\right) \ln[1 - Z(\tau)] \tag{g}$$

Figure E6.3.1 shows the reaction operating curve for different values of K_m/C_0.

b. Once the reaction operating curve is known, the species composition is readily determine using Eq. 2.7.4,

$$\frac{N_A(\tau)}{(N_{tot})_0} = 1 - Z(\tau)$$

$$\frac{N_B(\tau)}{(N_{tot})_0} = Z(\tau)$$

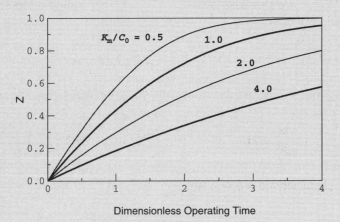

Dimensionless Operating Time

Figure E6.3.1   Reaction curve.

c. Using the relation between the conversion and extent for a single reaction (Eq. 2.6.2), for 80% conversion

$$Z = -\frac{y_A(0)}{s_A} f_A = 0.80$$

Substituting $Z = 0.80$ in (g) for $K_m/C_0 = 2$, we obtain $\tau = 4.019$. To determine the operating time t, use the definition of the dimensionless time, (e):

$$t = \tau t_{cr} = 80.4 \text{ min}$$

We continue the analysis of ideal, isothermal, constant-volume batch reactors with single reactions and consider now chemical reactions involving more than one reactant. Consider the general reaction form

$$A + bB \longrightarrow \text{Products} \tag{6.2.13}$$

that represents reactions between two reactants where $b = s_B/s_A$. The stoichiometric coefficients of the reaction are $s_A = -1$ and $s_B = -b$. We consider here chemical reactions whose rate expressions are of the form $r = kC_A^\alpha C_B^\beta$. We select the initial state as the reference state and express the species concentrations using Eq. 6.1.12:

$$C_A(\tau) = C_0[y_A(0) - Z(\tau)] \tag{6.2.14}$$

$$C_B(\tau) = C_0[y_B(0) - bZ(\tau)] \tag{6.2.15}$$

where $y_A(0) = C_A(0)/C_0$ and $y_B(0) = C_B(0)/C_0$. Substituting these equations into the rate expression,

$$r = kC_0^{\alpha+\beta}[y_A(0) - Z]^\alpha[y_B(0) - bZ]^\beta \tag{6.2.16}$$

Substituting Eq. 6.2.16 into Eq. 6.2.4, the design equation becomes

$$\frac{dZ}{d\tau} = kC_0^{\alpha+\beta}[y_A(0) - Z]^\alpha[y_B(0) - bZ]^\beta\left(\frac{t_{cr}}{C_0}\right) \tag{6.2.17}$$

We select characteristic reaction time

$$t_{cr} = \frac{1}{kC_0^{\alpha+\beta-1}} \tag{6.2.18}$$

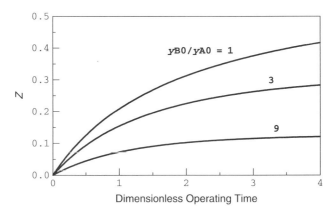

Figure 6.5 Reaction operating curves a reaction of the form $A + B \rightarrow$ products.

(the same t_{cr} as obtained by the procedure in Section 3.5), and the design equation reduces to

$$\frac{dZ}{d\tau} = [y_A(0) - Z]^\alpha [y_B(0) - bZ]^\beta \qquad (6.2.19)$$

We solve Eq. 6.2.19 numerically, subject to the initial condition that at $\tau = 0$, $Z(0) = 0$. Figure 6.5 shows the reaction operating curves for reaction 6.2.13 with $\alpha = \beta = 1$, for different proportions of the reactants. Note that when the reactants are in stoichiometric proportion, $y_B(0) = by_A(0)$, Eq. 6.2.19 reduces to

$$\frac{dZ}{d\tau} = b^\beta [y_A(0) - Z]^{\alpha+\beta} \qquad (6.2.20)$$

Also, since reactant A is the limiting reactant, for all other cases $by_A(0)/y_B(0) \leq 1$.

Example 6.4 Consider the gas-phase reaction

$$A + B \longrightarrow C$$

carried out in a 100-L isothermal constant-volume batch reactor operated at 250°C. The reaction is first order with respect to A and B. A gas mixture consisting of 40% A and 60% B (by mole) is charged into the reactor whose initial pressure is 2 atm. At 250°C, the reaction rate constant is $k = 800 \, \text{L/mol min}$.

a. Derive and plot the dimensionless reaction and species operating curves.

b. Determine the operating time needed for 80% conversion.

c. Calculate the pressure of the reactor at 80% conversion.

d. Calculate the content of species C at 80% conversion.

Solution   The stoichiometric coefficients of the chemical reaction are

$$s_A = -1 \qquad s_B = -1 \qquad s_C = 1 \qquad \Delta = -1$$

We select the initial state as the reference state; hence, $C_0 = C_A(0) + C_B(0) + C_C(0)$, and $y_A(0) = 0.4$, $y_B(0) = 0.6$, and $y_C(0) = 0$. Using Eq. 6.1.12, the species concentrations are

$$C_A = C_0(0.4 - Z) \tag{a}$$

$$C_B = C_0(0.6 - Z) \tag{b}$$

$$C_C = C_0 Z \tag{c}$$

The rate expression is $r = kC_A C_B$, and using (a) and (b),

$$r = kC_0^2(0.4 - Z)(0.6 - Z) \tag{d}$$

Substituting (d) into Eq. 6.2.1, the design equation becomes

$$\frac{dZ}{d\tau} = kC_0^2(0.4 - Z)(0.6 - Z)\left(\frac{t_{cr}}{C_0}\right) \tag{e}$$

We define the characteristic reaction time by

$$t_{cr} = \frac{1}{kC_0} \tag{f}$$

and the design equation reduces to

$$\frac{dZ}{d\tau} = (0.4 - Z)(0.6 - Z) \tag{g}$$

We solve (h) numerically, subject to the initial condition that at $\tau = 0$, $Z = 0$, and plot the reaction operating curve, shown in Figure E6.4.1.

Once we know the reaction curve, we can determine readily the species curves using Eq. 2.7.4:

$$\frac{N_A(\tau)}{(N_{tot})_0} = 0.4 - Z(\tau) \tag{h}$$

$$\frac{N_B(\tau)}{(N_{tot})_0} = 0.6 - Z(\tau) \tag{i}$$

$$\frac{N_C(\tau)}{(N_{tot})_0} = Z(\tau) \tag{j}$$

Figure E6.4.2 shows the species curves.

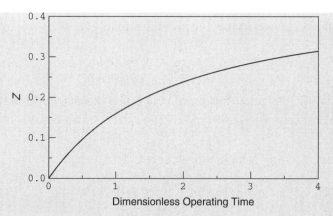

Figure E6.4.1   Reaction curve.

a. To determine the operating time required for 80% conversion, we have to calculate first the corresponding extent. Using Eq. 2.6.2,

$$Z = -\frac{y_A(0)}{s_A} f_A = 0.32$$

From the operating curve (or the numerical solution of the design equation), $Z = 0.32$ is reached at $\tau = 4.25$. To calculate the actual operating time, we first have to determine the characteristic reaction time. We calculate C_0, which for an ideal gas is

$$C_0 = \frac{P_0}{RT_0} = \frac{2 \text{ atm}}{(82.05 \times 10^{-3} \text{ L atm/mol K})(523 \text{ K})}$$

$$= 4.66 \times 10^{-2} \text{ mol/L}$$

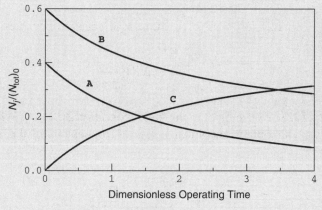

Figure E6.4.2   Species curves.

Using the definition of the dimensionless time, (f),

$$t = \frac{\tau}{kC_0} = 114.0 \text{ min}$$

b. Using Eq. 2.7.6, the reactor pressure at 80% conversion of reactant A is

$$P(t) = P_0[1 + \Delta Z(t)] = 1.36 \text{ atm}$$

c. Using (c), the concentration of product C is

$$C_C = C_0 Z = 1.49 \times 10^{-2} \text{ mol/L}$$

6.2.2 Gaseous, Variable-Volume Batch Reactors

In this section, we consider ideal batch reactors whose volume changes during the operation, like the one shown schematically in Figure 6.2. In general, when a gas-phase reaction results in a net change of the total number of moles, the volume of the reactor changes during the operation if the reactor is maintained at a constant pressure and temperature. For example, the combustion stroke of a car engine cycle (while the valves are closed) is carried out in a variable-volume reactor; in fact, the change in the volume is utilized to generate mechanical work. For such operations, we should incorporate the changes in the reactor volume into the design equation.

The variation of the reactor volume with the reaction extent is given by Eq. 6.1.14. For *isothermal-isobaric* operations with single reactions, and when the initial state is selected as the reference, Eq. 6.1.14 reduces to

$$\frac{V_R(t)}{V_{R0}} = 1 + \Delta Z(t) \tag{6.2.21}$$

Substituting Eq. 6.2.21 into Eq. 6.1.1, the design equation becomes

$$\frac{dZ}{d\tau} = r(1 + \Delta Z)\left(\frac{t_{\text{cr}}}{C_0}\right) \tag{6.2.22}$$

Equation 6.2.22 is solved subject to the initial condition $Z(0) = 0$. When r depends only on Z, we can obtain analytical solutions by separating the variables and integrating:

$$\tau = \left(\frac{C_0}{t_{\text{cr}}}\right) \int_0^{Z(\tau)} \frac{dZ}{r(1 + \Delta Z)} \tag{6.2.23}$$

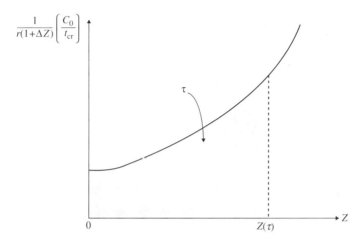

Figure 6.6 Graphical presentation of the design equation for variable-volume batch reactor.

Equation 6.2.23 is the integral form of the design equation, and it is represented graphically in Figure 6.6. Note that for the special case when $\Delta = 0$ (no change in the total number of moles), Eq. 6.2.22 reduces to Eq. 6.2.5, the design equations for a constant-volume batch reactor. Also, by multiplying both sides of Eq. 6.2.23 by t_{cr}, we express the design equation in terms of the actual operating time (rather than the dimensionless time):

$$t = C_0 \int_0^{Z(t)} \frac{dZ}{r(1 + \Delta Z)} \tag{6.2.24}$$

To solve the design equation, we have to express the reaction rate r in terms of Z and, to do so we relate the species concentrations to the dimensionless extent. From Eq. 6.1.11, for isothermal operations with single reactions, and when the reference state is the initial state:

$$C_j(\tau) = C_0 \frac{y_j(0) + s_j Z(\tau)}{1 + \Delta Z(\tau)} \tag{6.2.25}$$

Note that, when $\Delta \neq 0$, the species concentrations are *not* proportional to the reaction extent. To determine $Z(\tau)$ when $C_j(\tau)$ is given, we rearrange Eq. 6.2.25,

$$Z(\tau) = \frac{C_j(\tau) - y_j(0)C_0}{s_j C_0 - \Delta C_j(\tau)} \tag{6.2.26}$$

Below, we analyze the operation of variable-volume, gaseous batch reactors and describe how to apply the design equation for different cases.

First, we discuss the application of the design equation for a variable-volume batch reactor when the reaction rate is provided in the form of an algebraic expression. We start by considering chemical reactions of the form

$$A \longrightarrow \text{Products}$$

whose rate expression is $r = kC_A^{\alpha}$. For this case, $s_A = -1$, and when the initial state is the reference state, Eq. 6.2.25 is used to express the species concentrations. The rate expression becomes

$$r = kC_0^{\alpha} \left(\frac{y_A(0) - Z}{1 + \Delta Z} \right)^{\alpha} \tag{6.2.27}$$

Substituting Eq. 6.2.27 into Eq. 6.2.22, the design equation is

$$\frac{dZ}{d\tau} = kC_0^{\alpha} \left[(1 + \Delta Z) \left(\frac{y_A(0) - Z}{1 + \Delta Z} \right)^{\alpha} \right] \left(\frac{t_{cr}}{C_0} \right) \tag{6.2.28}$$

For αth-order reactions, the characteristic reaction time is

$$t_{cr} = \frac{1}{kC_0^{\alpha - 1}} \tag{6.2.29}$$

Substituting Eq. 6.2.29 into Eq. 6.2.28, the design equation reduces to

$$\frac{dZ}{d\tau} = \frac{[\, y_A(0) - Z]^{\alpha}}{(1 + \Delta Z)^{\alpha - 1}} \tag{6.2.30}$$

which can be solved subject to a specified initial condition, $Z(0)$. Separating the variables and integrating,

$$\tau = \int_{Z(0)}^{Z(\tau)} \frac{(1 + \Delta Z)^{\alpha - 1}}{(1 - Z)^{\alpha}} \, dZ \tag{6.2.31}$$

Equation 6.2.31 is the integral design equation for this case. Figure 6.7 shows the reaction operating curve of a second-order reaction ($\alpha = 2$) for different values of Δ. Note that for larger values of Δ, longer operating times are required for achieving a given extent. Also, the curve for $\Delta = 0$ represents the performance of a constant-volume batch reactor.

We continue the analysis of gaseous, variable-volume batch reactors and consider now chemical reactions with two reactants of the general form

$$A + bB \longrightarrow \text{Products}$$

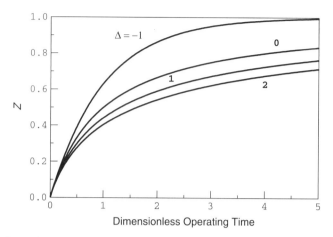

Figure 6.7 Operating curves of isothermal-isobaric batch reactor with second-order gaseous reaction A → products for different values of Δ.

whose rate expression is of the form $r = kC_A^{\alpha}C_B^{\beta}$. For this case, $s_A = -1$, $s_B = -b$, and the initial composition, $y_A(0)$ and $y_B(0)$, is specified. When the initial state is selected as the reference state, using Eq. 6.2.25, the rate expression becomes

$$r = kC_0^{\alpha+\beta} \frac{[y_A(0) - Z]^{\alpha}[y_B(0) - bZ]^{\beta}}{(1 + \Delta Z)^{\alpha+\beta}} \tag{6.2.32}$$

Substituting Eq. 6.2.32 into Eq. 6.2.22, the design equation becomes

$$\frac{dZ}{d\tau} = (1 + \Delta Z)\left[kC_0^{\alpha+\beta} \frac{[y_A(0) - Z]^{\alpha}[y_B(0) - bZ]^{\beta}}{(1 + \Delta Z)^{\alpha+\beta}}\right]\left(\frac{t_{cr}}{C_0}\right) \tag{6.2.33}$$

Using Eq. 3.5.4, for reactions whose overall order is $\alpha + \beta$, the characteristic reaction time is

$$t_{cr} = \frac{1}{kC_0^{\alpha+\beta-1}} \tag{6.2.34}$$

Substituting Eq. 6.2.34 into Eq. 6.2.33, the design equation reduces to

$$\frac{dZ}{d\tau} = \frac{[y_A(0) - Z]^{\alpha}[y_B(0) - bZ]^{\beta}}{(1 + \Delta Z)^{\alpha+\beta-1}} \tag{6.2.35}$$

Equation 6.2.35 is the differential design equation for this case and should be solved subject to the initial condition that at $\tau = 0$, $Z = 0$. Separating the variables and integrating, we obtain

$$\tau = \int\limits_{0}^{Z(\tau)} \frac{(1 + \Delta Z)^{\alpha+\beta-1}}{[y_A(0) - Z]^{\alpha}[y_B(0) - bZ]^{\beta}} \, dZ \qquad (6.2.36)$$

Equation 6.2.36 is the integral design equation for this case.

The design equation for gaseous variable-volume batch reactors was derived under two assumptions: (i) All the species are gaseous, and (ii) the mixture behaves as an ideal gas. In some operations, one or more of the species (especially heavier products generated by the reaction) may be saturated vapor. In this case, any additional amount generated will be in a condensed phase (liquid). While the ideal gas relation provides a reasonable approximation for the volume of species in the vapor phase, it cannot be applied for their volume in the liquid phase. Below, we modify the design equations for a variable-volume batch reactor with saturated vapors.

Recognizing that, with the exception of operations at very high pressures (or near critical conditions), the specific volume of a species in the liquid-phase is two to three orders of magnitude smaller than its specific volume in the vapor phase, the volume of the species that is in a liquid phase can be neglected. Hence, when considering variations in the reactor volume, we should account only for changes in the number of moles in the gas phase. Thus, the factor Δ used in the design equation should be calculated on the basis of the change in the number of moles in the gas phase only, Δ_{gas}, not Δ of the reaction. The calculation of Δ_{gas} depends on the reaction extent when the two phases are in equilibrium, as illustrated in the example below.

Example 6.5 The elementary gas-phase reaction

$$A + B \longrightarrow C$$

is carried out in an isothermal-isobaric (variable-volume) batch reactor. Its initial volume is 234.5 L, the pressure is 1.2 atm, and the temperature is 70°C. At this temperature, $k = 3.2 \, \text{L/mol min}$, and the vapor pressure of the product C is 0.3 atm. The reactor is filled with a stoichiometric mixture of A and B. Determine:

a. The reaction extent when C starts to condense.

b. The operating time when C starts to condense.

c. The operating time needed for 80% conversion.

Solution At the beginning of the operation, all the species are gaseous, the chemical reaction is

$$A(g) + B(g) \longrightarrow C(g)$$

and the stoichiometric coefficients are

$$s_A = -1 \qquad s_B = -1 \qquad s_C = 1 \qquad \Delta_{gas} = \Delta = -1$$

We select the initial state as the reference state, and the initial composition is $y_A(0) = y_B(0) = 0.5$ and $y_C(0) = 0$. Using Eqs. 2.7.4 and 2.7.6,

$$\frac{N_A(\tau)}{(N_{tot})_0} = \frac{N_B(\tau)}{(N_{tot})_0} = 0.5 - Z(\tau) \tag{a}$$

$$\frac{N_C(\tau)}{(N_{tot})_0} = Z(\tau) \tag{b}$$

$$\frac{N_{tot}(\tau)}{(N_{tot})_0} = 1 - Z(\tau) \tag{c}$$

Using Eq. 6.1.16, the concentrations of the species are

$$C_A = C_B = C_0 \frac{0.5 - Z}{1 - Z} \tag{d}$$

$$C_C = C_0 \frac{Z}{1 - Z} \tag{e}$$

Applying ideal gas law, the initial (reference) total number of moles is

$$(N_{tot})_0 = \frac{P_0 V_{R_0}}{RT_0} = 10 \text{ mol}$$

and the reference concentration is

$$C_0 \equiv \frac{(N_{tot})_0}{V_{R_0}} = 0.0426 \text{ mol/L}$$

a. Condensation of product C commences when its partial pressure in the reactor is 0.3 atm. Assuming ideal gas behavior,

$$P_C(\tau) = y_C(\tau)P(\tau) = \frac{N_C(\tau)}{N_{tot}(\tau)}P(\tau)$$

Using the stoichiometric relations (b) and (c),

$$P_C(\tau) = \frac{Z(\tau)}{1 - Z(\tau)}(1.2) = 0.3 \tag{f}$$

Solving (f), $Z(\tau) = 0.20$.

b. Using Eq. 6.2.22, the design equation for an isothermal-isobaric operation is

$$\frac{dZ}{d\tau} = r(1 - Z)\left(\frac{t_{cr}}{C_0}\right) \quad 0 \leq Z \leq 0.2 \tag{g}$$

Using (d) and (e), the rate expression is

$$r = kC_A C_B = kC_0^2 \frac{(0.5 - Z)^2}{(1 - Z)^2} \tag{h}$$

Using Eq. 3.5.4, for a second-order reaction, the characteristic reaction time is

$$t_{cr} = \frac{1}{kC_0} = 7.33 \text{ min} \tag{i}$$

and $\tau = kC_0 t$. Substituting (h) and (i) into (g), the design equation reduces to

$$\frac{dZ}{d\tau} = \frac{(0.5 - Z)^2}{1 - Z} \quad 0 \leq Z \leq 0.2 \tag{j}$$

The dimensionless operating time for $Z = 0.20$ is

$$\tau = \int_0^{0.2} \frac{1 - Z}{(0.5 - Z)^2} \, dZ = 1.177 \tag{k}$$

and, the operating time is

$$t = \frac{\tau}{kC_0} = 8.63 \text{ min} \tag{}$$

c. For $\tau > 1.178$ (or $Z > 0.2$), product C is formed in a liquid phase; hence the following reaction takes place in the reactor:

$$A(g) + B(g) \longrightarrow C(\text{liquid}) \tag{l}$$

For this reaction, $\Delta_{gas} = -2$. Hence, a portion of C is in gas phase and a portion in liquid phase (with negligible volume). For $\tau > 1.178$, the total number of moles in the gas phase (for $Z > 0.20$) is

$$(N_{tot})_{gas} = (N_{tot})_0[1 + (-1)0.2 + (-2)(Z - 0.2)] = (N_{tot})_0(1.2 - 2Z)$$

and, for isothermal-isobaric operation, the reactor volume varies according to

$$\frac{V_R}{V_{R_0}} = \frac{N_{\text{tot}}}{(N_{\text{tot}})_0} = 1.2 - 2Z \tag{m}$$

Using (m), for $\tau > 1.178$ the design equation is

$$\frac{dZ}{d\tau} = r(1.2 - 2Z)\left(\frac{t_{\text{cr}}}{C_0}\right) \quad Z > 0.2 \tag{n}$$

and the concentrations of the two reactants are

$$C_A = C_B = C_0 \frac{0.5 - Z}{1.2 - 2Z} \quad Z > 0.2 \tag{o}$$

Substituting (o) into (n), the design equation becomes

$$\frac{dZ}{d\tau} = \frac{(0.5 - Z)^2}{(1.2 - 2Z)} \quad Z > 0.2 \tag{p}$$

To determine the dimensionless operating time for 80% conversion ($Z = 0.4$), we separate the variables and integrate (numerically),

$$\tau = \int_{0.2}^{0.4} \frac{1.2 - 2Z}{(0.5 - Z)^2} dZ = 3.5305 \tag{q}$$

and the operating time for $Z > 0.2$ is

$$t = \frac{\tau}{kC_0} = 25.9 \text{ min}$$

The total operating time needed for 80% conversion is $t = 8.63 + 25.9 = 34.53$ min.

d. The reactor operation is described by two design equations, each applies for a certain period during the operation: (j) for $0 \leq Z \leq 0.2$ and (p) for $Z > 0.2$. Figure E6.5.1 shows the reaction curve. We use (a) and (b) to determine the species curves, shown in Figure E6.5.2. Note that in the second stage of the operation, the Δ factor is smaller than in the first stage; consequently, the total operating time is smaller than that calculated by ignoring the condensation of species C. If we integrated (k) between 0 and 0.4, we would have obtained $\tau = 5.61$ and the operating time is 41.15 min.

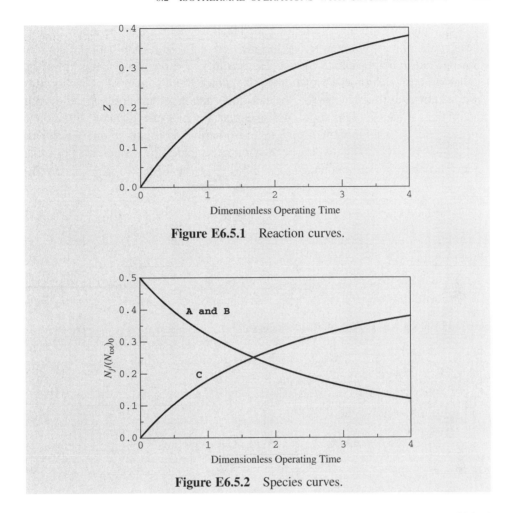

Figure E6.5.1 Reaction curves.

Figure E6.5.2 Species curves.

We conclude the discussion on the applications of the design equation of ideal batch reactors with two important comments. First, note that in all the cases discussed above, the design equation provides only the reactor operating time (either dimensionless or dimensional) needed to obtain a given extent level. The design equation does not indicate what reactor size (volume) should be used. The reactor size is determined by the required production rate and by the downtime between batches (for loading, downloading, and cleaning). Second, one may ask what extent level (or conversion) should be specified for a given operation. The optimal extent level is determined by the cost of the reactants, the value of the products, the cost of the equipment, and the operating expenses. These issues are discussed in Chapter 10.

6.2.3 Determination of the Reaction Rate Expression

In the preceding two sections, we discussed how to apply the design equation when the rate expression is known. In this section, we will describe methods to determine the reaction rate from operating data, and then determine the parameters of the rate expression.

The main difficulty in determining the reaction rate r is that the extent is not a measurable quantity. Therefore, we have to derive a relationship between the reaction rate and the appropriate measurable quantity. We do so by using the design equation and stoichiometric relations. Also, since the characteristic reaction time is not known a priori, we write the design equation in terms of operating time rather than dimensionless time. Assume that we measure the concentration of species j, $C_j(t)$, as a function of time in an isothermal, constant-volume batch reactor. To derive a relation between the reaction rate, r, and $C_j(t)$, we divide both sides of Eq. 6.2.4, by t_{cr} and obtain

$$\frac{dZ}{dt} = \frac{1}{C_0} r \qquad (6.2.37)$$

From Eq. 6.1.12,

$$C_j(t) = C_0[\, y_j(0) + s_j Z(t)] \qquad (6.2.38)$$

Differentiating Eq. 6.2.38 with respect to time,

$$\frac{dZ}{dt} = \frac{1}{C_0 s_j} \frac{dC_j}{dt} \qquad (6.2.39)$$

and substituting it into Eq. 6.2.37,

$$r = \frac{1}{s_j} \frac{dC_j}{dt} \qquad (6.2.40)$$

Since the determination of an r involves differentiating of experimental data with respect to time and applying the differential form of the design equation, it is commonly called the *differential method*.

Before we illustrate how to apply Eq. 6.2.40, a comment on numerical differentiation is in order. To achieve higher accuracy of the derivatives, we use second-order differentiation relations (see Appendix C). Therefore, when the data points are equally spaced, we calculate the slope of each point, except the two endpoints, using the central differentiation equation. For the first point, we use the backward differentiation equation, and, for the last point, we use the forward differentiation equation. When the points are not equally spaced, we use the central differentiation equation for the midpoint between any two adjacent data points. Hence, for n data points on species concentrations, we obtain $n - 1$ derivative values for the midpoints concentrations.

Next, we discuss how to determine the parameters of the rate expression. The determination of the rate expression is usually carried out in three steps:

- Determining the form of the rate expression [function $h(C_j$'s) in Eq. 3.3.1 and Eq. 6.1.5]

- Determining the value of the rate constant, $k(T)$, at a given operating temperature
- Determining the activation energy

The determination of the rate constant's parameters (activation energy and the frequency factor) is carried out by calculating the rate constant at a series of different temperatures and then applying the procedure described in Section 3.3. Below, we describe methods to determine the parameters of rate expressions that are power functions of the species concentrations. Similar procedures are used for other forms of the rate expression.

Consider a chemical reaction whose rate expression is powers of the concentrations:

$$r = k C_A{}^{\alpha} C_B{}^{\beta} \tag{6.2.41}$$

When the reaction rate is determined at different reactant concentrations, we rewrite Eq. 6.2.41 as

$$\ln r = \ln k + \alpha \ln C_A + \beta \ln C_B \tag{6.2.42}$$

Combining Eqs. 6.2.40 and 6.2.42,

$$\ln \left(\frac{1}{s_j} \frac{dC_j}{dt} \right) = \ln k + \alpha \ln C_A + \beta \ln C_B \tag{6.2.43}$$

To determine the reaction of individual orders (α, β, ...), we apply multivariable linear regression. To obtain reliable values for the orders, we need many data points of the reaction rate at different reactant concentrations. In many cases, we want a quick estimate of the orders. To obtain these, we conduct a few experiments while maintaining the concentration of some reactants constant. We write Eq. 6.2.41 for two runs and take the ratio between them:

$$\frac{r_1}{r_2} = \frac{k_1}{k_2} \left(\frac{C_{A_1}}{C_{A_2}} \right)^{\alpha} \left(\frac{C_{B_1}}{C_{B_2}} \right)^{\beta} \tag{6.2.44}$$

Thus if, for example, the two runs are conducted at the same temperature, ($k_1 = k_2$), and in both of them, C_B is the same ($C_{B1} = C_{B2}$), we can calculate the order of A, α, from

$$\alpha = \frac{\ln (r_1/r_2)}{\ln (C_{A1}/C_{A2})} \tag{6.2.45}$$

Example 6.6   Kinetic measurements of the reaction

$$A + B \longrightarrow C$$

were carried out in an isothermal, constant-volume batch reactor at 80°C. Based on the data below, determine the rate expression. *Data*:

Run	Concentration (mol/L)		Rate (mol/min L)
	C_A	C_B	
1	0.5	0.5	1.30
2	1.0	0.5	2.60
3	0.5	1.0	1.84

Solution   Noting that $C_{B1} = C_{B2}$, and since the temperature is constant, $k_1 = k_2$, we write Eq. 6.2.44 for runs 1 and 2,

$$\frac{1.30}{2.60} = \left(\frac{0.5}{1.0}\right)^{\alpha}$$

and find that $\alpha = 1$. Similarly, noting that $C_{A1} = C_{A3}$ and $k_1 = k_3$, we write Eq. 6.2.44 for runs 1 and 3,

$$\frac{1.30}{1.84} = \left(\frac{0.5}{1.0}\right)^{\beta}$$

and find that $\beta = 0.5$. The rate expression is $r = kC_A C_B^{0.5}$. Now that we know the orders, we can determine the value of the rate constant at 80°C. For each run,

$$k = \frac{r}{C_A^2 C_B^{0.5}}$$

The average value of k for the three runs is $6.13 \text{ L}^{1.5} \text{ mol}^{-1.5} \text{ min}^{-1}$.

Once the form of the rate expression is known, we can determine the reaction rate constant, $k(T)$, by using the integral form of the design equation. Consider the integral form of the design equation (Eq. 6.2.5):

$$\tau = \left(\frac{C_0}{t_{cr}}\right) \int_0^{Z(\tau)} \frac{dZ}{r} = \left(\frac{C_0}{t_{cr}}\right) G[Z(\tau)] \qquad (6.2.46)$$

The right-hand side of Eq. 6.2.46 is a function of $Z(\tau)$, denoted by $G[Z(\tau)]$, that depends on the extent at the end of the operation. For rate expressions that are powers of the concentrations, we can readily calculate the values of the function

$G[Z(\tau)]$ at different values of τ. For example, for reactions whose stoichiometry is A → Products and whose rate expression is of the form $r = kC_A{}^\alpha$, we derive $G[Z(\tau)]$ from Eq. 6.2.46:

$$G[Z(t)] = -\ln[1 - Z(t)] \qquad \text{for } \alpha = 1 \qquad (6.2.47a)$$

$$G[Z(t)] = \frac{1}{\alpha - 1}\left\{\left(\frac{Z(t)}{1 - Z(t)}\right)^{\alpha - 1} - 1\right\} \qquad \text{for } \alpha \neq 1 \qquad (6.2.47b)$$

Now, recalling the definition of the dimensionless operating time, $\tau = t/t_{cr}$, and, using Eq. 6.2.46, by plotting $G[Z(t)]$ versus the operating time t, we obtain a straight line that passes through the origin, whose slope is $1/t_{cr}$. Once we determine the value of t_{cr} from the slope, we calculate the value of the rate constant k, using Eq. 6.2.9. Because the integral form of the design equation is used, this method is commonly referred to as the *integral method*.

Two comments on the integral method are in order. First, since the extent is not a measurable quantity, to obtain $G[Z(t)]$, we have to express $Z(t)$ in terms of a measurable quantity (species concentration, pressure, etc.) using stoichiometric relations. Second, the integral method can be applied to determine the form of the rate expression. The determination of the rate expression goes as follows: first, we assume (hypothesize) the form of the rate expression and use Eq. 6.1.12 to derive the function $G[Z(t)]$. Next, we plot $G[Z(t)]$ versus the operating time t to check whether the experimental data fit the function $G[Z(t)]$. When the experimental data fit the derived straight line, we accept the hypothesized rate expression. Then, we use the slope of the line to determine the reaction rate constant. If the experimental data do not fit the straight line for the derived $G[Z(t)]$, we reject the hypothesized rate expression and test a different one. Note that each trial involves the integration of the rate expression, and the procedure is tedious and time-consuming. Hence, we usually apply the differential method to determine the form of the reaction, and then use the integral method to determine the value of the reaction rate constant.

Example 6.7 illustrates the determination of the reaction order, a, and the rate constant, $k(T)$, when the progress of the reaction is monitored by measuring the reactor pressure.

Example 6.7 Consider the decomposition of di-*tert*-butyl peroxide in a constant-volume batch reactor. The chemical reaction is

$$(CH_3)_3\,COOC(CH_3)_3 \longrightarrow C_2H_6 + 2CH_3COCH_3$$

Di-*tert*-butyl peroxide is charged into an isothermal, constant-volume batch reactor operated at 480°C. Based on the data below, determine:

a. The order of the reaction, using the differential method
b. The reaction rate constant, using the integral method

Data: Time (min)	0.0	2.5	5.0	10.0	15.0	20.0
Pressure (mm Hg)	7.5	10.5	12.5	15.8	17.9	19.4

Solution   We can describe the reaction as $A \rightarrow B + 2C$, whose stoichiometric coefficients are

$$s_A = -1 \qquad s_B = 1 \qquad s_C = 2 \qquad \Delta = 2$$

Since only reactant A is charged into the reactor, $y_A(0) = 1$ and $y_B(0) = y_C(0) = 0$.

a. From Eq. 6.2.37,

$$r = C_0 \frac{dZ}{dt} \tag{a}$$

where the rate expression is $r = kC_A{}^\alpha$. Using Eq. 6.1.12,

$$C_A(t) = C_0(1 - Z) \tag{b}$$

Substituting (b) into the rate expression and the latter in the design equation,

$$\frac{dZ}{dt} = kC_0{}^{\alpha-1}(1 - Z)^\alpha \tag{c}$$

Taking the log of both sides,

$$\ln\left(\frac{dZ}{dt}\right) = \ln\left(kC_0{}^{\alpha-1}\right) + \alpha \ln(1 - Z) \tag{d}$$

To determine α, we plot $\ln(dZ/dt)$ versus $\ln(1 - Z)$ and obtain α from the slope. In this case, we have data of $P(t)$, so we have to derive a relation between $P(t)$ and $Z(t)$. Selecting the initial state as the reference state and using Eq. 2.7.6. for an ideal gas in a constant-volume reactor

$$\frac{P(t)}{P(0)} = \frac{N_{\text{tot}}(t)}{N_{\text{tot}}(0)} = 1 + 2Z(t) \tag{e}$$

which reduces to

$$Z(t) = \frac{1}{2}\left[\frac{P(t)}{P(0)} - 1\right]$$ (f)

Differentiating (f) with time,

$$\frac{dZ}{dt} = \frac{1}{2P(0)}\frac{dP}{dt}$$ (g)

We substitute (g) and (f) into (d),

$$\ln\left(\frac{dP}{dt}\right) = (\text{constant}) + \alpha \ln\left[3 - \frac{P(t)}{P(0)}\right]$$ (h)

Thus, we plot $\ln(dP/dt)$ versus $\ln[3 - P(t)/P(0)]$ to determine the value of α. We calculate the derivatives numerically, using second-order forward, center, and backward numerical differentiation formulas, and obtain the following:

Run	t (min)	P (mm Hg)	Δt	dP/dt	$3 - P(t)/$ $P(0)$
1	0	7.5		1.40	2.000
2	2.5	10.5	2.5	1.00	1.600
3	5.0	12.5	2.5	0.69	1.333
4	10.0	15.8	5.0	0.54	0.893
5	15.0	17.9	5.0	0.42	0.613
6	20.0	19.4	5.0	0.30	0.413

The plot is shown in Figure E6.7.1. The slope of the line is

$$\alpha = \frac{\ln(1.0) - \ln(0.2)}{\ln(1.6) - \ln(0.31)} = 0.98 \approx 1.0$$

Note that the points are spread around the line. This is typical in plots based on numerical differentiation. It is common to round the value on the order to the nearest multiple of 0.5.

b. In this case, the rate expression is $r = kC_A$, and, from Eq. 3.5.4, for first-order reactions,

$$t_{cr} = \frac{1}{k}$$ (i)

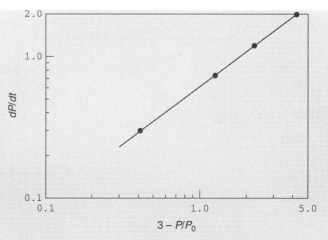

Figure E6.7.1   Determination of the reaction order.

Substituting $\alpha = 1$ into the rate expression and the latter into Eq. 6.2.4, and using (i), the design equation becomes

$$\frac{dZ}{d\tau} = 1 - Z \tag{j}$$

Separating the variables and integrating, using the initial condition that at $Z(0) = 0$,

$$\tau = k \cdot t = -\ln[1 - Z(t)] \tag{k}$$

To determine k, we plot $\ln(1 - Z)$ versus t and obtain a straight line whose slope is $-k$. Using (f),

$$1 - Z(t) = \frac{1}{2}\left[3 - \frac{P(t)}{P(0)}\right] \tag{l}$$

Substituting (l) into (k) and rearranging,

$$\ln\left[3 - \frac{P(t)}{P(0)}\right] = \ln 2 - kt \tag{m}$$

Hence, by plotting $\ln[3 - P(t)/P(0)]$ versus operating time, t, we obtain a straight line whose slope is $-k$. We calculate the needed quantities and show the plot in Figure E6.7.2. From the plot, the slope is

$$-k = \frac{\ln(2.0) - \ln(0.5)}{0 - 17.5} = -7.92 \times 10^{-2}\,\text{min}^{-1}$$

Therefore, the reaction rate constant at $480°C$ is $k = 7.92 \times 10^{-2}\,\text{min}^{-1}$.

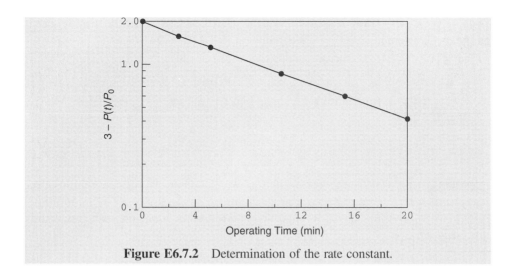

Figure E6.7.2 Determination of the rate constant.

We conclude the discussion with a brief description of several techniques to simplify the determination of the parameters of the rate expression. We can mix the reactants in proportions that are convenient for the determination of the individual orders or the overall orders. Commonly, one of the following mixtures is used:

Stoichiometric Proportion   When the two reactants are in stoichiometric proportion,

$$C_B(0) = \frac{s_B}{s_A} C_A(0) \tag{6.2.48}$$

it follows from Eq. 2.3.6 that for constant-volume reactors,

$$C_B(t) = \frac{s_B}{s_A} C_A(t) \tag{6.2.49}$$

Substituting Eq. 6.2.49, Eq. 5.2.41 becomes

$$r = k' C_A^{\alpha+\beta} \tag{6.2.50}$$

where $k' = k(s_B/s_A)$ is a new constant. When the concentration of species j, $C_j(t)$, is the measured quantity, from Eq. 6.2.50,

$$\ln\left(\frac{1}{s_j}\frac{dC_j}{dt}\right) = \ln k' + (\alpha + \beta)\ln C_A \tag{6.2.51}$$

By plotting $\ln\left[(1/s_j)(dC_j/dt)\right]$ versus $\ln C_A$, we obtain a straight line whose slope is $\alpha + \beta$. Hence, by conducting experiments with the reactants in stoichiometric proportion, we obtain the overall order of the reaction. To determine the orders of the individual species, we use one of the other methods below.

Large Excess of One Reactant   When $C_B(0) \gg (s_B/s_A)C_A(0)$, the relative change in C_B is small, $C_B(t) \approx C_B(0)$, and Eq. 6.2.41 becomes

$$r = k''C_A{}^{\alpha} \tag{6.2.52}$$

where $k'' = kC_B(0)$. Sunstituting Eq. 6.2.52 in Eq. 6.2.44

$$\ln\left(\frac{1}{s_A}\frac{dC_A}{dt}\right) = \ln k'' + \alpha \ln C_A \tag{6.2.53}$$

By plotting $\ln[1/s_j(dC_j/dt)]$ versus $\ln C_A$, we obtain α. Hence, by conducting experiments with a large excess of one reactant, we obtain the order of the limiting reactant. We determine the individual species orders by repeating the procedure with different mixtures, each time using a different limiting reactant.

Keeping the Concentration of One Reactant Constant   Assume we conduct an experiment while maintaining C_B constant. Equation 6.2.40 becomes

$$r = k'''C_A{}^{\alpha} \tag{6.2.54}$$

where $k''' = kC_B$. The method is similar to the one using the excess amount, except that experimentally, this one is more difficult to perform.

Initial Rate Method   For reversible reactions, we use a modified differential method—the initial rate method. In this case, a series of experiments are conducted at selected initial reactant compositions, and each run is terminated at low conversion. From the collected data, we calculate (by numerical differentiation) the reaction rate at the initial conditions. Since the reaction extent is low, the reverse reaction is negligible, and we can readily determine the orders of the forward reaction from the known initial compositions. The rate of the reversible reaction is determined by conducting a series of experiments when the reactor is charged with selected initial product compositions. The initial rate method is also used to determine the rates for complex reactions since it enables us to isolate the effect of different reactants.

6.3   ISOTHERMAL OPERATIONS WITH MULTIPLE REACTIONS

When more than one chemical reaction takes place in the reactor, we have to address several issues before we start the design procedure. We have to determine how many independent reactions take place in the reactor and select a set of independent reactions for the design formulation. Next, we have to identify all the reactions that actually take place (including the dependent reactions) and express their rates. To determine the reactor compositions and all other state quantities, we have to write Eq. 6.1.1 for each of the independent chemical reactions. To solve the design equations (obtain relationships between Z_m's and τ), we have to express the rates of the individual chemical reactions, r_m's and r_k's, in terms of Z_m's and τ. The procedure for designing batch reactors with multiple reactions goes as follows:

1. Identify all the chemical reactions that take place in the reactor and define the stoichiometric coefficients of each species in each reaction.
2. Determine the number of independent chemical reactions.
3. Select a set of independent reactions from among the reactions whose rate expressions are given.
4. For each dependent reaction, determine its α_{km} multipliers with *each* independent reaction, using Eq. 2.4.9.
5. Select a reference state [determine $(N_{tot})_0$, T_0, C_0, V_{R0}] and the initial species compositions, $y_j(0)$'s.
6. Write Eq. 6.1.1 for each independent chemical reaction.
7. Select a leading (or desirable) reaction and determine the expression of the characteristic reaction time, t_{cr}, and its numerical value.
8. Express the reaction rates in terms of the dimensionless extents of the independent reactions, Z_m's.
9. Solve the design equations (Z_m's as functions of τ) and obtain the reaction operating curves.
10. Calculate the species curves of all species, using Eq. 2.7.4.
11. Determine the reactor operating time based on the most desirable value of τ obtained from the dimensionless operating curves.

Below, we describe the design formulation of isothermal batch reactors with multiple reactions for various types of chemical reactions (reversible, series, parallel, etc.). In most cases, we solve the equations numerically by applying a numerical technique such as the Runge-Kutta method, but, in some simple cases, analytical solutions are obtained. Note that, for isothermal operations, we do not have to consider the effect of temperature variation, and we use the energy balance equation to determine the dimensionless heat-transfer number, HTN, required to maintain the reactor isothermal.

We start the analysis with single reversible reactions. When a reversible reaction takes place, there is only one independent reaction; hence, only one design equation should be solved. However, the rates of both forward and backward reactions should be considered. The design procedure is similar to the one discussed in Section 6.2. To illustrate the effect of the reverse reaction, consider the reversible elementary isomerization reaction $A \rightleftarrows B$ in a constant-volume batch reactor. We treat a reversible reaction as two chemical reactions:

$$\text{Reaction 1:} \quad A \longrightarrow B$$
$$\text{Reaction 2:} \quad B \longrightarrow A$$

We select the forward reaction as the independent reaction and the reverse reaction as the dependent reaction. Hence, the index of the independent reaction is $m = 1$, the index of the dependent reaction is $k = 2$. Since Reaction 2 is the reverse of Reaction 1, $\alpha_{21} = -1$. The stoichiometric coefficients of the independent reaction are

$$s_{A1} = -1 \qquad s_{B1} = 1 \qquad \Delta_1 = 0$$

Using Eq. 6.1.1, and selecting the initial state as the reference state, for a constant-volume reaction

$$\frac{dZ_1}{d\tau} = (r_1 - r_2)\left(\frac{t_{cr}}{C_0}\right) \tag{6.3.1}$$

Using Eq. 6.1.12 to express the species concentrations, the two reaction rates are

$$r_1 = k_1 C_0 [y_A(0) - Z_1]$$

$$r_2 = k_2 C_0 [y_B(0) + Z_1]$$

We define the characteristic reaction time on the basis of Reaction 1:

$$t_{cr} = \frac{1}{k_1}$$

Substituting these relations into Eq. 6.3.1, the design equation reduces to

$$\frac{dZ_1}{d\tau} = [y_A(0) - Z_1] - \left(\frac{k_2}{k_1}\right)[y_B(0) + Z_1] \tag{6.3.2}$$

To simplify the design equation, we rearrange Eq. 6.3.2 and notice that, at equilibrium, $(dZ_1/d\tau = 0)$,

$$Z_{1_{eq}} = \frac{y_A(0) - (k_2/k_1)y_B(0)}{1 + (k_2/k_1)} \tag{6.3.3}$$

where $Z_{1_{eq}}$ is the dimensionless extent at equilibrium. Using Eq. 6.3.3, the design equation becomes

$$\frac{dZ_1}{d\tau} = \left(\frac{k_1 + k_2}{k_1}\right)(Z_{1_{eq}} - Z_1) \tag{6.3.4}$$

Separating the variables and integrating, we obtain

$$Z(\tau) = Z_{1_{eq}}\left[1 - \exp\left(-\frac{k_1 + k_2}{k_1}\tau\right)\right] \tag{6.3.5}$$

Figure 6.8 shows the reaction operating curve for different values of k_2/k_1. Note that the design equation for batch reactors with single reversible reactions has two parameters (k_1 and k_2), whereas the design equation for reactors with an irreversible reaction has only one parameter. Also note that for an irreversible reaction, $k_2 = 0$, and, from Eq. 6.3.3, $Z_{1_{eq}} = y_A(0)$.

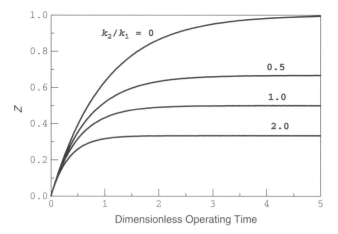

Figure 6.8 Reaction operating curves for single reversible reaction.

Example 6.8 The reversible gas-phase chemical reaction

$$A \rightleftharpoons 2B$$

takes place in an isothermal batch reactor. The forward reaction is first order, and the backward reaction is second order. We charge 100 mol of reactant A into the reactor that operates at 120°C. The initial pressure is 2 atm. At the operating conditions (120°C), $k_1 = 0.1 \text{ min}^{-1}$ and $k_2 = 0.322 \text{ L mol}^{-1} \text{ min}^{-1}$.

a. Derive the design equation and plot the reaction curve for a constant-volume batch reactor.

b. Determine the equilibrium composition at 120°C in a constant-volume reactor.

c. Determine the operating time to obtain 90% of the equilibrium conversion?

d. Derive the design equation and plot the reaction curve for an isobaric, variable-volume batch reactor.

e. Determine the equilibrium composition at 120°C in an isobaric variable-volume reactor.

f. What is the operating time to obtain 90% of the equilibrium conversion?

Solution We write the reversible reaction as two separate reactions:

$$\text{Reaction 1:} \quad A \longrightarrow 2B$$
$$\text{Reaction 1:} \quad 2B \longrightarrow A$$

whose stoichiometric coefficients are

$$s_{A_1} = -1 \quad s_{B_1} = 2 \quad \Delta_1 = 1$$
$$s_{A_2} = 1 \quad s_{B_2} = -2 \quad \Delta_2 = -1$$

We have here one independent reaction, and select Reaction 1 as the independent reaction; hence, the reaction indices are $m = 1$ and $k = 2$, and, for a reversible reaction, $\alpha_{21} = -1$. We select the initial state as the reference state, and the reference concentration is

$$C_0 = \frac{P_0}{RT_0} = \frac{2 \text{ atm}}{(82.06 \times 10^{-3} \text{ L atm/mol K})(393 \text{ K})} = 0.062 \text{ mol/L} \quad \text{(a)}$$

Since only reactant A is charged to the reactor, $y_A(0) = 1$ and $y_B(0) = 0$.

a. We use Eq. 6.1.1; for a constant-volume batch reactor the design equation is

$$\frac{dZ_1}{d\tau} = (r_1 - r_2)\left(\frac{t_{cr}}{C_0}\right) \quad \text{(b)}$$

We define the characteristic reaction time on the basis of reaction 1:

$$t_{cr} = \frac{1}{k_1} = 10 \text{ min} \quad \text{(c)}$$

Equation 6.1.12 expresses the species concentrations; the reaction rates of the two reactions are

$$r_1 = k_1 C_0 (1 - Z_1) \qquad r_2 = k_2 C_0^2 (2Z_1)^2$$

Substituting the rates and (c) into (b), the design equation reduces to

$$\frac{dZ_1}{d\tau} = 1 - Z_1 - \left(\frac{k_2 C_0}{k_1}\right) 4Z_1^2 \quad \text{(d)}$$

Once we solve the design equation, we use Eq. 2.7.4 to obtain the species curves:

$$\frac{N_A(\tau)}{(N_{tot})_0} = 1 - Z_1(\tau) \quad \text{(e)}$$

$$\frac{N_B(\tau)}{(N_{tot})_0} = 2Z_1(\tau) \quad \text{(f)}$$

Using Eq. 2.7.6

$$\frac{N_{tot}(\tau)}{(N_{tot})_0} = 1 + Z_1(\tau) \quad \text{(g)}$$

For the given data, $k_2 C_0 / k_1 = 0.2$, and we solve (d) subject to the initial condition that $Z_1(0) = 0$. Figure E6.8.1 shows the reaction operating curve

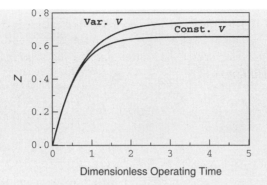

Figure E6.8.1 Reaction curves.

and compares it to the reaction curve obtained on a variable-volume batch reactor. Using (e) and (f), we determine the species curves shown in Figure E6.8.2.

b. At equilibrium, $dZ_1/d\tau = 0$, and, the equilibrium extent is $Z_{1eq} = 0.6563$. Using (e), (f), and (g), the species molar fractions at equilibrium are

$$y_{A_{eq}} = \frac{1 - Z_{1_{eq}}}{1 + Z_{1_{eq}}} = 0.2075 \qquad y_{B_{eq}} = \frac{2Z_{1_{eq}}}{1 + Z_{1_{eq}}} = 0.7925$$

c. A level of 90% equilibrium conversion corresponds to $Z = 0.5907$, and, from the reaction curve, it is reached at $\tau = 1.26$. Using (c), the operating time is

$$t = t_{cr}\tau = 12.6 \text{ min}$$

d. For an isobaric, variable-volume batch reactor, the design equation Eq. 6.1.1 is

$$\frac{dZ_1}{d\tau} = (r_1 - r_2)(1 + \Delta_1 Z_1)\left(\frac{t_{cr}}{C_0}\right) \qquad \text{(h)}$$

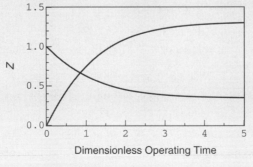

Figure E6.8.2 Species curves.

Using Eq. 6.1.16 to express the species concentrations, the reaction rates are

$$r_1 = k_1 C_0 \left(\frac{1 - Z_1}{1 + Z_1} \right) \qquad r_2 = k_2 C_0^2 \left(\frac{2Z_1}{1 + Z_1} \right)^2 \tag{i}$$

Substituting the rates and (c) into (h), the design equation becomes

$$\frac{dZ_1}{d\tau} = 1 - Z_1 - \left(\frac{k_2 C_0}{k_1} \right) \frac{4Z_1^2}{1 + Z_1} \tag{j}$$

We solve (j) subject to the initial condition that $Z_1(0) = 0$. The reaction curve is shown in Figure E6.8.1.

e. At equilibrium, $dZ_1/d\tau = 0$, and, from (j), the equilibrium extent is $Z_{1_{eq}} = 0.7455$. Using (e), (f), and (g), the species molar fractions at equilibrium are

$$y_{A_{eq}} = \frac{1 - Z_{1_{eq}}}{1 + Z_{1_{eq}}} = 0.1458 \qquad y_{B_{eq}} = \frac{2 \cdot Z_{1_{eq}}}{1 + Z_{1_{eq}}} = 0.8542$$

f. A level of 90% equilibrium conversion corresponds to $Z = 0.671$ and, from the operating curve, it is reached at $\tau = 1.56$; using (c), the operating time is 15.6 min.

Next, we consider series (consecutive) chemical reactions. These are reactions where the product of one reaction reacts to form undesirable species. In such cases, it is important to consider the amount of desirable and undesirable products formed in addition to the conversion of the reactant. In many instances, the yield of the desirable product provides a measure of the reactor performance.

Example 6.9   A valuable product B is produced in a batch reactor where the following liquid-phase reactions take place:

$$\text{Reaction 1:} \qquad A \longrightarrow 2B$$
$$\text{Reaction 2:} \qquad B \longrightarrow C + D$$

Each chemical reaction is first order. The reactor is charged with 200 L of an organic solution with a concentration of 4 mol A/L. The reactor is operated at 120°C. At the reactor temperature, $k_1 = 0.05\ \text{min}^{-1}$ and $k_2 = 0.025\ \text{min}^{-1}$.

a. Derive the design equations and plot the reaction and species curves.
b. Derive an expression for the yield of product B, and plot it as a function of the operating time.

c. Determine the operating time when the production of product B is maximized.

d. Determine the conversion of reactant A and the yield and selectivity of product B at optimal operating time.

Solution This is an example of series (sequential) chemical reactions. The stoichiometric coefficients of the chemical reactions are

$$s_{A_1} = -1 \qquad s_{B_1} = 2 \qquad s_{C_1} = 0 \qquad s_{D_1} = 0 \qquad \Delta_1 = 1$$
$$s_{A_2} = 0 \qquad s_{B_2} = -1 \qquad s_{C_2} = 1 \qquad s_{D_2} = 1 \qquad \Delta_2 = 1$$

Since each reaction has a species that does not participate in the other, the two reactions are independent, and there is no dependent reaction.

a. We write Eq. 6.1.1 for each independent reaction:

$$\frac{dZ_1}{d\tau} = r_1\left(\frac{t_{cr}}{C_0}\right) \tag{a}$$

$$\frac{dZ_2}{d\tau} = r_2\left(\frac{t_{cr}}{C_0}\right) \tag{b}$$

We select the initial state as the reference sate; hence, $C_0 = C_A(0)$, and

$$(N_{tot})_0 = V_R(0)C_0 = 800 \text{ mol}$$

Since only reactant A is charged into the reactor, $y_A(0) = 1$, and $y_B(0) = y_C(0) = y_D(0) = 0$. We use Eq. 6.1.12 to express the species concentrations, and the rates of the two reactions are

$$r_1 = k_1 C_0 (1 - Z_1) \tag{c}$$

$$r_2 = k_2 C_0 (2Z_1 - Z_2) \tag{d}$$

We select reaction 1, the desirable reaction, to define the characteristic reaction time

$$t_{cr} = \frac{1}{k_1} = 20 \text{ min} \tag{e}$$

Substituting (c), (d), and (e) into (a) and (b), the two design equations reduce to

$$\frac{dZ_1}{d\tau} = 1 - Z_1 \tag{f}$$

$$\frac{dZ_2}{d\tau} = \left(\frac{k_2}{k_1}\right)(2Z_1 - Z_2) \tag{g}$$

After we solve the design equations, we use Eq. 2.7.4, to determine the species curves,

$$\frac{N_A}{(N_{\text{tot}})_0} = 1 - Z_1 \tag{h}$$

$$\frac{N_B}{(N_{\text{tot}})_0} = 2Z_1 - Z_2 \tag{i}$$

$$\frac{N_C}{(N_{\text{tot}})_0} = \frac{N_D}{(N_{\text{tot}})_0} = Z_2 \tag{j}$$

In this case, the two design equations are not coupled, and we can obtain analytical solutions. We solve (f) subject to the initial condition that $Z_1(0) = 0$, and obtain

$$Z_1(\tau) = 1 - e^{-\tau} \tag{k}$$

We substitute (k) into (g), and the second design equation reduces to

$$\frac{dZ_2}{d\tau} + \left(\frac{k_2}{k_1}\right)Z_2 = 2\left(\frac{k_2}{k_1}\right)(1 - e^{-\tau}) \tag{l}$$

This is a first-order linear differential equation that can be solved analytically by using an integrating factor. We solve (l) subject to the initial condition that $Z_2(0) = 0$, and obtain

$$Z_2(\tau) = 2\left[1 - \left(\frac{k_1}{k_1 - k_2}\right)e^{-(k_2/k_1)\tau} + \left(\frac{k_2}{k_1 - k_2}\right)e^{-\tau}\right] \tag{m}$$

For the given data, $k_2/k_1 = 0.5$, we plot $Z_1(\tau)$ and $Z_2(\tau)$ in Figure E6.9.1. Once we have $Z_1(\tau)$ and $Z_2(\tau)$, we use (h) through (j) to plot the species curves, shown in Figure E6.9.2.

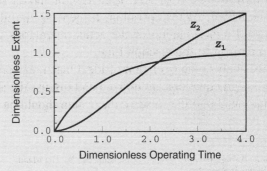

Figure E6.9.1 Reaction curves.

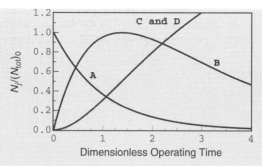

Figure E6.9.2   Species curves.

b. The desirable chemical reaction is Reaction 1, and using Eq. 2.6.12, the yield of product B is

$$\eta_B(\tau) = -\left(\frac{s_A}{s_B}\right)_1 \left[\frac{N_B(\tau) - N_B(0)}{N_A(0)}\right]$$

and, using (i),

$$\eta_B(\tau) = \frac{1}{2}[2Z_1(\tau) - Z_2(\tau)] = Z_1(\tau) - 0.5Z_2(\tau) \qquad \text{(n)}$$

Figure E6.9.3 shows the yield curve.

c. To determine the optimal operating time, we use either the curve of product B or the analytical solutions. From the graph, maximum N_B is reached at $\tau = 1.39$. Alternatively, using (i), (k), and (m),

$$\frac{N_B(\tau)}{N_{tot}(0)} = 2(1 - e^{-\tau}) - 2\left[1 - \left(\frac{k_1}{k_1 - k_2}\right)e^{-(k_2/k_1)\tau} + \left(\frac{k_2}{k_1 - k_2}\right)e^{-\tau}\right] \qquad \text{(o)}$$

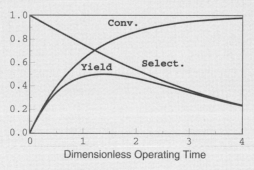

Figure E6.9.3   Yield and selectivity.

To determine the maximum, we take the derivative of (o) with respect to τ and equate it to zero

$$\tau_{\max_B} = \left(\frac{k_1}{k_1 - k_2}\right) \ln\left(\frac{k_1}{k_1 - k_2}\right) \tag{q}$$

For $k_2/k_1 = 0.5$, $\tau_{\max_B} = 1.39$. Using (e), the optimal operating time is

$$t = \tau t_{\mathrm{cr}} = 27.6 \text{ min}$$

d. For $\tau = 1.39$, the solutions of (l) and (n) are $Z_1 = 0.751$ and $Z_2 = 0.502$. Using (i) and (j), the species amounts are: $N_A = 199.2$ mol. Using (h), the conversion of reactant A is

$$f_A(\tau) = \frac{N_A(0) - N_A(\tau)}{N_A(0)} = 0.751$$

Using (n), the yield of product B is 0.5. To calculate the selectivity of product B, we use Eq. 2.6.17, (h), and (i),

$$\sigma_B(\tau) = -\left(\frac{-1}{2}\right) \frac{2Z_1(\tau) - Z_2(\tau)}{Z_1(\tau)} = 0.666$$

The selectivity curve is shown in Figure E6.9.3.

Next, we consider parallel chemical reactions. These are reactions in which a reactant of the desirable reaction also reacts in another reaction to form undesirable species. In such cases, it is important to consider not only the conversion of the reactant but also the amount of desirable and undesirable products formed. The yield of the desirable product provides a measure of the reactor performance.

Example 6.10 The catalytic oxidation of propylene on bismuth molybdate to produce acrolein is investigated in a constant-volume batch reactor. The following chemical reactions take place in the reactor:

Reaction 1:	$C_3H_6 + O_2 \longrightarrow C_3H_4O + H_2O$
Reaction 2:	$C_3H_4O + 3.5O_2 \longrightarrow 3O_2 + 2 H_2O$
Reaction 3:	$C_3H_6 + 4.5O_2 \longrightarrow 3CO_2 + 3H_2O$

The rate of each reaction is second order (first order with respect to each reactant), and the reaction rate constants at $460°C$ are $k_1 = 0.5\,\text{L mol}^{-1}\,\text{s}^{-1}$, $k_2/k_1 = 0.25$, $k_3/k_1 = 0.10$. A gas mixture consisting of 60% propylene and 40% oxygen is charged into a 4-L isothermal reactor operated at $460°C$. The initial pressure is 1.2 atm.

a. Derive the design equations, and plot the reaction and species operating curves.

b. Determine the operating time needed for 75% conversion of oxygen.

c. Determine the reactor composition at 75% conversion of oxygen.

Solution   The reactor design formulation of these chemical reactions was discussed in Example 4.2. Recall that there are two independent reactions and one dependent reaction, and, following the heuristic rule, we select Reactions 1 and 2 as a set of independent reactions; hence, $m = 1, 2$, and $k = 3$. The stoichiometric coefficients of the independent reactions are

$$(s_{C_3H_6})_1 = -1 \quad (s_{O_2})_1 = -1 \quad (s_{C_3H_4O})_1 = 1 \quad (s_{H_2O})_1 = 1 \quad (s_{CO_2})_1 = 0 \quad \Delta_1 = 0$$
$$(s_{C_3H_6})_2 = 0 \quad (s_{O_2})_2 = -3.5 \quad (s_{C_3H_4O})_2 = -1 \quad (s_{H_2O})_2 = 2 \quad (s_{CO_2})_2 = 3 \quad \Delta_2 = 0.5$$

Recall from Example 4.2 that the multipliers α_{km}'s of the dependent reaction (Reaction 3) and the two independent reactions (Reactions 1 and 2) are $\alpha_{31} = 1$ and $\alpha_{32} = 1$. We select the initial state as the reference state; hence,

$$C_0 = \frac{P_0}{RT_0} = 0.02\,\text{mol/L} \tag{a}$$

$$(N_{\text{tot}})_0 = V_R(0)C_0 = 0.08\,\text{mol}$$

The initial composition is

$$y_{C_3H_6}(0) = 0.6 \quad y_{O_2}(0) = 0.4 \quad y_{C_3H_4O}(0) = y_{H_2O}(0) = y_{CO_2}(0) = 0$$

a. We write Eq. 6.1.1 with $V_R(t) = V_R(0) = V_{R_0}$ for each the independent reaction,

$$\frac{dZ_1}{d\tau} = (r_1 + r_3)\left(\frac{t_{\text{cr}}}{C_0}\right) \tag{b}$$

$$\frac{dZ_2}{d\tau} = (r_2 + r_3)\left(\frac{t_{\text{cr}}}{C_0}\right) \tag{c}$$

Using Eq. 6.1.12 to express the species concentrations, the rates of the three chemical reactions are

$$r_1 = k_1 C_0{}^2 (0.6 - Z_1)(0.4 - Z_1 - 3.5Z_2) \tag{d}$$

$$r_2 = k_2 C_0{}^2 (Z_1 - Z_2)(0.4 - Z_1 - 3.5Z_2) \tag{e}$$

$$r_3 = k_3 C_0{}^2 (0.6 - Z_1)(0.4 - Z_1 - 3.5Z_2) \tag{f}$$

We select Reaction 1 as the leading reaction, and, using Eq. 3.5.4, the characteristic reaction time is

$$t_{cr} = \frac{1}{k_1 C_0} = 100 \text{ s} \tag{g}$$

Substituting (d), (e), (f), and (g) into (b) and (c), the design equations become

$$\frac{dZ_1}{d\tau} = \left(1 + \frac{k_3}{k_1}\right)(0.6 - Z_1)(0.4 - Z_1 - 3.5Z_2) \tag{h}$$

$$\frac{dZ_2}{d\tau} = \left(\frac{k_2}{k_1}\right)(Z_1 - Z_2)(0.4 - Z_1 - 3.5Z_2) +$$

$$\left(\frac{k_3}{k_1}\right)(0.6 - Z_1)(0.4 - Z_1 - 3.5Z_2) \tag{i}$$

Once we solve the design equations, we use Eq. 2.7.4 to obtain the species curves:

$$\frac{N_{C_3H_6}(\tau)}{(N_{tot})_0} = 0.6 - Z_1(\tau) \tag{j}$$

$$\frac{N_{O_2}(\tau)}{(N_{tot})_0} = 0.4 - Z_1(\tau) - 3.5Z_2(\tau) \tag{k}$$

$$\frac{N_{C_3H_4O}(\tau)}{(N_{tot})_0} = Z_1(\tau) - Z_2(\tau) \tag{l}$$

$$\frac{N_{H_2O}(\tau)}{(N_{tot})_0} = Z_1(\tau) + 2Z_2(\tau) \tag{m}$$

$$\frac{N_{CO_2}(\tau)}{(N_{tot})_0} = 3Z_2(\tau) \tag{n}$$

Using Eq. 2.7.6, the total molar content in the reactor is

$$\frac{N_{tot}(\tau)}{(N_{tot})_0} = 1 + 0.5Z_2(\tau) \tag{o}$$

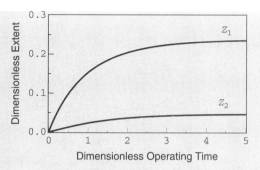

Figure E6.10.1   Reaction curves.

We substitute the numerical values $k_2/k_1 = 0.25$ and $k_3/k_1 = 0.1$ into (h) and (i), and solve them numerically subject to the initial conditions that at $\tau = 0$, $Z_1 = Z_2 = 0$. Figure E6.10.1 shows the reaction curves. We the apply (j) through (n) to determine the species curves, shown in Figure E6.10.2.

b. Using the definition of the conversion, at 75% conversion of oxygen,

$$N_{O_2} = (1 - 0.75)N_{O_2}(0) = (1 - 0.75)0.4(N_{tot})_0$$

hence, $N_{O_2}/(N_{tot})_0 = 0.1$. From the species operating curve (or tabulated values of the numerical solution), this is reached at $\tau = 1.78$. Using (g), the operating time is

$$t = \tau t_{cr} = 178 \text{ s}$$

c. From the reaction curves at $\tau = 1.78$, $Z_1 = 0.204$ and $Z_2 = 0.0275$. Using (j) through (n) and (o), the mole fractions of the individual species are

$$y_{C_3H_6} = 0.391 \qquad y_{O_2} = 0.099 \qquad y_{C_3H_4O} = 0.174$$

$$y_{H_2O} = 0.255 \qquad y_{CO_2} = 0.081$$

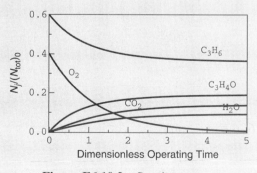

Figure E6.10.2   Species curves.

Example 6.11   Selective oxidation of ammonia is investigated in an isothermal constant-volume batch reactor. The following gaseous chemical reactions take place in the reactor:

Reaction 1:	$4NH_3 + 5O_2 \longrightarrow 4NO + 6H_2O$
Reaction 2:	$4NH_3 + 3O_2 \longrightarrow 2N_2 + 6H_2O$
Reaction 3:	$2NO + O_2 \longrightarrow 2NO_2$
Reaction 4:	$4NH_3 + 6NO \longrightarrow 5N_2 + 6H_2O$

The desired product is NO. A gas mixture consisting of 50% NH_3 and 50% O_2 (mol %) is charged into a 4-L reactor, and the initial pressure is 2 atm. The reactor operates at 609 K.

a. Derive the design equations and plot the reaction and species operating curves.

b. Determine the operating time at which the maximum amount of NO is reached.

c. Determine the reactor composition at that time.

The rate expressions of the reactions are

$$r_1 = k_1 C_{NH_3} C_{O_2}{}^2 \qquad r_2 = k_2 C_{NH_3} C_{O_2} \qquad r_3 = k_3 C_{O_2} C_{NO}{}^2$$
$$r_4 = k_4 C_{NH_3}{}^{2/3} C_{NO}$$

Data: At 609 K,

$$k_1 = 20 \ (L/mol)^2 min^{-1} \qquad k_2 = 0.04 \ (L/mol)min^{-1}$$
$$k_3 = 40 \ (l/mol)^2 min^{-1} \qquad k_4 = 0.0274 \ (L/mol)^{2/3} min^{-1}$$

Solution   The reactor design formulation of these chemical reactions was discussed in Example 4.3. Recall that there are three independent reactions and one dependent reaction, and, following the heuristic rule, we select Reactions 1, 2, and 3 as a set of independent reactions and Reaction 4 is a dependent reaction. Hence, $m = 1, 2, 3, k = 4$. The stoichiometric coefficients of the independent reactions are

$(s_{NH_3})_1 = -4 \ (s_{O_2})_1 = -5 \ (s_{NO})_1 = 4 \quad (s_{H_2O})_1 = 6 \ (s_{N_2})_1 = 0 \ (s_{NO_2})_1 = 0 \ \Delta_1 = 1$
$(s_{NH_3})_2 = -4 \ (s_{O_2})_2 = -3 \ (s_{NO})_2 = 0 \quad (s_{H_2O})_2 = 6 \ (s_{N_2})_2 = 2 \ (s_{NO_2})_2 = 0 \ \Delta_2 = 1$
$(s_{NH_3})_3 = 0 \quad (s_{O_2})_3 = -1 \ (s_{NO})_3 = -2 \ (s_{H_2O})_3 = 0 \ (s_{N_2})_3 = 0 \ (s_{NO_2})_3 = 2 \ \Delta_3 = -1$

Also, from Example 4.3 the multipliers α_{k_m}'s of the dependent reaction and the three independent reactions are $\alpha_{43} = 0$, $\alpha_{42} = 2.5$, and $\alpha_{41} = -1.5$. We select the initial state as the reference state; hence, the reference concentration is

$$C_0 = \frac{P_0}{RT_0} = 0.04 \, \text{mol/L} \tag{a}$$

The total molar content of the reference state is

$$(N_{\text{tot}})_0 = V_{R_0} C_0 = 0.16 \, \text{mol}$$

The initial composition is

$$y_{\text{NH}_3}(0) = 0.5 \quad y_{\text{O}_2}(0) = 0.5 \quad y_{\text{NO}}(0) = y_{\text{H}_2\text{O}}(0) = y_{\text{N}_2}(0) = y_{\text{NO}_2}(0) = 0$$

a. To design the reactor, we write Eq. 6.1.1 for each independent reaction,

$$\frac{dZ_1}{d\tau} = (r_1 - 1.5r_4)\left(\frac{t_{\text{cr}}}{C_0}\right) \tag{b}$$

$$\frac{dZ_2}{d\tau} = (r_2 + 2.5r_4)\left(\frac{t_{\text{cr}}}{C_0}\right) \tag{c}$$

$$\frac{dZ_3}{d\tau} = r_3\left(\frac{t_{\text{cr}}}{C_0}\right) \tag{d}$$

We use Eq. 6.1.12 to express the species concentrations, and the rates of the four chemical reactions are

$$r_1 = k_1 C_0^3 [y_{\text{NH}_3}(0) - 4Z_1 - 4Z_2][y_{\text{O}_2}(0) - 5Z_1 - 3Z_2 - Z_3]^2 \tag{e}$$

$$r_2 = k_2 C_0^2 [y_{\text{NH}_3}(0) - 4Z_1 - 4Z_2][y_{\text{O}_2}(0) - 5Z_1 - 3Z_2 - Z_3] \tag{f}$$

$$r_3 = k_3 C_0^3 [y_{\text{O}_2}(0) - 5Z_1 - 3Z_2 - Z_3][y_{\text{NO}}(0) + 4Z_1 - 2Z_3]^2 \tag{g}$$

$$r_4 = k_4 C_0^{5/3} [y_{\text{NH}_3}(0) - 4Z_1 - 4Z_2]^{2/3}[y_{\text{NO}}(0) + 4Z_1 - 2Z_3] \tag{h}$$

We select Reaction 1, the leading reaction, and using Eq. 3.5.4, the characteristic reaction time is

$$t_{\text{cr}} = \frac{1}{k_1 C_0^2} = \frac{1}{(20)(0.04)^2} = 31.25 \, \text{min} \tag{i}$$

Substituting (e) through (h) and (i) into (b), (j), and (c), the design equations are

$$\frac{dZ_1}{d\tau} = [y_{NH_3}(0) - 4Z_1 - 4Z_2][y_{O_2}(0) - 5Z_1 - 3Z_2 - Z_3]^2$$

$$- 1.5\left(\frac{k_4}{k_1 C_0^{4/3}}\right)[y_{NH_3}(0) - 4Z_1 - 4Z_2]^{2/3}[y_{NO}(0) + 4Z_1 - 2Z_3] \quad (j)$$

$$\frac{dZ_2}{d\tau} = \left(\frac{k_2}{k_1 C_0}\right)[y_{NH_3}(0) - 4Z_1 - 4Z_2][y_{O_2}(0) - 5Z_1 - 3Z_2 - Z_3]$$

$$+ 2.5\left(\frac{k_4}{k_1 C_0^{4/3}}\right)[y_{NH_3}(0) - 4Z_1 - 4Z_2]^{2/3}[y_{NO}(0) + 4Z_1 - 2Z_3] \quad (k)$$

$$\frac{dZ_3}{d\tau} = \left(\frac{k_3}{k_1}\right)[y_{O_2}(0) - 5Z_1 - 3Z_2 - Z_3][y_{NO}(0) + 4Z_1 - 2Z_3]^2 \quad (l)$$

Once we solve the design equations, we use Eq. 2.7.4, to obtain the species curves:

$$\frac{N_{NH_3}}{(N_{tot})_0} = y_{NH_3}(0) - 4Z_1 - 4Z_2 \quad (m)$$

$$\frac{N_{O_2}}{(N_{tot})_0} = y_{O_2}(0) - 5Z_1 - 3Z_2 - Z_3 \quad (n)$$

$$\frac{N_{NO}}{(N_{tot})_0} = y_{NO}(0) - 4Z_1 - 2Z_3 \quad (o)$$

$$\frac{N_{H_2O}}{(N_{tot})_0} = y_{H_2O}(0) + 6Z_1 + 6Z_2 \quad (p)$$

$$\frac{N_{N_2}}{(N_{tot})_0} = y_{N_2}(0) + 2Z_2 \quad (q)$$

$$\frac{N_{NO_2}}{(N_{tot})_0} = y_{NO_2}(0) + 2Z_3 \quad (r)$$

Using Eq. 2.7.6,

$$\frac{N_{tot}}{(N_{tot})_0} = 1 + Z_1 + Z_2 - Z_3 \quad (s)$$

The parameters of the design equations are

$$\left(\frac{k_2}{k_1 C_0}\right) = 0.05 \quad \left(\frac{k_3}{k_1}\right) = 2 \quad \left(\frac{k_4}{k_1 C_0^{4/3}}\right) = 0.1$$

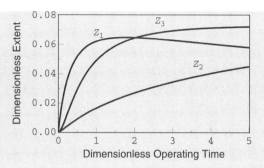

Figure E6.11.1   Reaction curves.

We substitute these values into (j), (k), and (l), and solve them numerically subject to the initial conditions that at $\tau = 0$, $Z_1 = Z_2 = Z_3 = 0$. Figure E6.11.1 shows the solutions of the design equations—the reaction operating curves. Next, we use (m) through (r) to determine the species curves, shown in Figure E6.11.2.

b. From the species curves (or the table of calculated data), the maximum amount of NO is achieved at $\tau = 1.275$, and, using (i), the required operating time is

$$t = \tau t_{cr} = 39.84 \text{ min}$$

c. At $\tau = 1.275$, $Z_1 = 0.0395$, $Z_2 = 0.0232$, and $Z_3 = 0.0095$. Substituting these values into (m) through (r) and using (s), the species molar fractions are

$$y_{NH_3} = 0.237 \qquad y_{O_2} = 0.21 \qquad y_{NO} = 0.132 \qquad y_{H_2O} = 0.357$$
$$y_{N_2} = 0.044 \qquad y_{NO_2} = 0.018$$

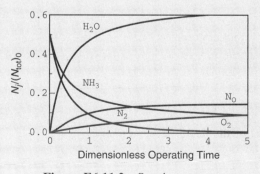

Figure E6.11.2   Species curves.

6.4   NONISOTHERMAL OPERATIONS

The design formulation of nonisothermal batch reactors with multiple reactions follows the procedure outlined in the previous section—we write Eq. 6.1.1 for each independent reaction. However, since the reactor temperature may vary during the operation, we should solve the design equations simultaneously with the energy balance equation. Note that the energy balance equation (Eq. 6.1.17) contains another variable—the temperature of the heating (or cooling) fluid, θ_F, that may also vary during the operation (θ_F is constant only when the fluid either evaporates or condenses). In general, we have to write the energy balance equation for the heating/cooling fluid to express changes in θ_F, but, in most batch reactor applications, θ_F is assumed constant. Because of the complex geometry of the heat-transfer surface (shell or coil), the average of the inlet temperature and the outlet temperature is usually used.

The procedure for setting up the energy balance equation goes as follows:

1. Define the *reference state*, and identify T_0, C_0, $(N_{tot})_0$, and its species compositions, y_{j_0}'s. Recall that, in most applications, the initial state is selected as reference state.
2. Determine the specific molar heat capacity of the reference state, $\hat{c}_{p_0}$.
3. Determine the dimensionless activation energies, γ_i's, of *all* chemical reactions.
4. Determine the dimensionless heat of reactions, DHR_m's, of each *independent* reaction.
5. Determine the correction factor of the heat capacity, $CF(Z_m, \theta)$.
6. Determine (or specify) the dimensionless temperature of the heating/cooling fluid, θ_F.
7. Specify the dimensionless heat-transfer number, HTN (using Eq. 6.1.26).
8. Determine (or specify) the initial dimensionless temperature, $\theta(0)$.
9. Solve the energy balance equation simultaneously with the design equations to obtain Z_m's and θ as functions of the dimensionless operating time, τ.

The design formulation of nonisothermal batch reactors consists of $n_I + 1$ non-linear first-order differential equations whose initial values are specified. The solutions of these equations provide Z_m's and θ as functions of τ. The examples below illustrate the design of nonisothermal ideal batch reactors.

Example 6.12   The first-order, liquid-phase endothermic reactions

Reaction 1: $A \longrightarrow 2B$
Reaction 2: $B \longrightarrow 2C + D$

take place in aqueous solution. A 200-L solution with a concentration of $C_A(0) = 0.8 \, \text{mol/L}$ is charged into a batch reactor. Initially, the solution is at 47°C, and the average temperature of the heating fluid is 67°C.

a. Derive the reaction and species curves for isothermal operation. Determine the operating time needed to maximize the production of product B, and the amount produced.

b. Determine the heating (or cooling) load during the isothermal operation.

c. Estimate the average HTN for the isothermal operation.

d. Derive the reaction, temperature, and species curves for adiabatic operation. Determine the operating time needed to maximize the production of product B, and the amount produced.

e. Examine the effect of HTN on the reactor temperature, the two reactions, and the production of product B.

Data: At 47°C,

$$k_1 = 0.04 \, \text{min}^{-1} \qquad k_2 = 0.03 \, \text{min}^{-1}$$

$$\Delta H_{R1}(T_0) = 40 \, \text{kcal/mol B} \qquad \Delta H_{R2}(T_0) = 60 \, \text{kcal/mol B}$$

$$E_{a1} = 7000 \, \text{cal/mol} \qquad E_{a2} = 6000 \, \text{cal/mol}$$

$$\rho = 1.0 \, \text{kg/L} \qquad \bar{c}_p = 1.0 \, \text{kcal/kgK}$$

Solution   The stoichiometric coefficients of the chemical reactions are:

$$
\begin{array}{lllll}
s_{A_1} = -1 & s_{B_1} = 2 & s_{C_1} = 0 & s_{D_1} = 0 & \Delta_1 = 1 \\
s_{A_2} = 0 & s_{B2} = -1 & s_{C_2} = 2 & s_{D_2} = 1 & \Delta_2 = 2
\end{array}
$$

The two reactions are independent, and there are no dependent reactions. We select the initial state as the reference state; hence, $T_0 = T(0) = 320 \, \text{K}$, $C_0 = C_A(0) = 0.8 \, \text{mol/L}$, and the total molar content of the reference state is

$$(N_{\text{tot}})_0 = V_{R_0} C_0 = 160 \, \text{mol} \tag{a}$$

The initial composition is $y_A(0) = 1.0$, and $y_B(0) = y_C(0) = y_D(0) = 0$. Using Eq. 6.1.23, for liquid-phase reactions, the specific molar heat capacity of the reference state is

$$\hat{c}_{p_0} = \frac{M}{(N_{\text{tot}})_0} \bar{c}_p = 1.25 \, \frac{\text{kcal}}{\text{mol K}}$$

The dimensionless activation energies of the two reactions are

$$\gamma_1 = \frac{E_{a_1}}{RT_0} = 14.15 \qquad \gamma_2 = \frac{E_{a_2}}{RT_0} = 9.44$$

The heat of reaction of Reaction 1 (for the chemical formula used) is

$$\Delta H_{R_1}(T_0) = \left(40\,\frac{\text{kcal}}{\text{mol B}}\right)\left(2\,\frac{\text{mol B}}{\text{mol extent}}\right) = 80\,\text{kcal/mol extent}$$

The dimensionless heat of reactions of the two chemical reactions are

$$\text{DHR}_1 = \frac{\Delta H_{R_1}(T_0)}{\hat{c}_{p_0}T_0} = 0.20 \qquad \text{DHR}_2 = \frac{\Delta H_{R_2}(T_0)}{\hat{c}_{p_0}T_0} = 0.15$$

For liquid-phase reactions (assuming constant heat capacity), $\text{CF}(Z_m, \theta) = 1$. We write Eq. 6.1.1 for each independent reaction,

$$\frac{dZ_1}{d\tau} = r_1\left(\frac{t_{\text{cr}}}{C_0}\right) \tag{a}$$

$$\frac{dZ_2}{d\tau} = r_2\left(\frac{t_{\text{cr}}}{C_0}\right) \tag{b}$$

Using Eq. 6.1.12 to express the species concentrations, the reaction rates are

$$r_1 = k_1(T_0)C_0(1 - Z_1)e^{\gamma_1(\theta-1)/\theta} \tag{c}$$

$$r_2 = k_2(T_0)C_0(2Z_1 - Z_2)e^{\gamma_2(\theta-1)/\theta} \tag{d}$$

We select Reaction 1 as the leading reaction, for first-order reaction, the characteristic reaction time is

$$t_{\text{cr}} = \frac{1}{k_1(T_0)} = 25\,\text{min} \tag{e}$$

Substituting (c), (d), and (e) into (a) and (b), the design equations become

$$\frac{dZ_1}{d\tau} = (1 - Z_1)e^{\gamma_1(\theta-1)/\theta} \tag{f}$$

$$\frac{dZ_2}{d\tau} = \left(\frac{k_2(T_0)}{k_1(T_0)}\right)(2Z_1 - Z_2)e^{\gamma_2(\theta-1)/\theta} \tag{g}$$

Using Eq. 6.1.17, the energy balance equation is

$$\frac{d\theta}{d\tau} = \frac{1}{CF(Z, \theta)} \left[HTN(\theta_F - \theta) - DHR_1\frac{dZ_1}{d\tau} - DHR_2\frac{dZ_2}{d\tau} \right] \qquad (h)$$

We have to solve (f), (g), and (h) simultaneously, subject to the initial conditions that at $\tau = 0$, $Z_1(0) = Z_2(0) = 0$, and $\theta(0) = 1$. Once we solve the design equation, we use Eq. 2.7.4 to obtain the species curves:

$$\frac{N_A(\tau)}{(N_{tot})_0} = 1 - Z_1(\tau) \qquad (i)$$

$$\frac{N_B(\tau)}{(N_{tot})_0} = 2Z_1(\tau) - Z_2(\tau) \qquad (j)$$

$$\frac{N_C(\tau)}{(N_{tot})_0} = 2Z_2(\tau) \qquad (k)$$

$$\frac{N_D(\tau)}{(N_{tot})_0} = Z_2(\tau) \qquad (l)$$

a. For isothermal operation, $\theta = 1$, and (f) and (g) reduce to

$$\frac{dZ_1}{d\tau} = (1 - Z_1) \qquad (m)$$

$$\frac{dZ_2}{d\tau} = \left[\frac{k_2(T_0)}{k_1(T_0)} \right](2Z_1 - Z_2) \qquad (n)$$

We solve (m) and (n) numerically subject to the initial condition $Z_1(0) = Z_2(0) = 0$. Figure E6.12.1 shows the reaction curves. We then use (i) through (l) to determine the species curves, shown in Figure E6.12.2. From the curve of product B, the highest $N_B/(N_{tot})_0$ is 0.8437 and is reached

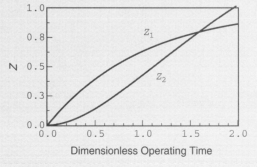

Figure E6.12.1   Reaction operating curves—isothermal operation.

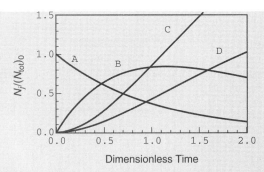

Figure E6.12.2   Species operating curves—isothermal operation.

at $\tau = 1.16$, $Z_1 = 0.6865$, $Z_2 = 0.5293$. Using (e), the required operating time is

$$t = t_{cr}\tau = 29 \text{ min}$$

The amount of product B produced is 135 mol.

b. For isothermal operations, $d\theta/d\tau = 0$, and using Eq. 6.1.20, the instantaneous dimensionless heat-transfer rate is

$$\frac{d}{d\tau}\left[\frac{Q}{(N_{tot})_0 \hat{c}_{p_0} T_0}\right] = DHR_1 \frac{dZ_1}{d\tau} + DHR_2 \frac{dZ_2}{d\tau} \tag{o}$$

Integrating (o) from 0 to τ,

$$Q(\tau) = [(N_{tot})_0 \hat{c}_{p_0} T_0][DHR_1 Z_1(\tau) + DHR_2 Z_2(\tau)] = 13.7 \times 10^3 \text{ kcal}$$

The positive value indicates that heat is added to the reactor.

c. Using Eq. 6.1.24, the HTN for isothermal operation ($\theta = 1$) is

$$HTN_{iso}(\tau) = \frac{1}{\theta_F - 1}\left(DHR_1 \frac{dZ_1}{dt} + DHR_2 \frac{dZ_2}{dt}\right) \tag{p}$$

Figure E6.12.3 shows the HTN curve during the operation. Using Eq. 6.1.25, for $\tau_{op} = 1.16$, the value of HTN_{ave} is 2.87.

d. For adiabatic operation, $HTN = 0$, and (h) reduces to

$$\frac{d\theta}{dt} = \frac{1}{CF(Z, \theta)}\left(-DHR_1 \frac{dZ_1}{dt} - DHR_2 \frac{dZ_2}{dt}\right) \tag{q}$$

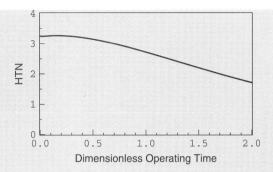

Figure E6.12.3   Instantaneous isothermal HTN curve.

We solve (f), (g), and (q) simultaneously, subject to the initial condition that $\tau = 0$, $Z_1(0) = Z_2(0) = 0$, and $\theta = 1$. Figure E6.12.4 shows the reaction curves, and Figure E6.12.5 shows the temperature curve. We then use (i) through (l) to determine the species curves, shown in Figure E6.12.6. From the curve of product B, the highest $N_B/(N_{\text{tot}})_0$ is 0.7224 and is reached at

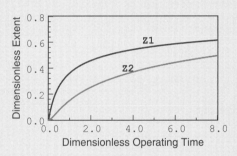

Figure E6.12.4   Reaction operating curves—adiabatic operation.

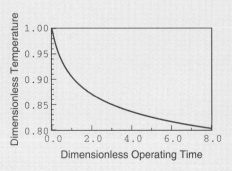

Figure E6.12.5   Temperature curve—adiabatic operation.

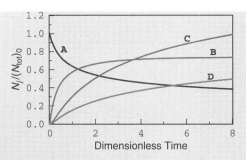

Figure E6.12.6   Species curves—adiabatic operation.

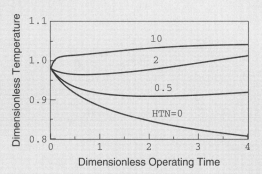

Figure E6.12.7   Effect of HTN on the reactor temperature.

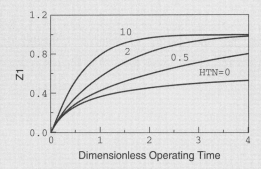

Figure E6.12.8   Effect of HTN on Reaction 1.

$\tau = 4.64$, $Z_1 = 0.5597$, $Z_2 = 0.3971$, and $\theta = 0.8285$. Using (e), the required operating time is

$$t = t_{cr}\tau = 1.116 \text{ min}$$

The amount of product B produces is 115.5 mol.

e. To examine the effect of HTN on the reactor operation, we solve (f), (g), and (h) for different values of HTN. Figure E6.12.7 shows the effect on the reactor temperature, Figure E6.12.8 shows the effect on the progress of Reaction

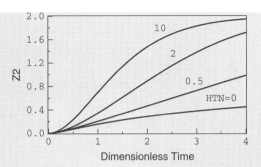

Figure E6.12.9 Effect of HTN on Reaction 2.

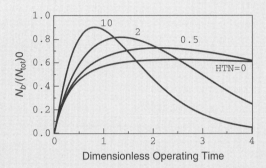

Figure E6.12.10 Effect of HTN on the production of product B.

1, Figure E6.12.9 shows the effect on the progress of Reaction 2, and Figure E6.12.10 shows the effect on the production of product B. Note that when HTN $\to \infty$, the temperature of the reactor is the same as the jacket temperature (67°C).

Example 6.13 The gas-phase elementary chemical reactions

$$\text{Reaction 1:} \quad A + B \longrightarrow C$$
$$\text{Reaction 2:} \quad C + B \longrightarrow D$$

take place in an isobaric (variable-volume) batch reactor. A gas mixture consisting of 50% reactant A and 50% of reactant B (mol %) is charged into the reactor, whose initial volume is 10 L. The pressure is 2 atm and at the beginning of the operation, the temperature is 150°C. The temperature of the cooling medium is 120°C. Based on the data below:

a. Derive the reaction operating curve for isothermal operation and determine the operating time needed to achieve the highest production of product C.

b. Determine the heating (or cooling) load during the isothermal operation.

c. Determine the instantaneous and average HTN for the isothermal operation.

d. Derive the reaction and the temperature curves when the reactor is operated adiabatically and determine the operating time needed to achieve the highest production of product C.

e. Examine the effect of HTN on the reactor temperature, the chemical reactions, and the production of product C.

The reaction rate expressions are $r_1 = k_1(T)C_A C_B$ and $r_2 = k_2(T)C_B C_C$.

Data: At 150°C, $k_1 = 0.02\,\mathrm{L\,mol^{-1}s^{-1}}$, $k_2 = 0.02\,\mathrm{L\,mol^{-1}s^{-1}}$

$$\Delta H_{R_1} = -4500\,\mathrm{cal/mol} \qquad \Delta H_{R_2} = -6000\,\mathrm{cal/mol}$$

$$E_{a_1} = 10{,}090\,\mathrm{cal/mol} \qquad E_{a_2} = 12{,}607\,\mathrm{cal/mol}$$

$$\hat{c}_{p_A} = 16 \qquad \hat{c}_{p_B} = 8 \qquad \hat{c}_{p_C} = 20 \qquad \hat{c}_{p_D} = 26\,\mathrm{cal/mol\,K}$$

Solution   Since each reaction has a species that does not appear in the other reaction, the two reactions are independent, and there is no dependent reaction. The stoichiometric coefficients of the chemical reactions are

$$
\begin{array}{lllll}
s_{A_1} = -1 & s_{B_1} = -1 & s_{C_1} = 1 & s_{D_1} = 0 & s_{\Delta_1} = -1 \\
s_{A_2} = 0 & s_{B_2} = -1 & s_{C_2} = -1 & s_{D_2} = 1 & s_{\Delta_2} = -1
\end{array}
$$

We select the initial state as the reference state; hence, $T_0 = T(0)$, and the reference concentration is

$$C_0 = \frac{P_0}{RT_0} = 0.0576\,\mathrm{mol/L}$$

and the molar content of the reference state is

$$(N_{\mathrm{tot}})_0 = V_{R_0} C_0 = 0.576\,\mathrm{mol}$$

The initial composition is $y_A(0) = 0.5$, $y_B(0) = 0.5$, and $y_C(0) = y_D(0) = 0$. Using Eq. 5.2.29, the specific molar heat capacity of the reference state is

$$\hat{c}_{p_0} = \sum_j^J y_{j_0} c_{p_j}(T_0) = (0.5)c_{p_A} + (0.5)c_{p_B} = 12\,\mathrm{cal/mol\,K}$$

The dimensionless activation energies are

$$\gamma_1 = \frac{E_{a_1}}{RT_0} = 12 \qquad \gamma_2 = \frac{E_{a_2}}{RT_0} = 15$$

The dimensionless heat of reactions are

$$DHR_1 = \frac{\Delta H_{R_1}(T_0)}{\hat{c}_{p_0} T_0} = -0.886$$

$$DHR_2 = \frac{\Delta H_{R_2}(T_0)}{\hat{c}_{p_0} T_0} = -1.1816$$

Using Eq. 5.2.34, the correction factor of the heat capacity is

$$CF(Z_m, \theta) = 1 + \frac{1}{\hat{c}_{p_0}} \sum_j^J \hat{c}_{p_j}(\theta) \sum_m^{n_I} (s_j)_m Z_m$$

$$= 1 + \frac{1}{\hat{c}_{p_0}} [\hat{c}_{p_A}(-Z_1) + \hat{c}_{p_B}(-Z_1 - Z_2) + \hat{c}_{p_C}(Z_1 - Z_2) + \hat{c}_{p_D} Z_2]$$

$$CF(Z_m, \theta) = \frac{6 - 2Z_1 - Z_2}{6} \tag{a}$$

We write Eq. 6.1.1 for each independent reaction, noting that $V_R/V_{R_0} = (1 - Z_1 - Z_2)\theta$,

$$\frac{dZ_1}{d\tau} = r_1 (1 - Z_1 - Z_2)\theta \left(\frac{t_{cr}}{C_0}\right) \tag{b}$$

$$\frac{dZ_2}{d\tau} = r_2 (1 - Z_1 - Z_2)\theta \left(\frac{t_{cr}}{C_0}\right) \tag{c}$$

We use Eq. 6.1.16, to express the species concentrations, and the rates of the two reactions are

$$r_1 = k_1(T_0) e^{\gamma_1(\theta-1)/\theta} C_0^2 \frac{(0.5 - Z_1)(0.5 - Z_1 - Z_2)}{[(1 - Z_1 - Z_2)\theta]^2} \tag{d}$$

$$r_2 = k_2(T_0) e^{\gamma_2(\theta-1)/\theta} C_0^2 \frac{(Z_1 - Z_2)(0.5 - Z_1 - Z_2)}{[(1 - Z_1 - Z_2)\theta]^2} \tag{e}$$

We select the characteristic reaction time on the basis of Reaction 1:

$$t_{cr} = \frac{1}{k_1(T_0)C_0} = 868 \text{ s} = 14.46 \text{ min} \tag{f}$$

Substituting (d) through (f) into (b) and (c), the two design equations become

$$\frac{dZ_1}{d\tau} = \frac{(0.5 - Z_1)(0.5 - Z_1 - Z_2)}{(1 - Z_1 - Z_2)\theta} e^{\gamma_1(\theta-1)/\theta} \tag{g}$$

$$\frac{dZ_2}{d\tau} = \left[\frac{k_2(T_0)}{k_1(T_0)}\right]\frac{(Z_1 - Z_2)(0.5 - Z_1 - Z_2)}{(1 - Z_1 - Z_2)\theta}e^{\gamma_2(\theta-1)/\theta} \tag{h}$$

Substituting (a) into Eq. 6.1.17, the energy balance equation for this case is

$$\frac{d\theta}{d\theta} = \left(\frac{6}{6 - 2Z_1 - Z_2}\right)\left[\text{HTN}(\theta_F - \theta) - \text{DHR}_1\frac{dZ_1}{d\tau} - \text{DHR}_2\frac{dZ_2}{d\tau}\right] \tag{i}$$

Once we solve the design equations, we use Eq. 2.7.4 to obtain the species curves:

$$\frac{N_A}{(N_{\text{tot}})_0} = 0.5 - Z_1 \tag{j}$$

$$\frac{N_B}{(N_{\text{tot}})_0} = 0.5 - Z_1 - Z_2 \tag{k}$$

$$\frac{N_C}{(N_{\text{tot}})_0} = Z_1 - Z_2 \tag{l}$$

$$\frac{N_D}{(N_{\text{tot}})_0} = Z_2 \tag{m}$$

a. For isothermal operation ($\theta = 1$), (g) and (h) reduce to

$$\frac{dZ_1}{d\tau} = \frac{(0.5 - Z_1)(0.5 - Z_1 - Z_2)}{1 - Z_1 - Z_2} \tag{n}$$

$$\frac{dZ_2}{d\tau} = (0.5)\frac{(Z_1 - Z_2)(0.5 - Z_1 - Z_2)}{1 - Z_1 - Z_2} \tag{o}$$

We solve (n) and (o) numerically, subject to the initial conditions that, at $\tau = 0$, $Z_1 = Z_2 = 0$. Figure E6.13.1 shows the reaction curves, and Figure E6.13.2 shows the species curves calculated by (j) through (m). From the curve of product C, the highest $N_C/(N_{\text{tot}})_0$ is 0.1839 and it is

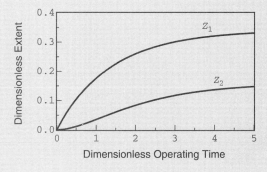

Figure E6.13.1   Reaction operating curves—isothermal operation.

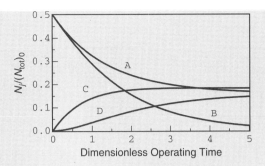

Figure E6.13.2 Species operating curves—isothermal operation.

reached at $\tau = 3.7$. At that operating time, $Z_1 = 0.3166$ and $Z_2 = 0.1332$. Hence, using (f), the operating time is

$$t = t_{cr}\tau = (14.46 \text{ min})(3.7) = 53.5 \text{ min}$$

b. For an isothermal operation, $d\theta/d\tau = 0$, substituting Eq. 6.1.20 into (i),

$$\frac{d}{d\tau}\left[\frac{Q}{(N_{tot})_0 \hat{c}_{p_0} T_0}\right] = \text{DHR}_1 \frac{dZ_1}{d\tau} + \text{DHR}_2 \frac{dZ_2}{d\tau} \qquad (p)$$

Integrating (l) over operating time τ_{op},

$$Q(\tau_{op}) = [(N_{tot})_0 \hat{c}_{p_0} T_0][\text{DHR}_1 Z_1(\tau_{op}) + \text{DHR}_2 Z(\tau_{op})] = -1281 \text{ cal}$$

The negative sign indicates that heat is removed from the reactor.

c. Using Eq. 6.1.24, the HTN at time τ is

$$\text{HTN}_{iso}(\tau) = \frac{1}{\theta_F - 1}\left(\text{DHR}_1 \frac{dZ_1}{d\tau} + \text{DHR}_2 \frac{dZ_2}{d\tau}\right)\frac{1}{1 - Z_1 - Z_2} \qquad (q)$$

Figure E6.13.3 shows the HTN curve during the operation. Using Eq. 6.1.23, for $\tau_{op} = 3.7$, $\text{HTN}_{ave} = 4.35$.

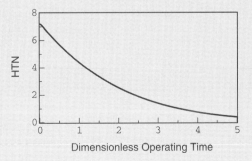

Figure E6.13.3 Isothermal HTN curve.

d. For adiabatic operation, HTN $= 0$, and the energy balance equation reduces to

$$\frac{d\theta}{d\theta} = \left(\frac{6}{6 - 2Z_1 - Z_2}\right)\left(-\text{DHR}_1\frac{dZ_1}{d\tau} - \text{DHR}_2\frac{dZ_2}{d\tau}\right) \tag{r}$$

We solve (g), (h), and (r) numerically, subject to the initial condition that at $\tau = 0$, $Z_1 = Z_2 = 0$ and $\theta = 1$. Figure E6.13.4 shows the reaction curves, Figure E6.13.5 shows the temperature curve, and Figure E6.13.6 shows the species curves, calculated by (j) through (m). From the curve of product C, the highest $N_C/(N_{\text{tot}})_0$ is 0.1194 and it is reached at $\tau = 0.25$. At that operating time, $Z_1 = 0.203$, $Z_2 = 0.0836$, and $\theta = 1.437$. Hence, using (f), the operating time is

$$t = t_{\text{cr}}\tau = 3.61 \text{ min}$$

e. To examine the effect of HTN on the reactor operation, we solve (f), (g), and (h) for different values of HTN. Figure E6.13.7 shows the effect on the reactor temperature, Figure E6.13.8 shows the effect on the progress of

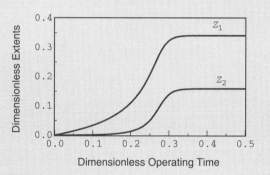

Figure E6.13.4   Reaction operating curves—adiabatic operation.

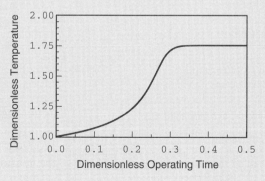

Figure E6.13.5   Temperature curve—adiabatic operation.

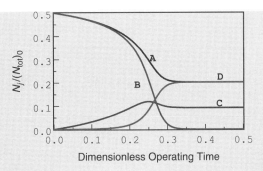

Figure E6.13.6   Species operating curves—adiabatic operation.

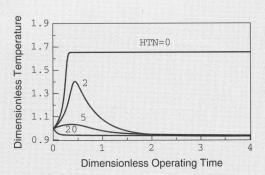

Figure E6.13.7   Effect of HTN on the reactor temperature.

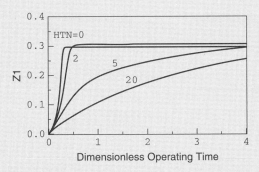

Figure E6.13.8   Effect of HTN on Reaction 1.

Reaction 1, Figure E6.13.9 shows the effect on the progress of Reaction 2, and Figure E6.13.10 shows the effect on the production of product C. Note that when HTN $\to \infty$, the temperature of the reactor is the same as the jacket temperature (120°C).

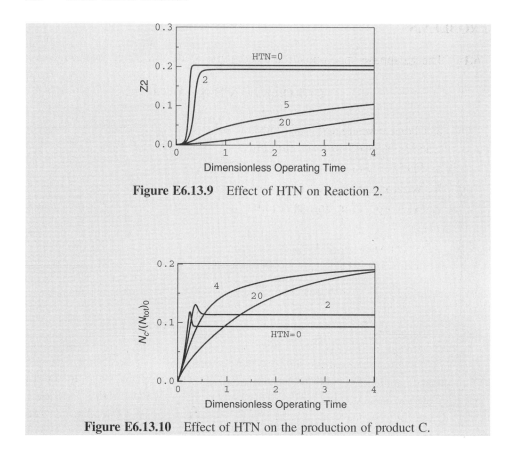

Figure E6.13.9   Effect of HTN on Reaction 2.

Figure E6.13.10   Effect of HTN on the production of product C.

6.5   SUMMARY

In this chapter, we analyzed the operation of ideal batch reactors. We covered the following topics:

1. The underlying assumptions for ideal batch reactor operations and when they are applied in practice.
2. Derived the dimensionless design equation for isothermal operation with single reactions and obtained the reaction operating curve.
3. Discussed the operation of constant-volume batch reactors.
4. Discussed the operation of gaseous, variable-volume batch reactors.
5. Discussed methods to determine the rate expression and its parameters.
6. Discussed the design of isothermal operations with multiple chemical reactions.
7. Discussed the design of nonisothermal operations with multiple chemical reactions.

PROBLEMS*

6.1₂ The gas-phase decomposition of ethylene oxide, described by the reaction

$$C_2H_4O(g) \longrightarrow CH_4(g) + CO(g)$$

has been investigated in a constant-volume batch reactor at $400°C$. The reaction is assumed to be first order. Based on the data below, determine:

a. Is the first-order assumption valid?

b. What is the reaction rate constant at $400°C$?

c. What operating time is required for 50% conversion?

Data:

Time (min)	0	7	12	18
Pressure (mm Hg)	119	130.7	138.2	146.4

6.2₂ Reactant A decomposes according to the reaction

$$A \longrightarrow 2B + C$$

in an aqueous solution containing a homogeneous catalyst. The measured relation between r and C_A is given in the table below. We want to run this reaction in a batch reactor at the same catalyst concentration as used in obtaining the data. Develop a graphical method to design the reactor without deriving the rate expression. Determine:

a. The time needed to lower the concentration of A from $C_A(0) = 10\,\text{mol}/\text{L}$ to $C_A(t) = 2\,\text{mol/L}$.

b. The final concentration after 10 h of operation if the reactor is charged with a solution of $C_A(0) = 7\,\text{mol/L}$.

Data:

$C_A(t)$ (mol/L)	1	2	4	6	7	9	12
r (mol/L h)	0.06	0.1	0.25	1.0	2.0	1.0	0.5

6.3₂ The chemical reaction

$$A \longrightarrow \text{Products}$$

is investigated in a constant-volume batch reactor. After 8 min in a batch reactor, 80% of reactant A is converted. After 18 min, 90% of A is

---

*Subscript 1 indicates simple problems that require application of equations provided in the text. Subscript 2 indicates problems whose solutions require some more in-depth analysis and modifications of given equations. Subscript 3 indicates problems whose solutions require more comprehensive analysis and involve application of several concepts. Subscript 4 indicates problems that require the use of a mathematical sofware or the writing of a computer code to obtain numerical solutions.

converted. If $C_A(0) = 1 \, \text{mol/L}$, determine the rate expression of this reaction.

6.4₂   Dvorko and Shilov (*Kinetics and Catalysis*, **4**, 212, 1964) studied the iodine-catalyzed addition reaction between HI and cyclohexene in benzene solution,

$$HI + C_6H_{10} \longrightarrow C_6H_{11}I$$

The reaction is believed to be first order with respect to each reactant. An experiment was conducted at 20°C in a benzene solution with an iodine concentration of $4.22 \ 10^{-4} \, \text{mol/L}$. The initial concentration of cyclohexene was $0.123 \, \text{mol/L}$. Based on the data below, determine:

a. Is the assumed rate expression valid?

b. What is the reaction rate constant at 20°C?

Data:

Time (s)	C_{HI} (mol/L)
0	0.106
150	0.090
480	0.087
870	0.076
1500	0.062
2280	0.050

6.5₂   Huang and Dauerman (*Ind. Eng. Chem. Process. Des. Develop.*, **8**, 227, 1969) studied the reaction between benzyl chloride and sodium acetate in dilute aqueous solution at 102°C:

$$NaAc + C_6H_5CH_2Cl \longrightarrow C_6H_5CH_2Ac + Na^+ + Cl^-$$

Initially, the concentration of both sodium acetate and benzyl chloride is $0.757 \, \text{mol/L}$. The fraction of unconverted benzyl as a function of time is given below. Assuming the reaction is first order with respect to each reactant, determine:

a. Is the assumed rate expression valid?

b. The reaction rate constant at 102°C.

Data:

Time (ks)	$C_B(t)/C_B(0)$
10.80	0.945
24.48	0.912
46.08	0.846
54.72	0.809
69.48	0.779
88.56	0.730
109.44	0.678
126.72	0.638
133.74	0.619
140.76	0.590

6.6$_2$ The dimerization second-order reaction $2A \rightarrow R$ takes place in a liquid solution. When the reactor is charged with a solution with $C_A(0) = 1$ mol/L, 50% conversion is reached after 1 h. What will the conversion be after 1 h if the initial concentration of A is 10 mol/L?

6.7$_2$ Enzyme E catalyzes the chemical reaction

$$A \longrightarrow P + R$$

in aqueous solution. The reaction rate expression is

$$(-r_A) = \frac{200 C_A C_E}{2 + C_A} \text{ (mol/L min)}$$

If we charge a solution with $C_E = 0.001$ mol/L and $C_A(0) = 10$ mol/L into a batch reactor, find the time it takes for the concentration of A to drop to 0.025 mol/L. Note that the concentration of the enzyme remains unchanged during the operation. Plot the dimensionless operating curve.

6.8$_2$ The reaction of cyclohexanol and acetic acid in dioxane solution as catalyzed by sulfuric acid was studied by McCracken and Dickson (*Ind. Eng. Chem. Proc. Des. and Dev.*, **6**, 286, 1967). The esterification reaction can be represented by the following stoichiometric reaction:

$$A + B \longrightarrow C + W$$

(aceticacid + cyclohexanol $\longrightarrow$ cyclohexylacetate + water)

The reaction is carried out in a well-stirred batch reactor at 40°C. Under these conditions, the esterification reaction can be considered as irreversible at conversions less than 70%. The following data were obtained using identical sulfuric acid concentrations in both runs.

Run 1: $C_A(0) = C_B(0) = 2.5 \text{ kmol/m}^3$

$C_A(t)$ (kmol/m^3)	Time, t (ks)
2.070	7.2
1.980	9.0
1.915	10.8
1.860	12.6
1.800	14.4
1.736	16.2
1.692	18.0
1.635	19.8
1.593	21.6
1.520	25.2
1.460	28.8

Run 2: $C_A(0) = 1$; $C_B(0) = 8 \text{ kmol/m}^3$

$C_A(t)$ (kmol/m^3)	Time, t (ks)
0.885	1.8
0.847	2.7
0.769	4.5
0.671	7.2
0.625	9.0
0.544	12.6
0.500	15.3
0.463	18.0

Determine the order of the reaction with respect to each reactant and the rate constant for the forward reaction under the conditions of the two runs. The rate constant should differ between runs, but the conditions are such that the order will not differ. The individual and overall orders are integers.

6.9$_2$ The elementary, gas-phase, dimerization reaction

$$2A \longrightarrow P$$

is carried out in an isobaric, isothermal batch reactor. Pure A is charged into the reactor, which operates at 5 atm and 300°C. Based on the data below, determine:

a. The time needed for 80% conversion of A.
b. The volume of the reactor at that time.
c. Plot the performance curve.

Data: The reaction rate constant at 300°C, $k = 1 \text{ min}^{-1} (\text{mol/L})^{-1}$
Initial volume of the reactor: 500 L.

6.10$_2$ The following data are typical of the pyrolysis of dimethylether at 504°C:

$$CH_3OCH_3 \longrightarrow CH_4 + H_2 + CO$$

The reaction takes place in the gas phase in an isothermal constant-volume reactor. Determine the order of the reaction and the reaction rate constant. The order may be assumed to be an integer.

Data:

Time (s)	P (kPa)
0	41.6
390	54.4
777	65.1
1195	74.9
3155	103.9
∞	124.1

6.11₁ For the chemical reaction

$$2A + 3B \longrightarrow 2C$$

the following data were obtained:

Run	Concentration (mol/L)		Rate
	A	B	(mol/L min)
1	0.50	1.00	0.1
2	0.50	2.00	0.3
3	1.00	1.00	0.4

Determine the form of the rate expression. What is the value of the reaction rate constant?

6.12₂ For the following data:

Concentration (mol/L)			Rate
A	B	C	(mol/L min)
0.01	0.20	0.10	2.8
0.01	0.40	0.10	5.6
0.01	0.80	0.05	5.6
0.02	0.10	0.10	2.8

Determine:
  a. The order of the reaction with respect to A, B, and C.
  b. The value of the reaction rate constant.

6.13₂ At 800 K, di-methyl ether decomposes according to the reaction

$$CH_3OCH_3(g) \longrightarrow CH_4(g) + CO(g) + H_2(g)$$

Ether is charged into a constant-volume batch reactor, and the progress of the reaction is monitored measuring the reactor pressure.
Based on the data below, determine:

a. The order of the reaction (use the differential method).

b. The reaction rate constant at 800 K (use the integral method).

c. The operating time for 50% conversion.

Data:

Time (min)	0	390	780	1200	3160
Pressure (mm Hg)	310	410	490	560	780

6.14$_2$ The gas-phase dimerization of trifluorochloroethylene may be represented by

$$2C_2F_3Cl \longrightarrow C_4F_6Cl_2$$

The following data are typical of this reaction at 440°C as it occurs in a constant-volume reactor:

Time, t (s)	Total Pressure (kPa)
0	82.7
100	71.1
200	64.0
300	60.4
400	56.7
500	54.8

Determine the order of the reaction and the reaction rate constant under these conditions. Assume the order is an integer.

6.15$_4$ The elementary liquid-phase reactions

$$A + B \longrightarrow C$$

$$C + B \longrightarrow D$$

take place in a 200-L batch reactor. A solution [$C_A(0) = 2$ mol/L, $C_B(0) = 2$ mol/L] at 80°C is charged into the reactor. Based on the data below, derive the reaction operating curves and the temperature curve for each of the operations below. For each case, determine the operating time required for maximum production of C and the amount of C and D formed at that time.

a. Isothermal operation at 80°C.

b. The heating/cooling load on the reactor in (a).

c. Determine the instantaneous and average isothermal HTN ($T_F = 70°C$).

d. Adiabatic operation. What is the reactor temperature at the end of the operation?

e. Non-isothermal operation with the value of HTN 30% of the average isothermal HTN.

Data: At 80°C, $k_1 = 0.1$ L mol^{-1} min^{-1}, $E_{a_1} = 6000$ cal/mol
At 80°C, $k_2 = 0.2$ L mol^{-1} min^{-1} $E_{a_2} = 8000$ cal/mol
At 80°C, $\Delta H_{R_1} = -15,000$ cal/mol extent $\Delta H_{R_2} = -10,000$ cal/mol extent
Density of the solution $= 1000$ g/L Heat capacity of the solution $= 1$ cal/g°C

6.16$_2$ The irreversible gas-phase reactions (both are first order):

$$A \longrightarrow 2V$$

$$V \longrightarrow 2W$$

are carried out in a constant-volume batch reactor. Species A is charged into a 100-L reactor, and initially $P = 3$ atm and $T = 731$ K ($C_{A0} = 0.05$ mol/L). Based on the data below, calculate,

a. For isothermal operation, the time needed for 40% conversion of A.

b. The production rate of V in (a).

c. The heating/cooling load in (a).

d. For adiabatic operation, the time needed for 40% conversion of A.

e. The production rate of V in (d).

f. Determine the instantaneous and average isothermal HTN ($T_F = 750$ K).

Data: At 731 K,
$k_1 = 2$ min^{-1} $k_2 = 0.5$ min^{-1}
$E_{a_1} = 8000$ cal/mol $E_{a_2} = 12,000$ cal/mol
At 731 K, $\Delta H_{R_1} = 3000$ cal/mol of V $\Delta H_{R_2} = 4.500$ cal/mol of W
$\hat{c}_{pA} = 65$ cal/mol K $\hat{c}_{pV} = 40$ cal/mol K $\hat{c}_{pW} = 25$ cal/mol K

6.17$_4$ The first-order gas-phase reaction

$$A \longrightarrow B + C$$

takes place in a constant-volume batch reactor. Species A is charged into a 200-L reactor, and the initial conditions are 731 K and 10 atm. Based on the data below, derive the reaction operating curves and the temperature curve for each of the operation below. For each case, determine the operating time required for 80% conversion.

a. Isothermal operation at 731 K.

b. The heating/cooling load in (a).

c. Estimate the isothermal HTN ($T_F = 750$ K).

d. Adiabatic operation.

e. Operation with HTN 25% of the value of the isothermal HTN, and $T_F = 750$ K.

Data: At 731 K, $k = 0.2$ s^{-1}, $E_a = 12,000$ cal/mol

$$\Delta H_R = 10,000 \text{ cal/mol extent}$$

$$\hat{c}_{p_A} = 25 \text{ cal/mol K} \quad \hat{c}_{p_B} = 15 \text{ cal/mol K} \quad \hat{c}_{p_C} = 18 \text{ cal/mol K}$$

BIBLIOGRAPHY

Information on the mechanical design of well-mixed reactors is available in:

J. Y. Oldshue, *Fluid Mixing Technology*, McGraw Hill, New York, 1983.

G. B. Tatterson, *Fluid Mixing and Gas Dispersion in Agitating Tanks*, McGraw Hill, New York, 1991.

G. B. Tatterson, *Scale-up and Design of Industrial Mixing Processes*, McGraw Hill, New York, 1994.

7

PLUG-FLOW REACTOR

The plug-flow reactor (PFR) is a mathematical model that depicts a certain type of continuous reactor operation. The model is based on three assumptions:

- The reactor is operated at steady state.
- The fluid moves in a flat (pistonlike or "plug") velocity profile.
- There is no spatial variation in species concentrations or temperature at any cross section in the reactor.

Chemical reactions take place along the reactor and, consequently, species compositions, and temperature, vary from point to point along the reactor.

In practice, the fluid velocity profile is rarely flat, and spatial gradients of concentration and temperature do exist, especially in large-diameter reactors. Hence, the plug-flow reactor model (Fig. 7.1) does not describe exactly the conditions in industrial reactors. However, it provides a convenient mathematical means to *estimate* the performance of some reactors. As will be discussed below, it also provides a measure of the most efficient flow reactor—one where no mixing takes place in the reactor. The plug-flow model adequately describes the reactor operation when one of the following two conditions is satisfied:

- Tubular reactors whose lengths are much larger than their diameter. The acceptable length-to-diameter ratio (L/D) depends on the flow conditions in the reactor. Typically, for turbulent flows, $L/D > 20$, and, for laminar flows, $L/D > 200$.

Principles of Chemical Reactor Analysis and Design, Second Edition. By Uzi Mann
Copyright © 2009 John Wiley & Sons, Inc.

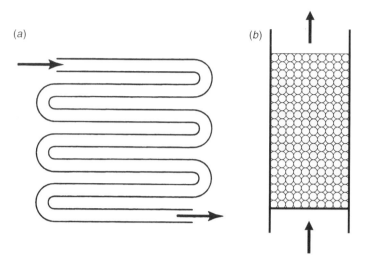

Figure 7.1 Schematic description of plug-flow reactor: (*a*) tubular reactor and (*b*) packed-bed reactor.

- Packed-bed reactors (the packing disperses the fluid laterally across the bed resulting in a near uniform flow).

7.1 DESIGN EQUATIONS AND AUXILIARY RELATIONS

The design equation of a plug-flow reactor was derived in Chapter 4. The design equation, written for the mth-independent reaction, is

$$\frac{dZ_m}{d\tau} = \left(r_m + \sum_k^{n_D} \alpha_{km} r_k \right) \left(\frac{t_{cr}}{C_0} \right) \qquad (7.1.1)$$

where Z_m is the dimensionless extent, defined by Eq. 2.7.2:

$$Z_m \equiv \frac{\dot{X}_m}{(F_{tot})_0} \qquad (7.1.2)$$

and τ is the dimensionless space time (dimensionless volume), defined by Eq. 4.4.8:

$$\tau \equiv \frac{V_R}{v_0 t_{cr}} = \frac{t_{sp}}{t_{cr}} \qquad (7.1.3)$$

where t_{cr} is a conveniently selected characteristic reaction time, defined by Eq. 3.5.1, and C_0 is the concentration of a conveniently selected reference stream, defined by

$$C_0 \equiv \frac{(F_{\text{tot}})_0}{v_0} \qquad (7.1.4)$$

where $(F_{\text{tot}})_0$ and v_0 are, respectively, the molar flow rate and the volumetric flow rate of the reference stream. As discussed in Chapter 4, to describe the operation of a reactor with multiple reactions, we have to write Eq. 7.1.1 for *each* independent chemical reaction. The solutions of the design equations (the relationships between Z_m's and τ) provide the reaction operating curves that describe the progress of the chemical reactions along the volume of the reactor.

To solve the design equations, we have to express the rates of *all* the chemical reactions in terms of Z_m's and τ. Below, we derive the auxiliary relations that relate the species concentrations to Z_m's and τ.

The volume-based rate of the ith chemical reaction (from Eq. 3.3.1 and Eq. 3.3.6) is

$$r_i = k_i(T_0)e^{\gamma_i(\theta-1)/\theta}h_i(C_j\text{'s}) \qquad (7.1.5)$$

where $k_i(T_0)$ is the reaction rate constant at reference temperature T_0, γ_i is the dimensionless activation energy ($\gamma_i = Ea_i/RT_0$), and $h_i(C_j\text{'s})$ is a function of the species concentrations, given by the rate expression.

To express the rates of the chemical reactions in terms of the reaction extents, we have to relate the species concentrations to Z_m's and τ. For plug-flow reactors, the concentration of species j at a given point in the reactor is

$$C_j = \frac{F_j}{v} \qquad (7.1.6)$$

where F_j is the local molar flow rate of species j, and v is the local volumetric flow rate. Using Eq. 2.7.7,

$$F_j = (F_{\text{tot}})_0\left\{\left[\frac{(F_{\text{tot}})_{\text{in}}}{(F_{\text{tot}})_0}\right]y_{j_{\text{in}}} + \sum_m^{n_I}(s_j)_m Z_m\right\} \qquad (7.1.7)$$

where $(F_{\text{tot}})_{\text{in}}$ and $y_{j_{\text{in}}}$ are, respectively, the total molar flow rate and the molar fraction of species j at the reactor inlet. Substituting Eq. 7.1.7 into Eq. 7.1.6, the local concentration of species j is

$$C_j = \frac{(F_{\text{tot}})_0}{v}\left\{\left[\frac{(F_{\text{tot}})_{\text{in}}}{(F_{\text{tot}})_0}\right]y_{j_{\text{in}}} + \sum_m^{n_I}(s_j)_m Z_m\right\} \qquad (7.1.8)$$

When the inlet stream is selected as the reference stream, as done in most applications, $(F_{tot})_0 = (F_{tot})_{in}$ and $y_{j_{in}} = y_{j_0}$, and Eq. 7.1.8 reduces to

$$C_j = \frac{(F_{tot})_0}{v}\left(y_{j_0} + \sum_{m}^{n_I}(s_j)_m Z_m\right) \tag{7.1.9}$$

For most *liquid-phase* reactions, the density of the reacting fluid is almost constant; hence, $v = v_{in}$, and Eq. 7.1.8 becomes

$$C_j = \frac{(F_{tot})_0}{v_{in}}\left\{\left(\frac{(F_{tot})_{in}}{(F_{tot})_0}\right)y_{j_{in}} + \sum_{m}^{n_I}(s_j)_m Z_m\right\} \tag{7.1.10}$$

When the inlet stream is selected as the reference stream, $v_0 = v_{in}$, and Eq. 7.1.10 reduces to

$$C_j = C_0\left(y_{j_0} + \sum_{m}^{n_I}(s_j)_m Z_m\right) \tag{7.1.11}$$

Equation 7.1.11 is used in most applications with liquid-phase reactions (when the inlet stream is selected as the reference stream).

For *gas-phase* reactions, the volumetric flow rate may vary along the reactor due to changes in molar flow rate, temperature, and pressure. Assuming ideal-gas behavior, the local volumetric flow rate is

$$v = v_0\left[\frac{F_{tot}}{(F_{tot})_0}\right]\left(\frac{T}{T_0}\right)\left(\frac{P_0}{P}\right) \tag{7.1.12}$$

Using Eq. 2.7.9 to express the total molar flow rate in terms of the extents of the independent reactions,

$$F_{tot} = (F_{tot})_0\left[\frac{(F_{tot})_{in}}{(F_{tot})_0} + \sum_{m}^{n_I}\Delta_m Z_m\right]$$

Equation 7.1.12 becomes

$$v = v_0\left[\frac{(F_{tot})_{in}}{(F_{tot})_0} + \sum_{m}^{n_I}\Delta_m Z_m\right]\theta\left(\frac{P_0}{P}\right) \tag{7.1.13}$$

where θ is the dimensionless temperature, T/T_0. Substituting Eq. 7.1.13 into Eq. 7.1.9, the local concentration is

$$C_j = C_0\frac{[(F_{tot})_{in}/(F_{tot})_0]y_{j_{in}} + \sum_{m}^{n_I}(s_j)_m Z_m}{[(F_{tot})_{in}/(F_{tot})_0 + \sum_{m}^{n_I}\Delta_m Z_m]\theta}\left(\frac{P}{P_0}\right) \tag{7.1.14}$$

When the inlet stream is selected as the reference stream, Eq. 7.1.14 reduces to

$$C_j = C_0 \frac{y_{j_0} + \sum_m^{n_I} (s_j)_m Z_m}{(1 + \sum_m^{n_I} \Delta_m Z_m)\theta} \left(\frac{P}{P_0}\right) \tag{7.1.15}$$

Equation 7.1.15 is used in most plug-flow reactor applications with gas-phase reactions (when the inlet stream is selected as the reference stream).

Since Eqs. 7.1.5 and 7.1.15 contain another dependent variable, θ, the dimensionless temperature, we should solve the energy balance equation simultaneously with the design equations. For plug-flow reactors with negligible viscous work, the energy balance equation (derived in Section 5.2) is

$$\frac{d\theta}{d\tau} = \frac{1}{CF(Z_m, \theta)} \left[HTN(\theta_F - \theta) - \sum_m^{n_I} DHR_m \frac{dZ_m}{d\tau} \right] \tag{7.1.16}$$

where HTN is the dimensionless local heat-transfer number, defined by Eq. 5.2.22,

$$HTN \equiv \frac{U t_{cr}}{C_0 \hat{c}_{p0}} \left(\frac{S}{V}\right) \tag{7.1.17}$$

DHR_m is the dimensionless heat of reaction of the mth-independent chemical reaction, defined by Eq. 5.2.23,

$$DHR_m \equiv \frac{\Delta H_{R_m}(T_0)}{\hat{c}_{p0} T_0} \tag{7.1.18}$$

and $CF(Z_m, \theta)$ is the correction factor of the heat capacity of the reacting fluid, determined by Eq. 5.2.54. The first term inside the bracket of Eq. 7.1.16 is the dimensionless local heat-transfer rate:

$$\frac{d}{d\tau} \left[\frac{\dot{Q}}{(F_{tot})_0 \hat{c}_{p0} T_0} \right] = HTN(\theta_F - \theta) \tag{7.1.19}$$

The second term inside the bracket of Eq. 7.1.16 represents the heat generated (or consumed) by the chemical reactions.

The specific molar specific heat capacity of the reference state, $\hat{c}_{p0}$, is defined differently for gas-phase and liquid-phase reactions. For gas-phase reactions, it is defined by Eq. 5.2.58,

$$\hat{c}_{p0} \equiv \sum_j^J y_{j_0} \hat{c}_{p_j}(T_0) \tag{7.1.20}$$

and for liquid-phase reactions, it is defined by Eq. 5.2.60,

$$\hat{c}_{p0} \equiv \frac{\dot{m}\bar{c}_p}{(F_{tot})_0} \tag{7.1.21}$$

To solve Eq. 7.1.16, we have to specify the value of HTN. However, its value depends on the heat-transfer coefficient, U, which depends on the flow conditions in the reactor, the properties of the fluid, and the heat-transfer area per unit volume, (S/V). These parameters are not known a priori. Therefore, we develop a procedure to estimate the range of HTN. For isothermal operation $(d\theta/d\tau = 0)$, we can determine the local HTN from Eq. 7.1.16 (taking the reactor temperature as the reference temperature, $\theta = 1$):

$$\text{HTN}_{\text{iso}}(\tau) = \frac{1}{\theta_F - 1} \sum_{m}^{n_I} \text{DHR}_m \frac{dZ_m}{d\tau} \qquad (7.1.22)$$

Since $dZ_m/d\tau$ varies along the reactor, the value of HTN also varies from point to point. We define an *average* HTN for isothermal operation by

$$\text{HTN}_{\text{ave}} \equiv \frac{1}{\tau_{\text{tot}}} \int_{0}^{\tau_{\text{tot}}} \text{HTN}(\tau)\, d\tau \qquad (7.1.23)$$

where τ_{tot} is the total dimensionless space time of the reactor. Recall that for adiabatic operation HTN = 0. In practice in most cases, the heat-transfer number would be

$$0 < \text{HTN} \leq \text{HTN}_{\text{ave}} \qquad (7.1.24)$$

Equation 7.1.24 provides only an estimate on the range of the value of HTN. We select a specific value after examining the reactor performance with different values of HTN. When multiple reactions take place, it is important to examine the reactor performance for different values of HTN, since it is difficult to predict the effect of the heat transfer on the relative rates of the individual reactions. Once the physical reactor has been designed, it is necessary to verify that the flow conditions in the reactor actually provide the specified value of HTN.

For convenience, Tables A.3a and A.3b in Appendix A provide the design equations and auxiliary relations used in the design of plug-flow reactors. Table A.4 provides the energy balance equation.

In the remainder of the chapter, we discuss how to apply the design equations and the energy balance equation to determine various quantities related to the operations of plug-flow reactors. In Section 7.2, we examine isothermal operations with single reactions to illustrate how the rate expressions are incorporated into the design equation and how rate expressions are determined. In Section 7.3, we expand the analysis to isothermal operations with multiple chemical reactions. In Section 7.4, we consider nonisothermal operations with multiple reactions. In all these cases, we assume that the pressure drop along the reactor is negligible. In Section 7.5, we consider the effect of pressure drop on the operations of plug-flow reactors with gas-phase reactions.

7.2 ISOTHERMAL OPERATIONS WITH SINGLE REACTIONS

We start the analysis of plug-flow reactors by considering isothermal operations with single reactions. For isothermal operations, $d\theta/d\tau = 0$, and we have to solve only the design equations. The energy balance equation provides the heating (or cooling) load necessary to maintain isothermal conditions. Furthermore, for isothermal operations, the reaction rate depends only on the species concentrations, and Eq. 7.1.5 reduces to

$$r = k(T_0)h(C_j\text{'s}) \qquad (7.2.1)$$

When a single chemical reaction takes place in the reactor, the operation is described by one design equation, and Eq. 7.1.1 reduces to

$$\frac{dZ}{d\tau} = r\left(\frac{t_{cr}}{C_0}\right) \qquad (7.2.2)$$

where Z is the dimensionless extent of the reaction and r is its rate. The solution of the design equation [$Z(\tau)$ versus τ] provides the reaction operating curve that describes the progress of the chemical reaction along the reactor. Once $Z(\tau)$ is known, we can apply Eq. 7.1.7 to obtain the species operating curves, indicating the species molar flow rates at any point along the reactor. Also, if one prefers to express the design equation in terms of the reactor volume, rather than the dimensionless space time τ, using Eq. 7.1.3, the design equation becomes

$$\frac{dZ}{dV_R} = \frac{r}{(F_{tot})_0} \qquad (7.2.3)$$

We have to solve Eq. 7.2.2 subject to the initial condition that at $\tau = 0$, $Z(0)$ is specified (Z at the inlet of the reactor). In some cases, the design equation can be solved by separating the variables and integrating

$$\tau = \left(\frac{C_0}{t_{cr}}\right) \int_{Z_{in}}^{Z_{out}} \frac{dZ}{r} \qquad (7.2.4)$$

where Z_{in} and Z_{out} are, respectively, the extent at the reactor inlet and outlet. Equation 7.2.4 is the integral form of the design equation of plug-flow reactors with a single chemical reaction. Figure 7.2 shows graphically the relationship among Z_{in}, Z_{out}, and τ.

To solve Eq. 7.2.2 or 7.2.4, we have to express the reaction rate r in terms of the dimensionless extent Z. To do so, we express the species concentrations in terms of Z. Selecting the inlet stream as the reference stream, $Z_{in} = 0$, and for *liquid-phase* reactions, Eq. 7.1.11 reduces to

$$C_j = C_0(y_{j_0} + s_j Z) \qquad (7.2.5)$$

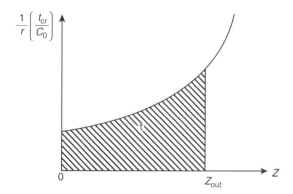

Figure 7.2   Graphical presentation of the PFR design equation for a single reaction.

For *gaseous* reactions, assuming negligible pressure drop, Eq. 7.1.15 reduces to

$$C_j = C_0 \left(\frac{y_{j_0} + s_j Z}{1 + \Delta Z} \right) \qquad (7.2.6)$$

Note that for given inlet conditions, Eqs. 7.2.2 and 7.2.4 have three variables: the dimensionless space time (or dimensionless reactor volume), τ, the reaction extent at the reactor outlet, Z_{out}, and the reaction rate, r. We apply the design equation to determine any one of these three variables when the other two are provided. In a typical design problem, we have to determine the reactor volume needed to obtain a specified extent (or conversion) for a given feed rate and given reaction rate. A second application of the design equation is to determine the extent (or conversion) at the reactor outlet for a given reactor volume and given reaction rate. The third application is to determine the reaction rate when the extent (or conversion) at the reactor outlet is provided for different feed rates.

Below, we analyze the operation of isothermal plug-flow reactors with single reactions for different types of chemical reactions. For convenience, we divide the analysis into two sections: (i) design and (ii) determination of the rate expression. In the former, we determine the size of the reactor for a known reaction rate, specified feed rate, and specified extent (or conversion). In the second section, we determine the rate expression and its parameters from reactor operating data.

7.2.1 Design

First, consider the application of the design equation when the reaction rate is provided in the form of an algebraic expression. We start with a first-order, gaseous, chemical reaction of the general form

$$A \longrightarrow \text{Products}$$

whose rate expression is $r = kC_A$. For this reaction, $s_A = -1$, and Δ is determined by the stoichiometry. We select the inlet stream as the reference stream; hence,

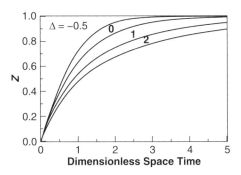

Figure 7.3 Reaction operating curves for a single first-order reaction of the form A $\rightarrow$ Products.

y_{A_0}, y_{B_0}, y_{C_0}, and so on are specified. For a gas-phase reaction, the species concentration is given by Eq. 7.2.6, and the reaction rate expression is

$$r = kC_0 \frac{y_{A_0} - Z}{1 + \Delta Z} \tag{7.2.7}$$

Substituting Eq. 7.2.7 into Eq. 7.2.2, the design equation becomes

$$\frac{dZ}{d\tau} = kC_0 \frac{y_{A_0} - Z}{1 + \Delta Z} \left(\frac{t_{cr}}{C_0} \right) \tag{7.2.8}$$

Using Eq. 3.5.4, for first-order reaction,

$$t_{cr} = \frac{1}{k}$$

(Note this characteristic reaction time is the same as the one obtained by the procedure described in Section 3.5.) When only reactant A is fed ($y_{A_0} = 1$), the design equation reduces to

$$\frac{dZ}{d\tau} = \frac{1 - Z}{1 + \Delta Z} \tag{7.2.9}$$

We solve this equation, subject to the initial condition that at $\tau = 0$, $Z = 0$. The solution of this design equation for different values of Δ is shown in Figure 7.3.

Example 7.1 The first-order, gas-phase, decomposition reaction

$$A \longrightarrow B + C$$

is carried out in an isothermal, plug-flow reactor. A stream of reactant A is fed to the reactor at a rate of 10 mol/min. At the feed conditions, the concentration of A is $C_{A_{in}} = 0.04$ mol/L. At the reactor operating temperature, the reaction rate constant is $k = 0.1$ s^{-1}.

a. Derive and plot the reaction and species curves.

b. Determine the volume of the reactor needed to achieve 80% conversion.

c. Determine the volume of the reactor needed to achieve 80% conversion if, by mistake, we take $\Delta = 0$.

d. Determine the conversion if the volume calculated in (c) is used.

e. Determine the feed flow rate to achieve 80% conversion in a reactor with the volume calculated in (c).

Solution   The stoichiometric coefficients of the chemical reaction are

$$s_A = -1 \qquad s_B = 1 \qquad s_C = 1 \qquad \Delta = 1$$

We select the inlet stream as the reference stream, and since only reactant A is fed to the reactor, $y_{A_0} = 1$, $y_{B_0} = y_{C_0} = 0$, $C_0 = C_{A_{in}} = 0.04$ mol/L, and 10 mol/min 10 mol/min. The volumetric flow rate of the reference stream is

$$v_0 = \frac{(F_{tot})_0}{C_0} = 250 \text{ L/min}$$

a. For a plug-flow reactor with a single chemical reaction, the design equation is

$$\frac{dZ}{d\tau} = r\left(\frac{t_{cr}}{C_0}\right) \tag{a}$$

For gas-phase reactions, C_A is expressed by Eq. 7.2.6, and the reaction rate is

$$r = kC_0 \frac{1 - Z}{1 + Z} \tag{b}$$

Substituting (b) into (a), the design equation becomes

$$\frac{dZ}{d\tau} = kC_0 \frac{1 - Z}{1 + Z}\left(\frac{t_{cr}}{C_0}\right) \tag{c}$$

Selecting the characteristic reaction time to be

$$t_{cr} = \frac{1}{k} = 10 \text{ s} = 0.167 \text{ min} \tag{d}$$

and substituting (d) into (c), the design equation reduces to

$$\frac{dZ}{d\tau} = \frac{1 - Z}{1 + Z} \tag{e}$$

We solve (e) numerically, subject to the initial condition that at $\tau = 0$, $Z = 0$.

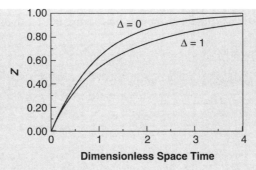

Figure E7.1.1 Reaction operating curves.

Figure E7.1.1 shows the reaction curve. Using Eq. 7.1.7, the species curves are

$$\frac{F_A(\tau)}{(F_{tot})_0} = 1 - Z(\tau)$$

$$\frac{F_B(\tau)}{(F_{tot})_0} = \frac{F_C(\tau)}{(F_{tot})_0} = Z(\tau)$$

Figure E7.1.2 shows the species curves.

b. Using Eq. 2.6.5, the relation between the conversion and extent for a single reaction,

$$Z = -\frac{y_A(0)}{s_A} f_A = -\frac{1}{(-1)} f_A = f_A$$

For $Z_{out} = 0.8$, from the reaction curve, $\tau = 2.42$. Note in this case we can determine τ by separating the variables and integrating (e),

$$\tau = \int_0^{0.8} \frac{1+Z}{1-Z} dZ = 2.42$$

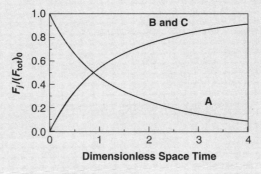

Figure E7.1.2 Species operating curves.

Using Eq. 7.1.3, the needed reactor volume is

$$V_R = \tau v_0 t_{cr} = 100.8\,\text{L} \tag{f}$$

c. If the variation in the volumetric flow rate along the reactor is not considered ($\Delta = 0$), the design equation is

$$\frac{dZ}{d\tau} = 1 - Z \tag{g}$$

We solve (g) by separating the variables and integrating. The solution is

$$Z(\tau) = 1 - e^{-\tau}$$

For $Z_{\text{out}} = 0.8$

$$\tau = \int_0^{0.8} \frac{1}{1-Z}\,dZ = 1.61 \tag{h}$$

Hence, the calculated reactor volume is

$$V_R = \tau v_0 t_{cr} = 67.18\,\text{L} \tag{i}$$

Note that by ignoring the effect of the volume expansion, we specify a reactor volume that is only 68% of the required volume.

d. For a reactor with a volume of 67.1 L, $\tau = 1.61$. Using the proper design equation, (e), the outlet extent, Z_{out}, is determined by

$$\int_0^{Z_{\text{out}}} \frac{1+Z}{1-Z}\,dZ = 1.61 \tag{j}$$

The solution is $Z_{\text{out}} = f_{A_{\text{out}}} = 0.680$. Hence, if the wrongly specified reactor volume is used, only 68% conversion is obtained.

e. To attain 80% conversion on the 67.1-L reactor, the feed flow rate should be reduced. From the design equation (e), $\tau = 2.42$. Using Eq. 7.1.3 to determine the feed rate,

$$v_0 = \frac{V_R}{\tau t_{cr}} = 166.3\,\text{L/min} \tag{k}$$

$$(F_{\text{tot}})_0 = v_0 C_0 = 6.65\,\text{mol/min}$$

Hence, when the wrongly specified reactor volume is used, the feed flow rate should be lowered by about 34% to maintain the specified conversion level of 80%.

Example 7.2 The second-order, gas-phase chemical reaction

$$2A \longrightarrow B$$

is carried out in a 1000 L isothermal plug-flow reactor. A feed stream consisting of 80% A and 20% inert (% mole) is fed to the reactor at a molar flow rate of 125 mol/min. The concentration of reactant A in the feed stream is $C_{A_{in}} = 0.05$ mol/L. The molar fraction of reactant A at the outlet of the reactor is 0.16. We want to modify the reactor such that it provides 90% conversion of A. What is the required additional volume?

Solution   The stoichiometric coefficients of the chemical reaction are

$$s_A = -2 \qquad s_B = 1 \qquad s_I = 0 \qquad \Delta = -1$$

In this case, the reaction rate constant is not provided, and we have to determine it from the operating data. First, we derive the design equation for this case. We select the inlet stream to the system as the reference stream. Hence, $(F_{tot})_0 = (F_{tot})_{in}$, $y_{A_0} = 0.8$, $y_{I_0} = 0.2$. The reference concentration is

$$C_0 = \frac{C_{A_0}}{y_{A_0}} = 0.0625 \text{ mol/L}$$

and the volumetric flow rate of the reference stream is

$$v_0 = \frac{(F_{tot})_0}{C_0} = 2000 \text{ L/min}$$

For a gas-phase reaction, the species concentration is given by Eq. 7.2.6, and the rate expression is

$$r = kC_0^2 \left(\frac{y_{A_0} + s_A Z}{1 + \Delta Z} \right)^2 \tag{a}$$

Using Eq. 3.5.4, for a second-order reaction, the characteristic reaction time, t_{cr}, is

$$t_{cr} = \frac{1}{kC_0} \tag{b}$$

Substituting (a) and (b) into Eq. 7.2.2, the design equation is

$$\frac{dZ}{d\tau} = \left(\frac{0.8 - 2Z}{1 - Z} \right)^2 \tag{c}$$

Separating the variables, the design equation for the reactor is

$$\tau = \int_0^{Z_{out}} \left(\frac{1-Z}{0.8-2Z} \right)^2 dZ \qquad (d)$$

where τ is the dimensionless space time of the reactor, defined by Eq. 7.1.3. To determine Z_{out}, we use Eq. 2.7.8 and Eq. 2.7.10 to express the molar fraction of reactant A at the reactor outlet:

$$y_{A_{out}} \equiv \frac{F_{A_{out}}}{F_{tot_{out}}} = \frac{0.8 - 2Z_{out}}{1 - Z_{out}} = 0.16 \qquad (e)$$

Solving (e), $Z_{out} = 0.348$. Substituting this value into (d), we obtain $\tau = 2.2$. Using Eq. 7.1.3, the characteristic reaction time is

$$t_{cr} = \frac{V_R}{v_0 \tau} = 0.2278 \, \text{min} \qquad (f)$$

Using (b), the reaction rate constant is

$$k = \frac{1}{C_0 t_{cr}} = 70.39 \, \text{L/mol min}$$

Now we design a reactor to provide 90% conversion. Using Eq. 2.6.5,

$$Z_{out} = -\frac{y_{A_{in}}}{s_A} f_{A_{out}} = -\frac{0.8}{-2}(0.9) = 0.36$$

Substituting in (d),

$$\tau_{new} = \int_0^{0.36} \left(\frac{1-Z}{0.8-2Z} \right)^2 dZ = 2.8 \qquad (g)$$

The required reactor volume is

$$V_{R_{new}} = v_0 t_{cr} \tau = (2000 \, \text{L/min})(0.2278 \, \text{min})(2.8) = 1275 \, \text{L} \qquad (h)$$

An additional 275 L should be added to the reactor.

Example 7.3   The elementary, gas-phase reaction

$$A + B \longrightarrow C$$

is carried out in an isothermal plug-flow reactor operated at 2 atm and 170°C. At this temperature, the reaction rate constant is $k = 90 \, \text{L}/(\text{mol min})$, and the vapor pressure of the product, C, is 0.3 atm. The reactor is fed with two gas streams: the first one consists of 80% A, 10% B, 10% inert (I), and is at 2.5 atm and 150°C; the second consists of 80% B, 20% I, and is at 3 atm and 180°C. The first stream is fed at a rate of 100 mol/min and the second at a rate of 120 mol/min. Determine:

a. The conversion of reactant A when C begins to condense.

b. The reactor volume where C starts to condense.

c. The reactor volume needed for 85% conversion of A.

Solution   In the first section of the reactor, all the species are gaseous, and the reaction is

$$A(g) + B(g) \longrightarrow C(g)$$

and its stoichiometric coefficients are

$$s_A = -1 \qquad s_B = -1 \qquad s_C = 1 \qquad \Delta = -1$$

First, we have to select a reference stream. Since the two streams are mixed at the reactor inlet, we select a fictitious reference stream that consists of the molar flow rates of the two streams, and is at 2 atm and 170°C (the reactor operating conditions). Hence,

$$F_{A_0} = 0.8 F_1 = 80 \, \text{mol/min}$$

$$F_{B_0} = 0.1 F_1 + 0.8 F_2 = 106 \, \text{mol/min}$$

$$F_{I_0} = 0.1 F_1 + 0.2 F_2 = 34 \, \text{mol/min}$$

$$(F_{\text{tot}})_0 = F_{A_0} + F_{B_0} + F_{I_0} = 220 \, \text{mol/min}$$

The composition of the reference stream is $y_{A_0} = 0.364$, $y_{B_0} = 0.482$, and $y_{I_0} = 0.154$. Assuming ideal-gas behavior, the concentration of the reference stream is

$$C_0 = \frac{P_0}{RT_0} = \frac{2 \, \text{atm}}{(0.08206 \, \text{L atm/mol K})(443 \, \text{K})} = 0.055 \, \text{mol/L}$$

Using Eq. 7.1.14, the volumetric flow rate of the reference stream is

$$v_0 = \frac{(F_{\text{tot}})_0}{C_0} = 4000 \, \text{L/min}$$

a. Species C starts to condense when its partial pressure in the reactor is 0.3 atm. At a given point in the reactor, the partial pressure of product C is

$$P_C = y_C P = \frac{F_C}{F_{tot}} P$$

Using Eqs. 2.7.8 and 2.7.10,

$$P_C = \frac{y_{C_0} + s_C Z}{1 + \Delta Z} P = \frac{Z}{1 - Z}(2 \text{ atm}) = 0.3 \text{ atm} \tag{a}$$

Solving (a), $Z = 0.130$, and using Eq. 2.6.5, the conversion is

$$f_A = -\frac{s_A}{y_{A_0}} Z = 0.358 \tag{b}$$

b. To determine the reactor volume for $Z = 0.130$, we use Eq. 7.2.2,

$$\frac{dZ}{d\tau} = r\left(\frac{t_{cr}}{C_0}\right) \tag{c}$$

Since the reaction is elementary, $r = k C_A C_B$, and, using Eq. 7.2.6,

$$r = k C_0^2 \frac{(y_{A_0} - Z)(y_{B_0} - Z)}{(1 + \Delta Z)^2} \tag{d}$$

Using Eq. 3.5.4, for a second-order reaction, the characteristic reaction time is

$$t_{cr} = \frac{1}{k C_0} = 0.202 \text{ min} \tag{e}$$

Substituting (d) and (e) into (c), the design equation reduces to

$$\frac{dZ}{d\tau} = \frac{(0.364 - Z)(0.482 - Z)}{(1 - Z)^2} \tag{f}$$

To determine the dimensionless space time for $Z = 0.130$, we solve (f) by separation of variables,

$$\tau = \int_0^{0.130} \frac{(1-Z)^2}{(0.364-Z)(0.482-Z)}dZ = 0.9306 \qquad \text{(g)}$$

Using Eq. 7.1.3, the volume of the reactor for $Z = 0.130$ is

$$V_R = \tau v_0 t_{cr} = (0.9306)(4000\,\text{L/min})(0.202\,\text{min}) = 751.9\,\text{L}$$

c. Any product C generated after dimensionless extent of 0.130 is reached cannot be in the vapor phase, as shown in Figure E7.3.1. Hence, the following reaction takes place downstream in the reactor:

$$A(g) + B(g) \longrightarrow C(\text{liq})$$

For this reaction, $\Delta_{gas} = -2$. For $Z > 0.130$, a portion of product C is in the gas phase and a portion in the liquid phase (with negligible volume). The total molar flow rate of the gas-phase is now (for $Z > 0.130$)

$$(F_{tot})_{gas} = (F_{tot})_0[1 + (-1)0.130 + (-2)(Z - 0.130)]$$
$$= (F_{tot})_0(1.13 - 2Z) \qquad \text{(h)}$$

Using Eq. 7.2.6, for $Z > 0.130$, the concentrations of the two reactants are

$$C_A = C_0\frac{0.364 - Z}{1.13 - 2Z} \qquad C_B = C_0\frac{0.482 - Z}{1.13 - 2Z} \qquad \text{(i)}$$

Substituting (i) and (d) into (c), for $Z > 0.130$, the design equation is

$$\frac{dZ}{d\tau} = \frac{(0.364 - Z)(0.482 - Z)}{(1.13 - 2Z)^2} \qquad \text{for } Z > 0.13 \qquad \text{(j)}$$

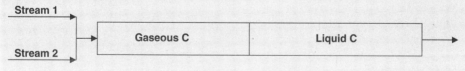

Figure E7.3.1 Zones of product C.

Using Eq. 2.6.5, for $f_A = 0.85$,

$$Z = -\frac{y_{A_0}}{s_A} f_A = 0.3094$$

To determine the dimensionless space time of the reactor section where product C is formed as a liquid, we integrate (j) between 0.130 and 0.3094:

$$\tau = \int_{0.130}^{0.3094} \frac{(1.13 - 2Z)^2}{(0.364 - Z)(0.482 - Z)} dZ = 2.52 \qquad (k)$$

Using Eq. 7.1.3, the volume of the reactor section where condensed C is formed is

$$V_R = \tau v_0 t_{cr} = 2036 \, L$$

The total volume of the reactor is $V_R = 751.9 + 2036 = 2788 \, L$.

d. The reactor operation is described by two design equations: (f) for $0 \le Z \le 0.13$ and (k) for $Z > 0.13$. Figure E7.3.2 shows the reaction operating curve. Using Eq. 2.7.8, the species curves are

$$\frac{F_A(\tau)}{(F_{tot})_0} = y_{A_0} - Z(\tau)$$

$$\frac{F_B(\tau)}{(F_{tot})_0} = y_{B_0} - Z(\tau)$$

$$\frac{F_C(\tau)}{(F_{tot})_0} = Z(\tau)$$

Figure E7.3.3 shows the species curves.

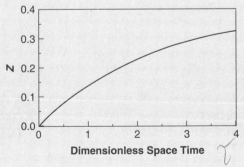

Figure E7.3.2 Reaction operating curve.

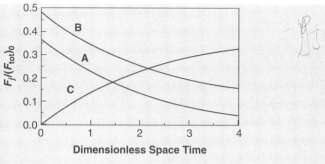

Figure E7.3.3   Species operating curves.

Example 7.4   Design a packed-bed reactor for the gas-phase, heterogeneous catalytic cracking reaction

$$A \longrightarrow B + C$$

The reactor should produce 20 metric ton per day of product B at 75% conversion of A. A process stream at 387°C and 2 atm, consisting of 90% A and 10% inert (I), is available in the plant. Based on the kinetic data below:

a. Determine the feed flow rate to the reactor.
b. Derive the design equation and plot the reaction and species curves.
c. Determine the volume of the reactor.
d. Determine the mass of the catalyst in the bed.

Data: Molecular mass of product B is 28 g/mol
Bulk density of the catalyst bed is 1.30 kg/L
The mass-based reaction rate expression is

$$r_w = \frac{k_w C_A}{1 + K C_A} \text{ mol (g catalyst)}^{-1} \text{ s}^{-1}$$

where $k_W = 0.192 \text{ cm}^3/\text{g-cat s}$, and $K = 60 \text{ L/mol}$.

Solution   This example shows the use of the design equation when the rate expression is given on the basis of mass, and it is not a power function of the species concentrations. The stoichiometric coefficients of the chemical reaction are

$$s_A = -1 \qquad s_B = 1 \qquad s_C = 1 \qquad s_I = 0 \qquad \Delta = 1$$

We select the feed stream as the reference stream; hence, $y_{A_0} = 0.9$, $y_{A_I} = 0.1$, and $y_{B_0} = y_{C_0} = 0$. The reference concentration is

$$C_0 = \frac{P_0}{RT_0} = \frac{2 \text{ atm}}{(0.08206 \text{ L atm/mol K})(660 \text{ K})} = 0.0369 \text{ mol/L}$$

a. To determine the feed flow rate, we use Eq. 2.7.8 to express the production rate of product B in terms of $(F_{tot})_0$. Selecting the inlet stream as the reference stream,

$$F_{B_{out}} = (F_{tot})_0 (y_{B_0} + s_B Z_{out}) \tag{a}$$

Using Eq. 2.6.5,

$$Z_{out} = -\frac{y_{A_0}}{s_A} f_{A_{out}} = 0.675$$

and the molar flow rate of product B at the reactor outlet is

$$F_{B_{out}} = \frac{\dot{m}_B}{MW_B} = 8.267 \text{ mol/s}$$

Hence, using (a), the molar flow rate of the reference stream is

$$(F_{tot})_0 = \frac{F_{B_{out}}}{s_B Z_{out}} = 12.25 \text{ mol/s}$$

The volumetric flow rate of the reference stream is

$$v_0 = \frac{(F_{tot})_0}{C_0} = 332.0 \text{ L/s} \tag{b}$$

b. Using Eq. 3.2.3, the volume-based rate expression is

$$r = \rho_{bed} \frac{k_w C_A}{1 + K C_A} = \frac{k C_A}{1 + K C_A} \tag{c}$$

where $k = \rho_{bed} k_w = 0.250 \text{ s}^{-1}$ is the volume-based reaction rate constant. We use Eq. 7.2.6 to express the species concentrations:

$$C_A = C_0 \frac{0.9 - Z}{1 + Z} \tag{d}$$

Substituting (c) and (d) in Eq. 7.2.2, the design equation becomes

$$\frac{dZ}{d\tau} = kC_0 \left(\frac{\dfrac{0.9 - Z}{1 + Z}}{1 + KC_0 \dfrac{0.9 - Z}{1 + Z}} \right) \left(\frac{t_{cr}}{C_0} \right) \tag{e}$$

We define the characteristic reaction time by

$$t_{cr} = \frac{1}{k} = 4\,\text{s} \tag{f}$$

and the design equation reduces to

$$\frac{dZ}{d\tau} = \frac{0.9 - Z}{1 + Z + KC_0(0.9 - Z)} \tag{g}$$

Substituting the numerical values of K and C_0, (g) becomes

$$\frac{dZ}{d\tau} = \frac{0.9 - Z}{2.99 - 1.21Z} \tag{h}$$

We solve (h) numerically subject to the initial condition that at $\tau = 0, Z = 0$. The reaction curve is shown in Figure E7.4.1. Using Eq. 2.7.8, the species curves are

$$\frac{F_A(\tau)}{(F_{\text{tot}})_0} = y_{A_0} - Z(\tau)$$

$$\frac{F_B(\tau)}{(F_{\text{tot}})_0} = \frac{F_C(\tau)}{(F_{\text{tot}})_0} = Z(\tau)$$

Figure E7.4.2 shows the species curves.

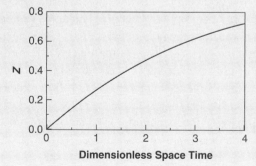

Figure E7.4.1 Reaction operating curve.

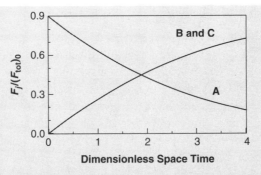

Figure E7.4.2 Species operating curves.

c. To determine the reactor volume, we use either the reaction operating curve or the integral form of the design equation:

$$\tau = \int_0^{0.675} \frac{2.99 - 1.21Z}{0.9 - Z} \, dZ = 3.458 \tag{i}$$

Using Eq. 7.1.3, the volume of the reactor is

$$V_R = \tau v_0 t_{cr} = 4584 \, \text{L} \tag{j}$$

d. The mass of the catalyst bed is

$$M_{bed} = \rho_{bed} V_R = 5959 \, \text{kg}$$

We conclude the discussion with two comments. First, note that the design equation provides us with the reactor volume needed to obtain a given extent (or conversion). However, it does not indicate whether we should use a long reactor with a small diameter or a short reactor with a large diameter (provided, of course, that the plug-flow assumption is valid). The reactor diameter is determined by other considerations such as the heat-transfer area needed to provide (or remove) heat to the reactor, and the pressure drop (pumping cost). The effect of heat-transfer on the performance of plug-flow reactors is discussed in Section 7.4. The effect of pressure drop on the performance of gaseous plug-flow reactors is discussed in Section 7.5. Second, in the examples above, we designed the reactor for a given specified extent. However, the question of what is the most suitable reaction extent for the operation has not been discussed. The cost of the reactants, the value of the products, the cost of the equipment, and the operating expenses (including separation costs) affect the optimal level of the extent. These points are covered in Chapter 10.

7.2.2 Determination of Reaction Rate Expression

In the preceding section, we described how to apply the design equation when the reaction rate expression is given. Now, we will discuss how to determine the rate expression from data obtained in plug-flow operations. The method is based on measuring the exit composition of the reactor at different space times, and then, by differentiating the data, we obtain the reaction rate at the exit conditions. Different space times are obtained by either withdrawing samples at different points along the reactor (different reactor volumes) or by varying the feed flow rate. The approach is similar to the differential method applied to batch reactors.

To derive a relation between the reaction rate and the extent, we rearrange Eq. 7.2.2 and use Eq. 7.1.3 to obtain

$$r = C_0 \frac{dZ}{dt_{sp}} \tag{7.2.10}$$

where $t_{sp} = V_R / v_0$ is the space time defined by Eq. 4.4.6. For the reactor outlet,

$$r_{out} = C_0 \frac{dZ_{out}}{dt_{sp}} \tag{7.2.11}$$

Hence, by plotting $C_0 Z_{out}$ versus t_{sp} and taking the derivatives, we can determine r_{out} as shown schematically in Figure 7.4. Note that r_{out} occurs at the outlet concentration.

The main difficulty in using Eq. 7.2.11 is that the extent is not a measurable quantity. Therefore, we have to derive a relationship between Z_{out} and an appropriate measured quantity. We do so by using the design equation and relevant stoichiometric relations. In most applications, we measure the concentration of a species at the reactor outlet and calculate the extent by either Eq. 7.2.5 for liquid-phase reactions or Eq. 7.2.6 for gas-phase reactions. We can then determine the orders of the individual species for power rate expressions.

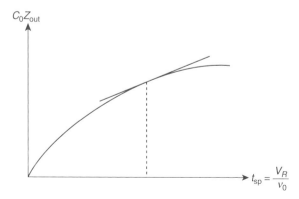

Figure 7.4 Determination of reaction rate from PFR operating data.

Example 7.5 A stream of gaseous reactant A at 3 atm and 30°C (C_{A_0} = 120 mmol/L) is fed into a 500-L plug-flow reactor where it decomposes according to the reaction:

$$A \longrightarrow 2B + C$$

The concentration of reactant A is measured at the outlet of the reactor for different feed flow rates. Based on the data below, determine:

a. The order of the reaction

b. The rate constant at 30°C

Data:

v_0 (L/min)	5.0	10.0	12.5	16.7	25.0	50.0	125	250	∞
$C_{A_{out}}$ (mmol/L)	7.1	13.5	16.2	20.0	26.8	40.6	62.6	79.5	120

Solution   The stoichiometric coefficients are:

$$s_A = -1 \qquad s_B = 2 \qquad s_C = 1 \qquad \Delta = 2$$

We select the inlet stream as the reference stream; hence, $y_{A_0} = 1$, $y_{B_0} = y_{C_0} = 0$, and $C_0 = C_{A_0} = 120$ mol/L. Using Eq. 7.2.6, we calculate Z_{out} for each outlet concentration:

$$Z_{out} = \frac{C_0 y_{A_0} - C_{A_{out}}}{\Delta\, C_{A_{out}} - s_A C_0} \tag{a}$$

The reactor space time is

$$t_{sp} = \frac{V_R}{v_0} \tag{b}$$

For each run, we calculate Z_{out} using (a), and t_{sp} using (b), and then calculate $C_0 Z_{out}$:

v_0 (L/min)	5.0	10.0	12.5	16.7	25.0	50.0	125	250	∞
$C_{A_{out}}$ (mmol/L)	7.1	13.5	16.2	20	26.8	40.6	62.6	79.5	120
Z_{out}	0.841	0.724	0.681	0.622	0.537	0.395	0.234	0.145	0
t_{sp} (min)	100	50	40	30	20	10	4	2	0
$C_0 Z_{out}$ (mmol/L)	101	86.9	81.7	74.6	64.4	47.4	28.1	17.4	0

a. To determine the reaction rate at the reactor outlet for each run (using Eq. 7.2.11), we differentiate the data. Since the data points are not equally spaced, we calculate the derivative at the midpoints of each of two adjacent points. The calculated values are given in the table below:

t_{sp} (min)	C_0Z_{out} (mmol/L)	$C_{A_{out}}$ (mmol/L)	$\Delta(C_0Z_{out})/\Delta t_{sp}$ (mmol/L min)	$(C_{A_{out}})_{ave}$ (mmol/L)
0	0.0	120		
2	17.4	79.5	8.70	99.7
4	28.1	62.6	5.35	71.0
10	47.4	40.6	3.21	51.6
20	64.4	26.8	1.70	33.7
30	74.7	20.2	1.03	23.5
40	81.7	16.2	0.70	18.2
50	86.9	13.5	0.52	14.9
100	101.0	7.1	0.28	10.3

Assuming the rate expression is of the form $r = kC_A^{\alpha}$, substituting Eq. 7.2.11, and taking the log,

$$\ln\left(C_0 \frac{dZ_{out}}{dt_{sp}} \right) = \ln(k) + \alpha \ln(C_{A_{out}}) \tag{c}$$

By plotting $\ln(C_0\,dZ_{out}/dt_{sp})$ versus $\ln(C_{A_{out}})$, we should get a straight line whose slope is α. The plot is shown in Figure E7.5.1. The slope is

$$\text{Slope} = \frac{\ln 10 - \ln 0.35}{\ln 100 - \ln 10} = 1.48 \tag{d}$$

b. Now that the order of the reaction is known ($\alpha = 1.5$), we use the integral form of the design equation (Eq. 7.2.4) to determine the value of k. For

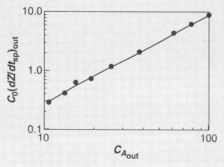

Figure E7.5.1   Determination of reaction order.

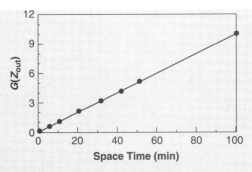

Figure E7.5.2   Determination of reaction rate constant.

$\alpha = 1.5$, Eq. 7.2.2 reduces to

$$\frac{dZ}{d\tau} = \left(\frac{1-Z}{1+2Z}\right)^{1.5} \tag{e}$$

Using Eq. 3.5.4, the characteristic reaction time is

$$t_{cr} = \frac{1}{k C_{A_0}^{0.5}} \tag{f}$$

Separating the variables and integrating (e),

$$\tau \equiv \frac{t_{sp}}{t_{cr}} = \int_0^{Z_{out}} \frac{(1+2Z)^{1.5}}{(1-Z)^{1.5}} dZ = G(Z_{out}) \tag{g}$$

The right-hand side of (g) is a function of Z_{out}, $G(Z_{out})$. Hence, by plotting $G(Z_{out})$ versus the space time, t_{sp}, we obtain a line whose slope is $1/t_{cr}$. The table below provides the calculated data, and the plot is shown in Figure E7.5.2.

t_{sp} (min)	100	50	40	30	20	10	4	2	0
Z_{out}	0.841	0.724	0.681	0.622	0.537	0.395	0.234	0.145	0
$G(Z_{out})$	10.01	5.00	4.00	3.00	2.00	1.00	0.40	0.20	0

The slope is

$$\text{Slope} = \frac{1}{t_{cr}} = k C_{A_0}^{0.5} = 0.1 \text{ min}^{-1} \tag{h}$$

and the rate constant at 30°C is $k = 9.13 \times 10^{-3}$ (L/mmol)$^{0.5}$ min^{-1}.

7.3   ISOTHERMAL OPERATIONS WITH MULTIPLE REACTIONS

When more than one chemical reaction takes place in a plug-flow reactor, we have to address several issues before we start the design. First, we have to determine how many independent reactions there are among the given reactions and then select a set of independent reactions. Next, we have to identify all the reactions that actually take place in the reactor (including dependent reactions) and express their rates. As discussed in Chapter 4, we have to write the design equation for each independent chemical reaction. To solve the design equations (obtain relationships between Z_m's and τ), we have to express the rates of the individual chemical reactions, r_m's and r_k's, in terms of the Z_m's and τ. The procedure for designing plug-flow reactors with multiple chemical reactions goes as follows:

1. Identify all the chemical reactions that take place in the reactor and define the stoichiometric coefficients of each species in each reaction.
2. Determine the number of independent chemical reactions.
3. Select a set of independent reactions from among the reactions whose rate expressions are given.
4. For each *dependent* reaction, determine its α_{km} multipliers with each of the independent reactions using Eq. 2.4.9.
5. Select a reference stream [determine $(F_{\text{tot}})_0$, C_0, v_0] and its species compositions, y_{j_0}'s.
6. Write Eq. 7.1.1 for each independent chemical reaction.
7. Select a leading (or desirable) chemical reaction and determine the expression and the value of its characteristic reaction time, t_{cr}.
8. Express the reaction rates in terms of the extents of the independent reactions, Z_m's.
9. Specify the inlet conditions.
10. Solve the design equations for Z_m's as functions of the dimensionless space time, τ, and obtain the reaction operating curves.
11. Determine the species operating curves using Eq. 2.7.8.
12. Determine the reactor volume using Eq. 7.1.3.

Below, we describe the design formulation of isothermal plug-flow reactors with multiple reactions for various types of chemical reactions (reversible, series, parallel, etc.). In most cases, we solve the design equations numerically by applying a numerical technique such as the Runge-Kutta method or using commercial mathematical software such as HiQ, Mathcad, Maple, and Mathematica. In some simple cases, we can obtain analytical solutions. Note that, for isothermal operations, $d\theta = 0$, and we do not have to solve the energy balance equation simultaneously with the design equations.

Example 7.6   Product B is produced in an isothermal tubular reactor where the following gas-phase, first-order chemical reactions take place:

$$\text{Reaction 1:} \quad A \longrightarrow 2B$$
$$\text{Reaction 2:} \quad B \longrightarrow C + D$$

A gaseous stream of reactant A ($C_{A_0} = 0.04\,\text{mol/L}$) is fed into a 200 L reactor at a rate of $100\,\text{L/min}$. At the reactor operating temperature, $k_1 = 2\,\text{min}^{-1}$ and $k_2 = 1\,\text{min}^{-1}$. The pressure drop along the reactor is negligible.
a. Derive the design equations and plot the reaction and species curves.
b. Derive and plot the reaction, yield, and selectivity curves.
c. Determine the conversion of reactant A and the production rate of products B, C, and D for the given reactor.
d. Determine the optimal reactor volume to maximize the production rate of product B.
e. Determine the conversion of A, the production rate of products B and C, and the yield of product B in the optimal reactor.

Solution   Since each reaction has a species that does not participate in the other, the two reactions are independent, and there is no dependent reaction. The stoichiometric coefficients are

$$s_{A_1} = -1 \quad s_{B_1} = 2 \quad s_{C_1} = 0 \quad s_{D_1} = 0 \quad \Delta_1 = 1$$
$$s_{A_2} = 0 \quad s_{B_2} = -1 \quad s_{C_2} = 1 \quad s_{D_2} = 1 \quad \Delta_2 = 1$$

We write Eq. 7.1.1 for each of the independent reactions ($m = 1, 2$),

$$\frac{dZ_1}{d\tau} = r_1\left(\frac{t_{cr}}{C_0}\right) \tag{a}$$

$$\frac{dZ_2}{d\tau} = r_2\left(\frac{t_{cr}}{C_0}\right) \tag{b}$$

We select the inlet stream as the reference stream; hence, $Z_{1_{in}} = Z_{2_{in}} = 0$. Since only reactant A is fed into the reactor, $C_0 = C_{A_0} = 4\,\text{mol/L}$, $y_{A_0} = 1$, and $y_{B_0} = y_{C_0} = y_{D_0} = 0$. Also,

$$(F_{\text{tot}})_0 = v_0 C_0 = 4\,\text{mol/min}$$

a. For gas-phase reactions, isothermal operation ($\theta = 1$), and negligible pressure drop, we use Eq. 7.1.15 to express the species concentrations, and the reaction rates are

$$r_1 = k_1 C_0 \frac{1 - Z_1}{1 + Z_1 + Z_2} \tag{c}$$

$$r_2 = k_2 C_0 \frac{2Z_1 - Z_2}{1 + Z_1 + Z_2} \tag{d}$$

We define the characteristic reaction time on the basis of Reaction 1; hence,

$$t_{\text{cr}} = \frac{1}{k_1} = 0.5 \, \text{min} \tag{e}$$

Substituting (c), (d), and (e) into (a) and (b), the design equations become

$$\frac{dZ_1}{d\tau} = \frac{1 - Z_1}{1 + Z_1 + Z_2} \tag{f}$$

$$\frac{dZ_2}{d\tau} = \left(\frac{k_2}{k_1} \right) \frac{2Z_1 - Z_2}{1 + Z_1 + Z_2} \tag{g}$$

We solve (f) and (g) numerically subject to the initial condition that at $\tau = 0$, and $Z_1 = Z_2 = 0$. The reaction curves are shown in the Figure E7.6.1. Once we have Z_1 and Z_2 as a function of the dimensionless space time, we use Eq. 2.7.8 to determine the species curves:

$$\frac{F_A}{(F_{\text{tot}})_0} = 1 - Z_1 \tag{h}$$

$$\frac{F_B}{(F_{\text{tot}})_0} = 2Z_1 - Z_2 \tag{i}$$

$$\frac{F_C}{(F_{\text{tot}})_0} = Z_2 \tag{j}$$

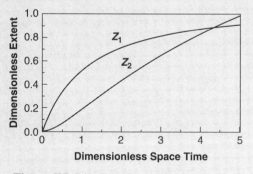

Figure E7.6.1   Reaction operating curves.

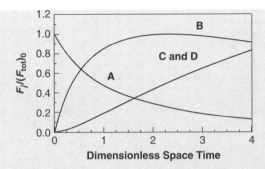

Figure E7.6.2 Species operating curves.

$$\frac{F_D}{(F_{tot})_0} = Z_2 \qquad (k)$$

Figure E7.6.2 shows the species curves.

b. Using Eq. 2.6.2,

$$f_A(\tau) = \frac{F_{A_{in}} - F_{A_{out}}}{F_{A_{in}}} = 1 - Z_1(\tau) \qquad (l)$$

The desirable reaction is Reaction 1. Using Eq. 2.6.15, the yield of product B is

$$\eta_B(\tau) = -\left(\frac{-1}{2}\right)[2Z_1(\tau) - Z_2(\tau)] \qquad (m)$$

Using Eq. 2.6.19, the selectivity of product B is

$$\sigma_B(\tau) = -\left(\frac{-1}{2}\right)\frac{2Z_1(\tau) - Z_2(\tau)}{Z_1(\tau)} \qquad (n)$$

Figure E7.6.3 shows the conversion, yield, and selectivity curves.

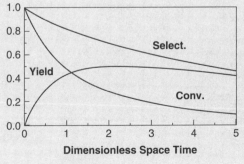

Figure E7.6.3 Conversion, yield, and selectivity curves.

c. For the given reactor and feed flow rate, using Eq. 7.1.3 and (i), the dimensionless space time is

$$\tau = \frac{V_R}{v_0} k_1 = 4$$

From the reaction curves (or tabulated calculated data), for $\tau = 4$, $Z_1 = 0.874$ and $Z_2 = 0.832$. Using (c) through (f), the species molar flow rates at the reactor outlet are $F_A = 0.504 \text{ mol/min}$, $F_B = 3.66 \text{ mol/min}$, $F_C = F_D = 3.33 \text{ mol/min}$. Using (l), $f_A = 0.874$, using (m), the yield of product B is 0.458, and using (n) the selectivity is 0.524.

d. To determine the reactor volume that provides the highest production rate of product B, we use the species curves (or tabulated calculated data) and find that the highest F_B is reached at $\tau = 2.30$. Using Eq. 7.1.3 and (e), the optimal reactor volume is

$$V_R = \frac{\tau v_0}{k_1} = 115 \text{ L}$$

From the reaction operating curves, at $\tau = 2.30$, the solutions of (f) and (g) are $Z_1 = 0.751$ and $Z_2 = 0.501$. Using (h) through (k), the species molar flow rates at the optimal reactor outlet are: $F_A = 0.996 \text{ mol/min}$, $F_B = 4.00 \text{ mol/min}$, $F_C = F_D = 2.00 \text{ mol/min}$. Using (l), the conversion of reactant A is 0.751. Using (m), the yield of product B is 0.500. Using (n), the selectivity of product B is 0.667. The table below provides a comparison between the performance of the given reactor and the optimal reactor.

	Given Reactor	Optimal Reactor
Reactor volume (L)	200	115
Volumetric feed rate (L/min)	100	100
Conversion of reactant A	0.874	0.751
Yield of product B	0.458	0.500
Selectivity of product B	0.524	0.667
Production rate of product B (mol/min)	3.66	3.99
Production rate of product C (mol/min)	3.33	2.00

We see that the volume of the optimal reactor is slightly more than half that of the given reactor. While the feed rate is the same, the conversion of reactant A is slightly lower, the production rate of product B is about 10% higher, and the production of the by-products (products C and D) is reduced by 40%.

Example 7.7   Diels-Alder reactions are organic reactions between a hydrocarbon with two double bonds and another unsaturated hydrocarbon to form a
cyclic molecule. However, in many cases, the species with the two double
bonds reacts with itself to form an undesired product. A gaseous mixture containing 50% butadiene (B) and 50% acrolein (A) is fed at a rate of 1 mol/min
into a plug-flow reactor, where the following reactions take place:

$$\text{Reaction 1:} \quad A + B \longrightarrow P$$
$$\text{Reaction 2:} \quad 2B \longrightarrow W$$

The rate expressions of the chemical reactions are: $r_1 = k_1 C_A C_B$ and $r_2 = k_2 C_B{}^2$.
The reactor operates at 330°C and 2 atm. At these conditions, $k_1 = 5.86$
L/mol min^{-1} and $k_2 = 0.72$ L/mol min^{-1}. Assume ideal-gas behavior.
a. Derive the design equations and plot the reaction curves.
b. Derive and plot the species curves.
c. Determine the reactor volume needed to achieve 80% conversion of
   reactant B.

Solution   The stoichiometric coefficients of the chemical reactions are

$$
\begin{array}{ccccc}
s_{A_1} = -1 & s_{B_1} = -1 & s_{P_1} = 1 & s_{W_1} = 0 & \Delta_1 = -1 \\
s_{A_2} = 0 & s_{B_2} = -2 & s_{P_2} = 0 & s_{W_2} = 1 & \Delta_2 = -1
\end{array}
$$

Since each chemical reaction has a species that does not appear in the other, the
two reactions are independent, and there is no dependent reaction in this case.
We select the inlet stream as the reference stream; hence, $Z_{1_{\text{in}}} = Z_{2_{\text{in}}} = 0$:

$$C_0 = \frac{P_0}{RT_0} = \frac{2 \text{ atm}}{(82.05 \times 10^{-3} \text{ L atm/mol K})(603 \text{ K})} = 4.04 \times 10^{-2} \text{ mol/L} \quad \text{(a)}$$

$$v_0 = \frac{(F_{\text{tot}})_0}{C_0} = 24.75 \text{ L/min}$$

The feed composition is $y_{A_0} = 0.5$ and $y_{B_0} = 0.5$.

a. We write Eq. 7.1.1 for each independent reaction:

$$\frac{dZ_1}{d\tau} = r_1 \left(\frac{t_{\text{cr}}}{C_0}\right) \quad \text{(b)}$$

$$\frac{dZ_2}{d\tau} = r_2 \left(\frac{t_{\text{cr}}}{C_0}\right) \quad \text{(c)}$$

We use Eq. 7.1.14 to express the species concentrations, and the reaction rates are

$$r_1 = k_1 C_0{}^2 \frac{(y_{A_0} - Z_1)(y_{B_0} - Z_1 - 2Z_2)}{(1 - Z_1 - Z_2)^2} \tag{d}$$

$$r_2 = k_2 C_0{}^2 \left(\frac{y_{B_0} - Z_1 - 2Z_2}{1 - Z_1 - Z_2} \right)^2 \tag{e}$$

We select Reaction 1 as the leading reaction; the characteristic reaction time is

$$t_{cr} = \frac{1}{k_1 C_0} = 4.22 \, \text{min} \tag{f}$$

Substituting (d), (e), and (f) into (b) and (c), the design equations reduce to

$$\frac{dZ_1}{d\tau} = \frac{(y_{A_0} - Z_1)(y_{B_0} - Z_1 - 2Z_2)}{(1 - Z_1 - Z_2)^2} \tag{g}$$

$$\frac{dZ_2}{d\tau} = \left(\frac{k_2}{k_1} \right) \left(\frac{y_{B_0} - Z_1 - 2Z_2}{1 - Z_1 - Z_2} \right)^2 \tag{h}$$

From the given data, $k_2/k_1 = 0.123$, and we solve (g) and (h) numerically subject to the initial conditions that at $\tau = 0$, $Z_1 = Z_2 = 0$. Figure E7.7.1 shows the reaction curves.

b. Now that the reaction curves are known, we obtain the species curves by using Eq. 2.7.8:

$$\frac{F_A}{(F_{\text{tot}})_0} = y_{A_0} - Z_1 \tag{i}$$

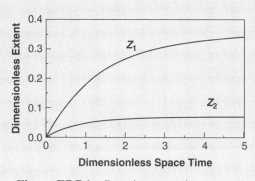

Figure E7.7.1 Reaction operating curves.

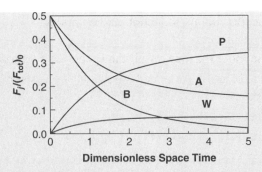

Figure E7.7.2   Species operating curves.

$$\frac{F_B}{(F_{tot})_0} = y_{B_0} - Z_1 - 2Z_2 \tag{j}$$

$$\frac{F_P}{(F_{tot})_0} = y_{P_0} + Z_1 \tag{k}$$

$$\frac{F_W}{(F_{tot})_0} = y_{W_0} + Z_2 \tag{l}$$

Figure E7.7.2 shows the species curves.

c. Using Eq. 2.6.2, for 80% conversion of reactant B, $F_B/F_{B_0} = 0.20$ and $F_B/(F_{tot})_0 = 0.10$. From the species operating curve of reactant B, $F_B/(F_{tot})_0 = 0.10$ is achieved at $\tau = 2.80$. Using Eq. 7.1.3 and (f), the required reactor volume is

$$V_R = \tau v_0 t_{cr} = 292.7 \, \text{L} \tag{m}$$

Example 7.8   The following simplified reversible reactions were proposed for gas-phase cracking of hydrocarbons:

$$
\begin{array}{lll}
\text{Reactions } 1 \& 2\text{:} & A \rightleftharpoons 2B \\
\text{Reactions } 3 \& 4\text{:} & A + B \rightleftharpoons C \\
\text{Reactions } 5 \& 6\text{:} & A + C \rightleftharpoons D
\end{array}
$$

Species C is the desired product. We want to design an isothermal, isobaric tubular reactor (plug-flow reactor) to be operated at 489°C and 5 atm. A stream of reactant A at a rate of 1 mol/s is available in the plant.

a. Derive the design equations and plot the reaction operating curves.
b. Plot the species operating curves.

c. Determine the maximum production rate of product C and the required reactor volume to achieve it.

d. Determine the mole fractions of all the species at equilibrium ($\tau = \infty$).

e. Repeat parts a, b, and d for a feed stream that consists only of species D.
The reaction rates are:

$$r_1 = k_1 C_A \qquad r_2 = k_2 C_B{}^2 \qquad r_3 = k_3 C_A C_B \qquad r_4 = k_4 C_C$$
$$r_5 = k_5 C_A C_C \qquad r_6 = k_6 C_D$$

Data: At 489°C

$$k_1 = 0.2 \, \text{min}^{-1} \qquad k_2 = 0.1 \, \text{L/mol min}^{-1} \qquad k_3 = 50 \, \text{L/mol min}^{-1}$$
$$k_4 = 0.8 \, \text{min}^{-1} \qquad k_5 = 125 \, \text{L/mol min}^{-1} \qquad k_6 = 4 \, \text{min}^{-1}$$

Solution This is an example of series–parallel reversible reactions. The reactor design formulation of these chemical reactions was discussed in Example 4.1. Here, we complete the design for an isothermal plug-flow reactor and obtain the reaction and species curves. Recall that we select Reactions 1, 3, and 5 as a set of independent reactions. Hence, the indices of the independent reactions are $m = 1$, 3, and 5, the indices of the dependent reactions are $k = 2$, 4, and 6, and we express the design equations in terms of Z_1, Z_3, and Z_5. The stoichiometric coefficients of the three independent reactions are

$$s_{A_1} = -1 \qquad s_{B_1} = 2 \qquad s_{C_1} = 0 \qquad s_{D_1} = 0 \qquad \Delta_1 = -1$$
$$s_{A_3} = -1 \qquad s_{B_3} = -1 \qquad s_{C_3} = 1 \qquad s_{D_3} = 0 \qquad \Delta_3 = -1$$
$$s_{A_5} = -1 \qquad s_{B_5} = 0 \qquad s_{C_5} = -1 \qquad s_{D_5} = 1 \qquad \Delta_5 = -1$$

The multipliers of the independent reactions to obtain the dependent reactions are

$$\alpha_{21} = -1 \qquad \alpha_{23} = 0 \qquad \alpha_{25} = 0$$
$$\alpha_{41} = 0 \qquad \alpha_{43} = -1 \qquad \alpha_{45} = 0$$
$$\alpha_{61} = 0 \qquad \alpha_{63} = 0 \qquad \alpha_{65} = -1$$

We select the inlet stream as the reference stream; hence, $Z_{1_{\text{in}}} = Z_{3_{\text{in}}} = Z_{5_{\text{in}}} = 0$:

$$C_0 = \frac{P_0}{RT_0} = \frac{5}{(82.05 \times 10^{-3} \, \text{L atm/mol K})(762 \, \text{K})} = 0.08 \, \text{mol/L} \qquad \text{(a)}$$

and

$$v_0 = \frac{(F_{\text{tot}})_0}{C_0} = 12.5 \, \text{L/s} = 750 \, \text{L/min}$$

We write Eq. 7.1.1 for each independent reaction:

$$\frac{dZ_1}{d\tau} = (r_1 - r_2)\left(\frac{t_{cr}}{C_0}\right) \tag{b}$$

$$\frac{dZ_3}{d\tau} = (r_3 - r_4)\left(\frac{t_{cr}}{C_0}\right) \tag{c}$$

$$\frac{dZ_5}{d\tau} = (r_5 - r_6)\left(\frac{t_{cr}}{C_0}\right) \tag{d}$$

We use Eq. 7.1.15 to express the species concentrations, and the rates of the six chemical reactions are

$$r_1 = k_1 C_0 \frac{y_{A_0} - Z_1 - Z_3 - Z_5}{1 + Z_1 - Z_3 - Z_5} \tag{e}$$

$$r_2 = k_2 C_0^2 \left(\frac{y_{B_0} + 2Z_1 - Z_3}{1 + Z_1 - Z_3 - Z_5}\right)^2 \tag{f}$$

$$r_3 = k_3 C_0^2 \frac{(y_{A_0} - Z_1 - Z_3 - Z_5)(y_{B_0} + 2Z_1 - Z_3)}{(1 + Z_1 - Z_3 - Z_5)^2} \tag{g}$$

$$r_4 = k_4 C_0 \frac{y_{C_0} + Z_3 - Z_5}{1 + Z_1 - Z_3 - Z_5} \tag{h}$$

$$r_5 = k_5 C_0^2 \frac{(y_{A_0} - Z_1 - Z_3 - Z_5)(y_{C_0} + Z_3 - Z_5)}{(1 + Z_1 - Z_3 - Z_5)^2} \tag{i}$$

$$r_6 = k_6 C_0 \frac{y_{D_0} + Z_5}{1 + Z_1 - Z_3 - Z_5} \tag{j}$$

We select Reaction 1 and define the characteristic reaction time by

$$t_{cr} = \frac{1}{k_1} = 5 \text{ min} \tag{k}$$

Substituting (e) through (j) and (k) into (b), (c), and (d), the design equations reduce to

$$\frac{dZ_1}{d\tau} = \frac{y_{A_0} - Z_1 - Z_3 - Z_5}{1 + Z_1 - Z_3 - Z_5} - \left(\frac{k_2 C_0}{k_1}\right)\left(\frac{y_{B_0} + 2Z_1 - Z_3}{1 + Z_1 - Z_3 - Z_5}\right)^2 \tag{l}$$

$$\frac{dZ_3}{d\tau} = \left(\frac{k_3 C_0}{k_1}\right)\frac{(y_{A_0} - Z_1 - Z_3 - Z_5)(y_{B_0} + 2Z_1 - Z_3)}{(1 + Z_1 - Z_3 - Z_5)^2} - \left(\frac{k_4}{k_1}\right)\frac{y_{C_0} + Z_3 - Z_5}{1 + Z_1 - Z_3 - Z_5} \tag{m}$$

$$\frac{dZ_5}{d\tau} = \left(\frac{k_5 C_0}{k_1}\right)\frac{(y_{A_0} - Z_1 - Z_3 - Z_5)(y_{C_0} + Z_3 - Z_5)}{(1 + Z_1 - Z_3 - Z_5)^2} - \left(\frac{k_6}{k_1}\right)\frac{y_{D_0} + Z_5}{1 + Z_1 - Z_3 - Z_5} \tag{n}$$

For the given kinetic data,

$$\frac{k_2 C_0}{k_1} = 0.04 \quad \frac{k_3 C_0}{k_1} = 20 \quad \frac{k_4}{k_1} = 4 \quad \frac{k_5 C_0}{k_1} = 50 \quad \frac{k_6}{k_1} = 20 \tag{o}$$

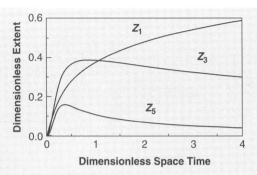

Figure E7.8.1 Reaction operating curves.

a. When a stream of reactant A is fed into the reactor, $y_{A_0} = 1$, $y_{B_0} = y_{C_0} = y_{D_0} = 0$. We substitute these values into (l), (m), and (n), and solve them numerically, subject to the initial conditions that at $\tau = 0$, $Z_1 = Z_3 = Z_5 = 0$. Figure E7.8.1 shows the solution—the reaction curves for the three independent chemical reactions.

b. Now that the reaction curves are known, we use Eq. 2.7.8 to obtain the species curves:

$$\frac{F_A}{(F_{\text{tot}})_0} = y_{A_0} - Z_1 - Z_3 - Z_5 \tag{p}$$

$$\frac{F_B}{(F_{\text{tot}})_0} = y_{B_0} + 2Z_1 - Z_3 \tag{q}$$

$$\frac{F_C}{(F_{\text{tot}})_0} = y_{C_0} + Z_3 - Z_5 \tag{r}$$

$$\frac{F_D}{(F_{\text{tot}})_0} = y_{D_0} + Z_5 \tag{s}$$

Figure E7.8.2 shows the species curves.

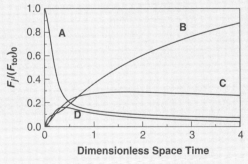

Figure E7.8.2 Species operating curves.

c. From the curve of product C (or tabulated calculated values), the maximum value of $F_{C_{out}}/(F_{tot})_0$ is 0.290, and it is reached at $\tau = 1.64$. The highest production rate of product C is

$$F_{C_{out}} = (0.29)(F_{tot})_0 = 0.29 \, \text{mol/s}$$

Using Eq. 7.1.3 and (k), the reactor volume required to maximize the production rate of product B is

$$V_R = \tau v_0 t_{cr} = 6150 \, \text{L}$$

d. We calculate the extents of the independent reactions at equilibrium by equating (l), (m), and (n) to zero. We obtain a set of nonlinear algebraic equations whose solutions are

$$Z_{1_{eq}} = 0.2451 \qquad Z_{3_{eq}} = 0.3693 \qquad Z_{5_{eq}} = 0.2100$$

Substituting these values into (p) through (s), we obtain

$$\frac{F_{A_{eq}}}{(F_{tot})_0} = 0.1755 \qquad \frac{F_{B_{eq}}}{(F_{tot})_0} = 0.1209 \qquad \frac{F_{C_{eq}}}{(F_{tot})_0} = 0.1593$$

$$\frac{F_{D_{eq}}}{(F_{tot})_0} = 0.210$$

Using Eq. 2.7.10, the total molar flow rate at equilibrium,

$$\frac{F_{tot}}{(F_{tot})_0} = 1 - Z_{1_{eq}} - Z_{3_{eq}} - Z_{5_{eq}} = 0.6657 \tag{t}$$

and the corresponding mole fractions are

$$y_{A_{eq}} = 0.2636 \qquad y_{B_{eq}} = 0.1816 \qquad y_{C_{eq}} = 0.2393 \qquad y_{D_{eq}} = 0.3155$$

e. The reactor design calculations when the feed stream consists of species D proceeds in the same way as in parts a and b. The only difference is that now $y_{A_0} = y_{B_0} = y_{C_0} = 0$ and $y_{D_0} = 1$. Substituting these values into (l), (m), and (n), we solve them numerically. Figure E7.8.3 shows the reaction curves for this case. Note that in this case, the extents of Reactions 3 and 5 are negative since they proceed in the reverse direction. Once we have the reaction curves, we use (p) through (s) to determine the species curves, shown in Figure E7.8.4. To determine the equilibrium composition, we equate (l), (m), and (n) to zero and obtain a set of nonlinear algebraic equations whose solutions are

$$Z_{1_{eq}} = 0.1128 \quad Z_{3_{eq}} = -0.0766 \quad Z_{5_{eq}} = -0.4749$$

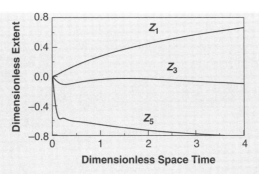

Figure E7.8.3 Reaction operating curves for feed of species D.

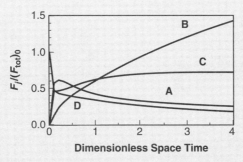

Figure E7.8.4 Species operating curves for feed of species D.

Substituting these values into (p) through (s), we obtain

$$\frac{F_{A_{eq}}}{(F_{tot})_0} = 0.4388 \quad \frac{F_{B_{eq}}}{(F_{tot})_0} = 0.3021 \quad \frac{F_{C_{eq}}}{(F_{tot})_0} = 0.3986 \quad \frac{F_{D_{eq}}}{(F_{tot})_0} = 0.5251$$

The corresponding mole fractions at equilibrium are

$$y_{A_{eq}} = 0.2637 \quad y_{B_{eq}} = 0.1815 \quad y_{C_{eq}} = 0.2393 \quad y_{D_{eq}} = 0.3155$$

Note that, as expected, the equilibrium composition is independent of the feed composition.

Example 7.9 Ammonia oxidation is carried out in an isothermal plug-flow reactor. The following gas-phase chemical reactions take place in the reactor:

Reaction 1:	$4NH_3 + 5O_2 \longrightarrow 4NO + 6H_2O$
Reaction 2:	$4NH_3 + 3O_2 \longrightarrow 2N_2 + 6H_2O$
Reaction 3:	$2NO + O_2 \longrightarrow 2NO_2$
Reaction 4:	$4NH_3 + 6NO \longrightarrow 5N_2 + 6H_2O$

The desired product is NO. A stream consisting of 50% NH_3 and 50% O_2 (mol%) at 609 K and 2 atm is fed into the reactor at a rate of 240 L/min. Based on the rate data below:

a. Derive the design equations and plot the reaction and species operating curves.

b. Determine the volume of the plug-flow reactor for optimal production of NO; what is the flow rate of each species at the reactor exit?

The rate expressions of the reactions are:

$$r_1 = k_1 C_{NH_3} C_{O_2}^2 \quad r_2 = k_2 C_{NH_3} C_{O_2} \quad r_3 = k_3 C_{O_2} C_{NO}^2 \quad r_4 = k_4 C_{NH_3}^{2/3} C_{NO}$$

Data: At 609 K:

$$k_1 = 20 \ (L/mol)^2 min^{-1} \qquad k_2 = 0.04 \ (L/mol)min^{-1}$$

$$k_3 = 40 \ (L/mol)^2 min^{-1} \qquad k_4 = 0.0274 \ (L/mol)^{2/3}min^{-1}$$

Solution The reactor design formulation of these chemical reactions was discussed in Example 4.3. Here, we complete the design for an isothermal plug-flow reactor and obtain the dimensionless reaction and species curves. Recall that there are three independent reactions and one dependent reaction, and, following the heuristic rule, we select a set of three reactions from the given reactions. We select Reactions 1, 2, and 3 as a set of independent reactions; hence, $m = 1, 2, 3$, and $k = 4$, and we express the design equations in terms of Z_1, Z_2, and Z_3. The stoichiometric coefficients of the independent reactions are

$$(s_{NH_3})_1 = -4 \quad (s_{O_2})_1 = -5 \quad (s_{NO})_1 = 4 \quad (s_{H_2O})_1 = 6 \quad (s_{N_2})_1 = 0$$

$$(s_{NO_2})_1 = 0 \quad \Delta_1 = 1$$

$$(s_{NH_3})_2 = -4 \quad (s_{O_2})_2 = -3 \quad (s_{NO})_2 = 0 \quad (s_{H_2O})_2 = 6 \quad (s_{N_2})_2 = 2$$

$$(s_{NO_2})_2 = 0 \quad \Delta_2 = 1$$

$$(s_{NH_3})_3 = 0 \quad (s_{O_2})_3 = -1 \quad (s_{NO})_3 = -2 \quad (s_{H_2O})_3 = 0 \quad (s_{N_2})_3 = 0$$

$$(s_{NO_2})_3 = 2 \quad \Delta_3 = -1$$

Recall from Example 4.3 that the multipliers α_{km}'s of the dependent reaction (Reaction 4) and the three independent reactions are $\alpha_{43} = 0$, $\alpha_{42} = 2.5$, and $\alpha_{41} = -1.5$. We select the feed stream as the reference stream; hence, $Z_{1_{in}} = Z_{2_{in}} = Z_{3_{in}} = 0$, and the reference concentration is

$$C_0 = \frac{P_0}{R \, T_0} = \frac{2}{(0.08206 \, L \, atm/mol \, K)(609 \, K)} = 0.04 \, mol/L \qquad \text{(a)}$$

The molar flow rate of the reference stream is

$$(F_{\text{tot}})_0 = v_0 C_0 = 9.6 \, \text{mol/min}$$

For the selected reference stream, $y_{\text{NH}_3}(0) = 0.5, y_{\text{O}_2}(0) = 0.5, y_{\text{NO}}(0)$ $y_{\text{N}_2}(0) = y_{\text{NO}_2}(0) = 0$ $y_{\text{N}_2}(0) = y_{\text{NO}_2}(0) = 0$.

a. To design a plug-flow reactor, we write Eq. 7.1.1 for each independent reaction:

$$\frac{dZ_1}{d\tau} = (r_1 - 1.5 r_4)\left(\frac{t_{\text{cr}}}{C_0}\right) \tag{b}$$

$$\frac{dZ_2}{d\tau} = (r_2 + 2.5 \, r_4)\left(\frac{t_{\text{cr}}}{C_0}\right) \tag{c}$$

$$\frac{dZ_3}{d\tau} = r_3 \left(\frac{t_{\text{cr}}}{C_0}\right) \tag{d}$$

We select Reaction 1 as the reference reaction, and define characteristic reaction time by

$$t_{\text{cr}} = \frac{1}{k_1 C_0^2} = 31.21 \, \text{min} \tag{e}$$

We use Eq. 7.1.15 to express the species concentrations, and the rates of the four chemical reactions are

$$r_1 = k_1 C_0{}^3 \left(\frac{0.5 - 4Z_1 - 4Z_2}{1 + Z_1 + Z_2 - Z_3}\right)\left(\frac{0.5 - 5Z_1 - 3Z_2 - Z_3}{1 + Z_1 + Z_2 - Z_3}\right)^2 \tag{f}$$

$$r_2 = k_2 C_0{}^2 \left(\frac{0.5 - 4Z_1 - 4Z_2}{1 + Z_1 + Z_2 - Z_3}\right)\left(\frac{0.5 - 5Z_1 - 3Z_2 - Z_3}{1 + Z_1 + Z_2 - Z_3}\right) \tag{g}$$

$$r_3 = k_3 C_0{}^3 \left(\frac{0.5 - 5Z_1 - 3Z_2 - Z_3}{1 + Z_1 + Z_2 - Z_3}\right)\left(\frac{4Z_1 - 2Z_3}{1 + Z_1 + Z_2 - Z_3}\right)^2 \tag{h}$$

$$r_4 = k_4 C_0{}^{5/3}\left(\frac{0.5 - 4Z_1 - 4Z_2}{1 + Z_1 + Z_2 - Z_3}\right)^{2/3}\left(\frac{4Z_1 - 2Z_3}{1 + Z_1 + Z_2 - Z_3}\right) \tag{i}$$

Substituting (f) through (i) and (e) into (b), (c), and (d), the design equations become

$$\frac{dZ_1}{d\tau} = \frac{(0.5 - 4Z_1 - 4Z_2)(0.5 - 5Z_1 - 3Z_2 - Z_3)^2}{(1 + Z_1 + Z_2 - Z_3)^3}$$

$$- 1.5\left(\frac{k_4}{k_1 C_0^{4/3}}\right)\frac{(0.5 - 4Z_1 - 4Z_2)^{2/3}(4Z_1 - 2Z_3)}{(1 + Z_1 + Z_2 - Z_3)^{5/3}} \tag{j}$$

$$\frac{dZ_2}{d\tau} = \left(\frac{k_2}{k_1 C_0}\right)\frac{(0.5 - 4Z_1 - 4Z_2)(0.5 - 5Z_1 - 3Z_2 - Z_3)}{(1 + Z_1 + Z_2 - Z_3)^2}$$

$$+ 2.5\left(\frac{k_4}{k_1 C_0^{4/3}}\right)\frac{(0.5 - 4Z_1 - 4Z_2)^{2/3}(4Z_1 - 2Z_3)}{(1 + Z_1 + Z_2 - Z_3)^{5/3}} \tag{k}$$

$$\frac{dZ_3}{d\tau} = \left(\frac{k_3}{k_1}\right)\frac{(0.5 - 5Z_1 - 3Z_2 - Z_3)(4Z_1 - 2Z_3)^2}{(1 + Z_1 + Z_2 - Z_3)^3} \tag{l}$$

For the given data, the parameters of the dimensionless design equations are

$$\left(\frac{k_2}{k_1 C_0}\right) = 0.05 \qquad \left(\frac{k_3}{k_1}\right) = 2 \qquad \left(\frac{k_4}{k_1 C_0^{4/3}}\right) = 0.1$$

We substitute these values into (j), (k), and (l), and solve them numerically for Z_1, Z_2, and Z_3, subject to the initial conditions that at $\tau = 0$, $Z_1 = Z_2 = Z_3 = 0$. Figure E7.9.1 shows the reaction curves. Now that we have the reaction curves, we use Eq. 2.7.8 to obtain the species curves:

$$\frac{F_{NH_3}}{(F_{tot})_0} = 0.5 - 4Z_1 - 4Z_2 \tag{m}$$

$$\frac{F_{O_2}}{(F_{tot})_0} = 0.5 - 5Z_1 - 3Z_2 - Z_3 \tag{n}$$

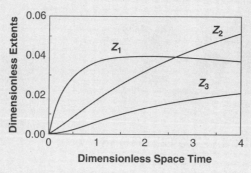

Figure E7.9.1   Reaction operating curves.

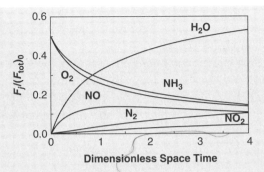

Figure E7.9.2   Species operating curves.

$$\frac{F_{NO}}{(F_{tot})_0} = 4Z_1 - 2Z_3 \tag{o}$$

$$\frac{F_{H_2O}}{(F_{tot})_0} = 6Z_1 + 6Z_2 \tag{p}$$

$$\frac{F_{N_2}}{(F_{tot})_0} = 2Z_2 \tag{q}$$

$$\frac{F_{NO_2}}{(F_{tot})_0} = 2Z_3 \tag{r}$$

Figure E7.9.2 shows the reaction curves.

b. From the NO curve (or the table of calculated data), the maximum production is $F_{NO}/(F_{tot})_0 = 0.1373$, and it is achieved at $\tau = 1.36$. Using Eq. 7.1.3 and (e), the required reactor volume is

$$V_R = \tau v_0 t_{cr} = 10,200\,\text{L}$$

At $\tau = 1.36$, $Z_1 = 0.0388$, $Z_2 = 0.0232$, and $Z_3 = 0.0089$. Substituting these values into (m) through (r), the species molar flow rates in the reactor outlet are

$$F_{NH_{3out}} = 2.42 \qquad F_{O_{2out}} = 2.19 \qquad F_{NO_{out}} = 1.32 \qquad F_{H_2O_{out}} = 3.57$$

$$F_{N_{2out}} = 0.445 \qquad F_{NO_{2out}} = 0.171\,\text{mol/min}$$

7.4   NONISOTHERMAL OPERATIONS

The design formulation of nonisothermal plug-flow reactors with multiple reactions follows the same procedure outlined in the previous section—we write design equation Eq. 7.1.1 for each independent reaction. Since the temperature varies along the reactor, we should solve the design equations simultaneously with the energy balance equation, Eq. 7.1.16.

The energy balance equation contains another variable—the temperature of the heating (or cooling) fluid, θ_F, which may also vary along the reactor. (Note that θ_F is constant only when the fluid either evaporates or condenses.) We distinguish between two configurations: co-current flow, shown in Figure 7.5a, and countercurrent flow, shown in Figure 7.5b.

To derive an equation for θ_F, we write the energy balance equation over the heating fluid in the shell element. For co-current flow,

$$d\dot{Q} = U(T_F - T)\,dS = -\dot{m}_F \bar{c}_{p_F}\,dT_F \qquad (7.4.1)$$

where $\dot{m}_F$ and $\bar{c}_{p_F}$ are, respectively, the mass flow rate and the heat capacity of the heating fluid. Using Eq. 5.2.51, $dS = (S/V)\,dV_R$, and Eq. 7.4.1 reduces to

$$\frac{dT_F}{dV_R} = -\left(\frac{S}{V}\right)\frac{U(T_F - T)}{\dot{m}_F \bar{c}_{p_F}} \qquad (7.4.2)$$

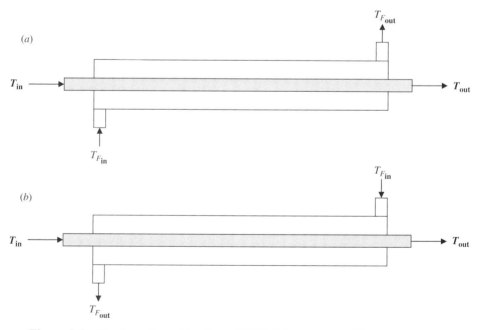

Figure 7.5 Configuration of heating of PFR; (a) co-current, (b) countercurrent.

To convert Eq. 7.4.2 to dimensionless form, we use the definition of the dimensionless space time, Eq. 7.1.3, and divide both sides by the reference thermal energy,

$$\frac{d\theta_F}{d\tau} = -\text{HTN}\left(\frac{(F_{\text{tot}})_0 \hat{c}_{p_0}}{\dot{m}_F \bar{c}_{p_F}}\right)(\theta_F - \theta) \tag{7.4.3}$$

where HTN is the heat-transfer number, defined by Eq. 7.1.17. For countercurrent flow, the energy balance over the heating fluid is

$$d\dot{Q} = U(T_F - T)\,dS = \dot{m}_m \hat{c}_{p_F}\,dT_F \tag{7.4.4}$$

Following a similar procedure to the one we used for co-current flow, Eq. 7.4.4 reduces to

$$\frac{d\theta_F}{d\tau} = \text{HTN}\left(\frac{(F_{\text{tot}})_0 \hat{c}_{p_0}}{\dot{m}_m \bar{c}_{p_F}}\right)(\theta_F - \theta) \tag{7.4.5}$$

Hence, when designing a plug-flow reactor with a heating/cooling fluid whose temperature varies, we have to solve either Eq. 7.4.3 or 7.4.5 simultaneously with the design equations and the energy balance equation of the reactor.

We adopt the following procedure for setting up the energy balance equation:

1. Select and define the temperature of the reference stream, T_0.
2. Determine the specific molar heat capacity of the reference stream, $\hat{c}_{p_0}$.
3. Determine the dimensionless activation energies, γ_i's, of *all* chemical reactions.
4. Determine the dimensionless heat of reactions, DHR$_m$'s, of the *independent* reactions.
5. Determine the correction factor of the heat capacity of the reacting fluid, CF(Z_m, θ).
6. Specify the dimensionless heat-transfer number, HTN (using Eq. 7.1.24).
7. Determine (or specify) the inlet temperature, θ_{in}.
8. Determine (or specify) the inlet dimensionless temperature of the heating/cooling fluid.
9. Solve the design equations simultaneously with the energy balance equation, and, if necessary, the energy balance equation of the heating/cooling fluid to obtain Z_m's, θ, and θ_F as functions of the dimensionless space time, τ.

The design formulation of nonisothermal plug-flow reactors consists of n_{I+2} nonlinear first-order differential equations. Note that usually the inlet temperature of the heating/cooling fluid, $T_{F_{\text{in}}}$, is known. Hence, the case of co-current configuration can be readily solved, but the countercurrent configuration is solved iteratively by trial and error. In the latter, we guess the value of $T_{F_{\text{out}}}$ as an initial value and solve the formulation equations. We then check if the value of $T_{F_{\text{in}}}$ at the end of the reactor is in agreement with the given value of $T_{F_{\text{in}}}$. If not, we guess another value of $T_{F_{\text{out}}}$ and repeat the calculation.

Example 7.10 The first-order, gas-phase reaction

$$A \longrightarrow B + C$$

takes place in a tubular (plug-flow) reactor. The reactor is made of a 20-cm tube, and it operates at 2 atm. A stream consisting of 80% reactant A and 20% inert I (by mole) is fed into the reactor at 450 K at the rate of 30 L/min.

a. Determine the needed reactor length for isothermal operation at 450 K.
b. Determine the overall heating (or cooling) load in (a).
c. Determine the local and estimate the average isothermal HTN.
d. Determine the needed reactor length for adiabatic operation.
e. Examine the effect of HTN on the temperature profile along the reactor and the production of product B.

Data: At 450 K,

$$k(T_0) = 15.0\,\text{hr}^{-1} \qquad \Delta H_R(T_0) = 8000\,\text{cal/mol extent} \qquad E_a/R = 2000\,\text{K}$$
$$\hat{c}_{p_A} = 20\,\text{cal/mol K} \qquad \hat{c}_{p_B} = 15\,\text{cal/mol K}$$
$$\hat{c}_{p_C} = 12\,\text{cal/mol K} \qquad \hat{c}_{p_I} = 10\,\text{cal/mol K}$$

The temperature of the reactor wall is maintained at 470 K.

Solution The stoichiometric coefficients are

$$s_A = -1 \qquad s_B = 1 \qquad s_C = 1 \qquad s_I = 0 \qquad \Delta = 1$$

Using Eq. 7.1.1, the design equation is

$$\frac{dZ}{d\tau} = r\left(\frac{t_{\text{cr}}}{C_0}\right) \tag{a}$$

We select the inlet stream as the reference stream, $Z_{\text{in}} = 0$, and $T_0 = 450$ K.

$$C_0 = \frac{P}{RT_0} = \frac{2}{(0.08206)(450)} = 0.0542\,\text{mol/L} \tag{b}$$
$$(F_{\text{tot}})_0 = v_0 C_0 = 1.626\,\text{mol/min}$$

For the given feed composition, $y_{A_0} = 0.80$, $y_{I_0} = 0.20$, and $y_{B_0} = y_{C_0} = 0$. We use Eq. 7.1.15 to express the species concentrations, and the reaction rate is

$$r = k(T_0)e^{\gamma(\theta-1)/\theta} C_0 \frac{0.8 - Z}{(1 + Z)\theta} \qquad (c)$$

Using Eq. 3.3.5, the dimensionless activation energy is

$$\gamma = \frac{E_a}{RT_0} = 4.44$$

We define a characteristic reaction time,

$$t_{cr} = \frac{1}{k(T_0)} = 4\,\text{min} \qquad (d)$$

Substituting (c) and (d) into (a), the design equation reduces to

$$\frac{dZ}{d\tau} = \frac{0.8 - Z}{(1 + Z)\theta} e^{\gamma(\theta-1)/\theta} \qquad (e)$$

Using Eq. 7.1.16, the energy balance equation becomes

$$\frac{d\theta}{d\tau} = \frac{1}{CF(Z_m, \theta)} \left[HTN(\theta_F - \theta) - DHR \frac{dZ}{d\tau} \right] \qquad (f)$$

We have to solve (e) and (f) simultaneously. Using Eq. 7.1.20, the specific molar heat capacity of the reference stream is

$$\hat{c}_{p0} = \sum_j^J y_{j0} \hat{c}_{pj}(T_0) = y_{A_0} \hat{c}_{pA}(T_0) + y_{I_0} \hat{c}_{pI}(T_0) = 18\,\text{cal/mol K} \qquad (g)$$

Using Eq. 7.1.18, the dimensionless heat of reaction is

$$DHR = \frac{\Delta H_R(T_0)}{T_0 \hat{c}_{p0}} = 0.9877 \qquad (h)$$

Using Eq. 5.2.61, the correction factor of the heat capacity of the reacting fluid is

$$CF(Z_m, \theta) = 1 + \frac{1}{\hat{c}_{p0}} [\hat{c}_{pA}(\theta)(-Z) + \hat{c}_{pB}(\theta)Z + \hat{c}_{pC}(\theta)Z] = \frac{18 + 7Z}{18} \qquad (i)$$

Once we solve the design equation, we use Eq. 2.7.8 to determine the species operating curves:

$$\frac{F_A}{(F_{tot})_0} = 0.8 - Z \qquad (j)$$

$$\frac{F_B}{(F_{tot})_0} = Z \qquad (k)$$

$$\frac{F_C}{(F_{tot})_0} = Z \qquad (m)$$

a. For isothermal operation, $\theta = 1$, and (e) reduces to

$$\frac{dZ}{d\tau} = \frac{0.8 - Z}{1 + Z} \qquad (n)$$

We solve (n) analytically. Using Eq. 2.6.5, for 60% conversion, the extent is

$$Z = -\frac{y_{A0}}{s_A}f_A = 0.48$$

From the solution, $Z = 0.48$ is reached at $\tau = 1.17$. Using (d) and Eq. 7.1.3, the required reactor volume is

$$V_R = v_0 \tau t_{cr} = 140.4\,\text{L} \qquad (o)$$

The length of the reactor is 4.46 m.

b. For isothermal operation, $d\theta/d\tau = 0$. Combining Eqs. 7.1.16 and 7.1.19, the local dimensionless heat-transfer rate along the reactor is

$$\frac{d}{d\tau}\left[\frac{\dot{Q}}{T_0(F_{tot})_0\hat{c}_{p0}}\right] = \left[\frac{\Delta H_R(T_0)}{T_0\hat{c}_{p0}}\right]\frac{dZ}{d\tau} \qquad (p)$$

Integrating (p), the total heating load is

$$\dot{Q} = (F_{tot})_0\,\Delta H_R(T_0)Z_{out} = 6.24\,\text{kcal/min}$$

c. Using 7.1.22, the local isothermal HTN is

$$\text{HTN}_{iso}(\tau) = \frac{1}{\theta_F - 1}\text{DHR}\frac{dZ}{d\tau} \qquad (q)$$

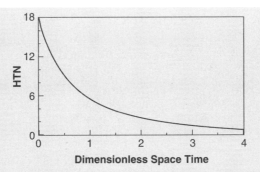

Figure E7.10.1 The local isothermal HTN.

where $\theta_F = 470/450 = 1.044$, and $dZ/d\tau$ is given by (n). Figure E7.10.1 shows the local isothermal HTN. Using Eq. 7.1.23, the average isothermal HTN for $0 \leq \tau \leq 1.17$ is 8.78.

d. For adiabatic operation, HTN = 0, and using (i), the energy balance equation, (f), reduces to

$$\frac{d\theta}{d\tau} = -\left(\frac{18}{18 + 7Z}\right) DHR \frac{dZ}{d\tau} \tag{r}$$

We solve (e) and (r) simultaneously, subject to the initial conditions that at $\tau = 0$, $Z = 0$ and $\theta = 1$. Figure E7.10.2 shows the reaction curve and the temperature curve, and Figure E7.10.3 shows the species curves determined by (j) through (m). From the calculated data, an extent of $Z = 0.48$ is reached at $\tau = 6.96$, and, at that space time, $\theta = 0.565$. Using (g) and Eq. 7.1.3, the required reactor volume is

$$V_R = v_0 \tau t_{cr} = 835\,L$$

The length of the reactor is 26.5 m. The exit temperature is

$$T = T_0 \theta = 295\,K$$

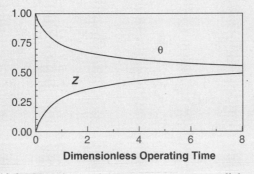

Figure E7.10.2 Reaction and temperature curves—adiabatic operation.

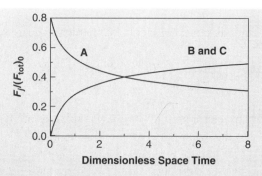

Figure E7.10.3 Species curves—adiabatic operation.

e. To examine the effect of HTN on the reactor operation, we solve (e) and (f)
for different values of HTN. Figure E7.10.4 shows the effect on the reactor
temperature, Figure E7.10.5 shows the effect of HTN on the reaction
extent (and from (k), on the production rate of product B). Note that when
HTN → ∞, the temperature of the reactor is the same as the wall temperature
(470 K, θ = 1.025).

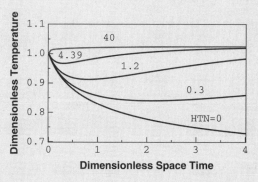

Figure E7.10.4 Effect of HTN on the temperature profile along the reactor.

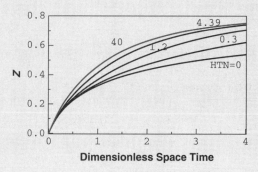

Figure E7.10.5 Effect of HTN on the reaction extent.

Example 7.11 The gas-phase elementary chemical reactions

$$\text{Reaction 1:} \quad A + B \longrightarrow V$$
$$\text{Reaction 2:} \quad V + B \longrightarrow W$$

take place in a tubular reactor (plug flow). A gas mixture of 40% A, 40% B, and 20% I (inert) is fed to the reactor at a rate of 400 L/min. The inlet temperature is 150°C, and the reactor operates at a constant pressure of 2 atm. It is desirable to maximize the production of product V. The temperature of the cooling fluid is 130°C.

a. Determine the reactor volume for maximum production of product V for isothermal operation.
b. Determine the heating/cooling load for isothermal operation.
c. Determine the local and average HTN for isothermal operation.
d. Determine the reactor volume for maximum production of product V for adiabatic operation.
e. Examine the effect of HTN on the temperature profile along the reactor, the two reactions, and the production of product V.

Data: At 150°C, $k_1 = 0.2\,\text{L/mol s}^{-1}$ $k_2 = 0.2\,\text{L/mol s}^{-1}$

$$\Delta H_{R_1} = -12{,}000\,\text{cal/mol extent} \qquad \Delta H_{R_2} = -9000\,\text{cal/mol extent}$$

$$E_{a_1} = 4950\,\text{cal/mol} \qquad E_{a_2} = 7682\,\text{cal/mol}$$

$$\hat{c}_{p_A} = 16\,\text{cal/mol K} \qquad \hat{c}_{p_B} = 8\,\text{cal/mol K} \quad \hat{c}_{p_V} = 20\,\text{cal/mol K}$$

$$\hat{c}_{p_W} = 26\,\text{cal/mol K} \qquad \hat{c}_{p_I} = 10\,\text{cal/mol K}$$

Solution The purpose of this example is to illustrate the effect of heat transfer on the production rate and product distribution. The two chemical reactions are independent, and there is no dependent reaction. The stoichiometric coefficients of the reactions are

$$s_{A_1} = -1 \qquad s_{B_1} = -1 \qquad s_{C_1} = 1 \qquad s_{D_1} = 0 \qquad \Delta = -1$$
$$s_{A_2} = 0 \qquad s_{B_2} = -1 \qquad s_{C_2} = -1 \qquad s_{D_2} = 1 \qquad \Delta = -1$$

We write Eq. 7.1.1 for each reaction:

$$\frac{dZ_1}{d\tau} = r_1 \left(\frac{t_{cr}}{C_0} \right) \tag{a}$$

$$\frac{dZ_2}{d\tau} = r_2 \left(\frac{t_{cr}}{C_0} \right) \tag{b}$$

We select the inlet stream as the reference stream; hence, $Z_{1_{in}} = Z_{2_{in}} = 0$, and $T_0 = 423$ K. Also,

$$C_0 = \frac{P_0}{RT_0} = 0.0576 \, \text{mol/L}$$

$$(F_{tot})_0 = v_0 C_0 = 23.04 \, \text{mol/min}$$

and $y_{A_0} = 0.40$, $y_{B_0} = 0.40$, $y_{I_0} = 0.20$, and $y_{V_0} = y_{W_0} = 0$. Using Eq. 5.2.58, the specific molar heat capacity of the reference stream is

$$\hat{c}_{p_0} = \sum_j^J y_{j_0} \hat{c}_{p_j}(T_0) = y_{A_0} \hat{c}_{p_A}(T_0) + y_{B_0} \hat{c}_{p_B}(T_0) + y_{I_0} \hat{c}_{p_I}(T_0) = 11.6 \, \text{cal/mol K}$$

The dimensionless heat of reactions of the two chemical reactions are

$$\text{DHR}_1 = \frac{\Delta H_{R_1}(T_0)}{\hat{c}_{p_0} T_0} = -2.45 \qquad \text{DHR}_2 = \frac{\Delta H_{R_2}(T_0)}{\hat{c}_{p_0} T_0} = -1.834$$

Using Eq. 5.2.61, the correction factor of the heat capacity of the reacting fluid is

$$\text{CF}(Z_m, \theta) = 1 + \frac{1}{\hat{c}_{p_0}} \sum_j^J \hat{c}_{p_j}(\theta) \sum_m^{n_I} (s_j)_m Z_m$$

$$= 1 + \frac{1}{\hat{c}_{p_0}} [\hat{c}_{p_A}(-Z_1) + \hat{c}_{p_B}(-Z_1 - Z_2) + \hat{c}_{p_C}(Z_1 - Z_2) + \hat{c}_{p_D} Z_2]$$

$$\text{CF}(Z_m, \theta) = \frac{11.6 - 4Z_1 - 2Z_2}{11.6} \tag{c}$$

Using 7.1.15 to express the species concentrations, the rates of the two reactions are

$$r_1 = k_1 C_A C_B = k_1(T_0) e^{\gamma_1 (\theta - 1)/\theta} C_0^2 \frac{(y_{A_0} - Z_1)(y_{B_0} - Z_1 - Z_2)}{[(1 - Z_1 - Z_2)\theta]^2} \tag{d}$$

$$r_2 = k_2 C_C C_B = k_2(T_0) e^{\gamma_2 (\theta - 1)/\theta} C_0^2 \frac{(Z_1 - Z_2)(y_{B_0} - Z_1 - Z_2)}{[(1 - Z_1 - Z_2)\theta]^2} \tag{e}$$

where the dimensionless activation energies of the two reactions are

$$\gamma_1 = \frac{E_{a_1}}{RT_0} = 5.89 \qquad \gamma_2 = \frac{E_{a_2}}{RT_0} = 9.14$$

We select a characteristic reaction time on the basis of Reaction 1:

$$t_{cr} = \frac{1}{k_1(T_0)C_0} = 1.45 \, \text{min} \tag{f}$$

Substituting (d), (e), and (f) into (a) and (b), the two design equations become

$$\frac{dZ_1}{d\tau} = \frac{(y_{A_0} - Z_1)(y_{B_0} - Z_1 - Z_2)}{[(1 - Z_1 - Z_2)\theta]^2} e^{\gamma_1(\theta-1)/\theta} \tag{g}$$

$$\frac{dZ_2}{d\tau} = \frac{k_2(T_0)}{k_1(T_0)} \frac{(Z_1 - Z_2)(y_{B_0} - Z_1 - Z_2)}{[(1 - Z_1 - Z_2)\theta]^2} e^{\gamma_2(\theta-1)/\theta} \tag{h}$$

Using Eq. 7.1.16, the energy balance equation becomes

$$\frac{d\theta}{d\tau} = \frac{1}{\text{CF}(Z_m, \theta)} \left[\text{HTN}(\theta_F - \theta) - \text{DHR}_1 \frac{dZ_1}{d\tau} - \text{DHR}_2 \frac{dZ_2}{d\tau} \right] \tag{i}$$

We have to solve (g), (h), and (i) simultaneously, subject to the initial condition that at $\tau = 0$, $Z_1 = Z_2 = 0$ and $\theta = 1$. Once we solve the design equations, we can determine the species curves, using Eq. 2.7.8:

$$\frac{F_A}{(F_{\text{tot}})_0} = 0.4 - Z_1 \tag{j}$$

$$\frac{F_B}{(F_{\text{tot}})_0} = 0.4 - Z_1 - Z_2 \tag{k}$$

$$\frac{F_V}{(F_{\text{tot}})_0} = Z_1 - Z_2 \tag{l}$$

$$\frac{F_W}{(F_{\text{tot}})_0} = Z_2 \tag{m}$$

a. For isothermal operation, $\theta = 1$, and the two design equations, (g) and (h), reduce to

$$\frac{dZ_1}{d\tau} = \frac{(0.4 - Z_1)(0.4 - Z_1 - Z_2)}{(1 - Z_1 - Z_2)^2} \tag{n}$$

$$\frac{dZ_2}{d\tau} = (0.5) \frac{(Z_1 - Z_2)(0.4 - Z_1 - Z_2)}{(1 - Z_1 - Z_2)^2} \tag{o}$$

We solve (n) and (o) numerically. Figure E7.11.1 shows the reaction curves. Once we have the solution, we use (j) through (m) to determine the species

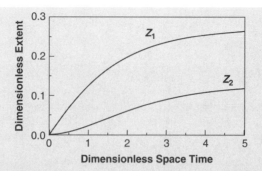

Figure E7.11.1   Reaction operating curves—isothermal operation.

curves, shown in Figure E7.11.2. From the curve of product V, the highest value of $F_V/(F_{tot})_0$ is 0.147, and it is reached at $\tau = 3.85$. At that space time, $Z_1 = 0.253$ and $Z_2 = 0.106$. Hence, using Eq. 7.1.3, the reactor volume is

$$V_R = v_0 t_{cr} \tau = 2233 \text{ L}$$

Using (l), the production rate of product V is

$$F_V = (23.04)(0.253 - 0.105) = 3.41 \text{ mol/min}$$

b. Combining Eqs. 7.1.16 and 7.1.19, the local dimensionless heat-transfer rate along the reactor is

$$\frac{d}{d\tau}\left[\frac{\dot{Q}}{(F_{tot})_0 \hat{c}_{p_0} T_0}\right] = \text{DHR}_1 \frac{dZ_1}{d\tau} + \text{DHR}_2 \frac{dZ_2}{d\tau} \qquad (p)$$

Integrating (p), the total heating/cooling load is

$$\dot{Q} = [(F_{tot})_0 \hat{c}_{p_0} T_0][\text{DHR}_1 Z_1 + \text{DHR}_2 Z_2] = -91.75 \text{ kcal/min}$$

The negative sign indicates that heat is removed from the reactor.

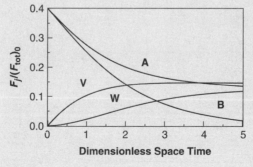

Figure E7.11.2   Species operating curves—isothermal operation.

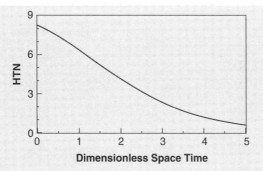

Figure E7.11.3   HTN curve.

c. Using Eq. 7.1.22, the local isothermal HTN is

$$\text{HTN}_{\text{iso}}(\tau) = \frac{1}{\theta_F - 1}\left(\text{DHR}_1 \frac{dZ_1}{d\tau} + \text{DHR}_2 \frac{dZ_2}{d\tau}\right) \qquad (q)$$

where $\theta_F = 0.953$. Figure E7.11.3 shows the local isothermal HTN. Using Eq. 7.1.23, for $\tau_{\text{op}} = 3.85$, the average isothermal HTN is 4.5.

d. For adiabatic operations, HTN = 0, and (h) reduces to

$$\frac{d\theta}{d\tau} = \left(\frac{11.6}{11.6 - 4Z_1 - 2Z_2}\right)\left(\text{DHR}_1 \frac{dZ_1}{d\tau} + \text{DHR}_2 \frac{dZ_2}{d\tau}\right) \qquad (r)$$

We solve (f), (g), and (r) numerically, subject to the initial condition that at $\tau = 0$, $Z_1 = Z_2 = 0$, and $\theta = 1$. Figure E7.11.4 shows the reaction curves, and Figure E7.11.5 shows the temperature curve. Once we have the solution, we use (j) through (m) to determine the species curves, shown in Figure E7.11.6. From the curve of product V, the highest value of $F_V/(F_{\text{tot}})_0 = 0.0891$, and it

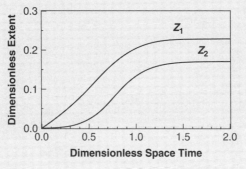

Figure E7.11.4   Reaction operating curves—adiabatic operation.

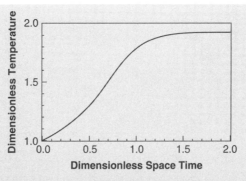

Figure E7.11.5   Temperature curve—adiabatic operation.

is reached at $\tau = 0.68$. At that space time, $Z_1 = 0.1468$, $Z_2 = 0.0577$, and $\theta = 1.4797$. Using Eq. 7.1.3, the reactor volume is

$$V_R = v_0 t_{cr} \tau = 394.4 \, \text{L}$$

Using (e), the production rate of product V is

$$F_V = (23.04)(0.1468 - 0.0577) = 2.05 \, \text{mol/min}$$

The exit temperature is $T = T_0 \theta = (423)(1.4797) = 625.9 \, \text{K}$

e. To examine the effect of HTN on the reactor operation, we solve (f), (g), and (h) for different values of HTN. Figure E7.11.7 shows the effect on the reactor temperature, Figure E7.11.8 shows the effect on the progress of Reaction 1, Figure E7.11.9 shows the effect on the progress of Reaction 2, and Figure E7.11.10 shows the effect on the production of product V. Note that when HTN $\rightarrow \infty$, the temperature of the reactor is the same as the jacket temperature (130°C, $\theta = 0.9527$). Also, since the activation energy of Reaction 2 is higher than that of Reaction 1, by operating at lower temperature, the reactor produces more product V, but a larger reactor is required.

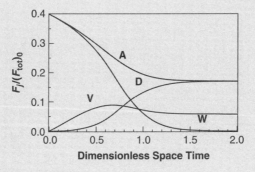

Figure E7.11.6   Species operating curves—adiabatic operation.

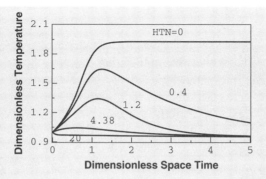

Figure E7.11.7   Effect of HTN on the temperature profile.

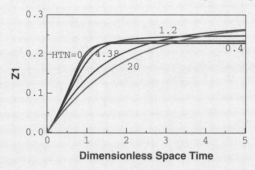

Figure E7.11.8   Effect of HTN on Reaction 1.

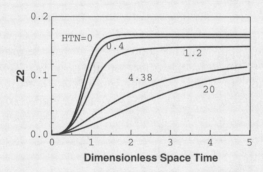

Figure E7.11.9   Effect of HTN on Reaction 2.

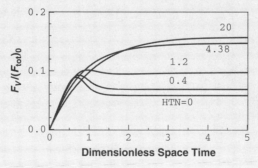

Figure E7.11.10   Effect of HTN on the production of product V.

7.5 EFFECTS OF PRESSURE DROP

In the preceding sections, we discussed the operation of plug-flow reactors with gas-phase reactions under the assumption that the pressure does not vary along the reactor. However, in some applications, the pressure significantly changes and, therefore, affects the reaction rates. In this section, we incorporate the variation in pressure into the design equations. For convenience, we divide the discussion into two parts: tubular tube with uniform diameter and packed-bed reactors.

We consider first a cylindrical reactor of uniform diameter D. To derive an expression for the pressure drop, we write the steady-state momentum balance equation for a reactor element of length dL and cross-sectional area A:

$$-A \, dP - F_f = \dot{m} \, du \tag{7.5.1}$$

where $-(A \, dP)$ is the pressure force, and F_f is the friction force. Expressing the friction force in terms of a friction factor, f, and the kinetic energy, for an empty cylindrical tube of diameter D,

$$F_f = f(\pi D \, dL)\frac{1}{2}\rho u^2$$

and Eq. 7.5.1 becomes

$$-\frac{dP}{dL} = 4f\,\frac{\dot{m}}{DA}\frac{1}{2}u + \frac{\dot{m}}{A}\frac{du}{dL} \tag{7.5.2}$$

The first term on the right indicates the pressure drop due to friction, and the second indicates the pressure drop due to change in velocity (kinetic energy). In many applications, the second term in Eq. 7.5.2 is small in comparison to the first, and noting that $u = v/A$, the momentum balance equation reduces to

$$-\frac{dP}{dL} \approx 2f\,\frac{\dot{m}}{DA^2}v \tag{7.5.3}$$

where v is the local volumetric flow rate. Using Eq. 7.1.12 and noting from Eq. 7.1.3 that $dL = (v_0 t_{\mathrm{cr}}/A) \, d\tau$, Eq. 7.5.3 becomes

$$-\frac{d(P/P_0)}{d\tau} \approx \left(2f\,\frac{\dot{m}v_0^2 t_{\mathrm{cr}}}{DA^3 P_0}\right)\left(1 + \sum_m^{n_I} \Delta_m Z_m\right)\theta\left(\frac{P_0}{P}\right) \tag{7.5.4}$$

Equation 7.5.4 provides an *approximate* relation for the changes in pressure along a plug-flow reactor, expressed in terms of dimensionless extents and temperature. It is applicable when the velocity does not exceed 80–90% of the sound velocity. For these situations, we solve Eq. 7.5.4 simultaneously with the design equation

(Eq. 7.1.1) and energy balance equation (Eq. 7.1.16), subject to specified initial conditions. Note that the first parenthesis on the right is a dimensionless friction number for the reference stream.

When the gas velocity approaches the sound velocity, the kinetic energy and viscous work terms in the energy balance equation are not negligible (as assumed in Chapter 5). For these cases, we write the general energy balance equation for a differential plug-flow reactor with length dL (see Eq. 5.2.44),

$$d\dot{Q} - d\dot{W}_{vis} = \dot{m} \, d\left(h + \frac{1}{2}u^2 + gz \right) \tag{7.5.5}$$

Assuming negligible change in the potential energy,

$$\frac{d\dot{Q}}{\dot{m}} - \frac{d\dot{W}_{vis}}{\dot{m}} = dh + u \, du$$

For flow in cylindrical conduits, the viscous work per unit mass of the fluid is expressed in terms of a friction factor and a specific kinetic energy by

$$\frac{d\dot{W}_{vis}}{\dot{m}} = 4f\frac{dL}{D}\frac{1}{2}u^2 \tag{7.5.6}$$

and the energy balance equation becomes

$$d\dot{Q} - 4f\dot{m}\frac{dL}{D}\frac{1}{2}u^2 = \dot{m} \, dh + \dot{m}u \, du \tag{7.5.7}$$

Multiplying Eq. 7.5.7 by ρ and then subtracting Eq. 7.5.2, we obtain

$$-\frac{dP}{dL} = v\left(\frac{d\dot{Q}}{dL} - \dot{m}\frac{dh}{dL} \right) \tag{7.5.8}$$

For cylindrical reactors,

$$d\dot{Q} = U(\pi D \, dL)(T_F - T) \tag{7.5.9}$$

and, differentiating Eq. 5.2.47, the differential enthalpy difference of the reacting fluid is

$$\dot{m} \, dh = d\dot{H} = \sum_{m}^{n_I} \Delta H_{R_m}(T_0) \, d\dot{X}_m + \left(\sum_{j}^{J} F_j\hat{c}_{p_j} \right) dT \tag{7.5.10}$$

Substituting these relationships into Eq. 7.5.7 and using Eqs. 5.2.54 and 7.1.11, we obtain

$$-\frac{dP}{dL} = \frac{(P/P_0)[U\pi D(T_F - T) - \sum_m^{n_I} \Delta H_{R_m}(T_0)(d\dot{X}_m/dL) - (F_{\text{tot}})_0 \hat{c}_{p_0}(\text{CF})(dT/dL)]}{v_0\left(1 + \sum_m^{n_I} \Delta_m Z_m\right)\theta}$$

$$(7.5.11)$$

From Eq. 7.1.3, $dL = (v_0 t_{cr}/A) \, d\tau$, using Eqs. 3.3.4, 7.1.2, 7.1.4, 5.2.22, and 5.2.23, and, noting that for ideal gas, $RC_0 = P_0/T_0$, Eq. 7.5.11 reduces to

$$-\frac{d(P/P_0)}{d\tau} = \frac{(\hat{c}_{p_0}/R)(P/P_0)[\text{HTN}(\theta_F - \theta) - \sum_m^{n_I} \text{DHR}_m(dZ_m/d\tau) - (\text{CF})(d\theta/d\tau)]}{\left(1 + \sum_m^{n_I} \Delta_m Z_m\right)\theta}$$

$$(7.5.12)$$

Equation 7.5.12 provides the changes in pressure along a plug-flow reactor, and it should be solved simultaneously with the design equations, and the energy balance equation. Note that the expression inside the parenthesis in the numerator on right-hand side of Eq. 7.5.12 is the energy balance equation derived under the assumption that the kinetic energy term and the friction work term are negligible. Indeed, under this assumption, $d(P/P_0)/d\tau = 0$. To design a plug-flow reactor with pressure drop, we have to derive the energy balance equation that includes terms for kinetic energy and the friction work.

To derive an expression for the temperature variation in a tubular plug-flow reactor, we rearrange the general energy balance equation Eq. 7.5.7:

$$\frac{d\dot{Q}}{dL} - 2f\dot{m}\frac{1}{D}u^2 = \dot{m}\frac{dh}{dL} + \dot{m}u\frac{du}{dL} \qquad (7.5.13)$$

For a reactor with uniform cross-sectional area A, $u = v/A$, and, using Eq. 7.1.12,

$$u = \frac{v_0}{A}\left(1 + \sum_m^{n_I} \Delta_m Z_m\right)\theta\left(\frac{P_0}{P}\right) \qquad (7.5.14)$$

Differentiating Eq. 7.5.14 with dL,

$$\frac{du}{dL} = \frac{v_0}{A}\left(1 + \sum_m^{n_I} \Delta_m Z_m\right)\left(\frac{P_0}{P}\right)\frac{d\theta}{dL} + \frac{v_0}{A}\theta\left(\frac{P_0}{P}\right)\sum_m^{n_I} \Delta_m \frac{dZ_m}{dL}$$

$$-\frac{v_0}{A}\left(1 + \sum_m^{n_I} \Delta_m Z_m\right)\left(\frac{P_0}{P^2}\right)\theta\frac{dP}{dL}$$

$$(7.5.15)$$

We substitute Eqs. 7.5.15 and 7.5.14 into Eq. 7.5.13 and note from Eq. 7.1.3 that $dL = (v_0 t_{cr}/A) \, d\tau$. We convert Eqs. 7.5.13 and 7.5.15 to dimensionless forms by using Eqs. 3.3.4, 7.1.2, 7.1.4, 5.2.22, and 5.2.23, and noting that, for an ideal gas, $RC_0 = P_0/T_0$. The first terms in Eq. 7.5.13 reduce to

$$\frac{d\dot{Q}}{dL} = \left(\frac{v_0 t_{cr}}{A}\right)\frac{d\dot{Q}}{d\tau} = \left(\frac{A}{v_0 t_{cr}}\right)\text{HTN}F_0 \hat{c}_{p_0} T_0 (\theta_F - \theta)$$

The second term in Eq. 7.5.13 reduces to

$$2f\dot{m}\frac{1}{D}u^2 = 2f\dot{m}\frac{1}{D}\frac{v_0^2}{A^2}\left(1 + \sum_m^{n_I}\Delta_m Z_m\right)^2 \theta^2 \left(\frac{P_0}{P}\right)^2$$

The third term in Eq. 7.5.13 reduces to

$$\dot{m}\frac{dh}{dL} = \left(\frac{A}{v_0 t_{cr}}\right)\frac{d\dot{H}}{d\tau} = \left(\frac{A}{v_0 t_{cr}}\right)F_0 \hat{c}_{p_0} T_0 \left(\sum_m^{n_I}\text{DHR}_m \frac{dZ_m}{d\tau} + (\text{CF})\frac{d\theta}{d\tau}\right)$$

The fourth term in Eq. 7.5.13 reduces to

$$\dot{m}u\frac{du}{dL} = \left(\frac{A}{v_0 t_{cr}}\right)\dot{m}u\frac{du}{d\tau} = \left(\frac{A}{v_0 t_{cr}}\right)\dot{m}\frac{v_0}{A}\left(1 + \sum_m^{n_I}\Delta_m Z_m\right)\theta\left(\frac{P_0}{P}\right)\frac{du}{d\tau}$$

$$= \left(\frac{\dot{m}v_0}{At_{cr}}\right)\left(1 + \sum_m^{n_I}\Delta_m Z_m\right)^2 \theta\left(\frac{P_0}{P}\right)^2 \frac{d\theta}{d\tau}$$

$$+ \left(\frac{\dot{m}v_0}{At_{cr}}\right)\left(1 + \sum_m^{n_I}\Delta_m Z_m\right)^2 \theta^2 \left(\frac{P_0}{P}\right)^2 \sum_m^{n_I}\Delta_m \frac{dZ_m}{d\tau}$$

$$- \left(\frac{\dot{m}v_0}{At_{cr}}\right)\left(1 + \sum_m^{n_I}\Delta_m Z_m\right)^2 \theta^2 \left(\frac{P_0}{P}\right)^3 \frac{d(P/P_0)}{d\tau}$$

Substituting those terms into Eq. 7.5.13 and using Eq. 7.5.12,

$$\frac{d\theta}{d\tau} = \frac{(\text{TERM})}{(\text{Denominator})} \qquad (7.5.16)$$

where

$$(\text{TERM}) = \text{HTN} \left[1 - \frac{\dot{m} v_0{}^2}{A^2 F_0 \hat{c}_{p_0} T_0} \left(1 + \sum_m^{n_I} \Delta_m Z_m \right)^2 \theta \left(\frac{P_0}{P} \right)^2 \right] (\theta_F - \theta)$$

$$- 2f \frac{\dot{m} v_0{}^3 t_{\text{cr}}}{D A^3 F_0 \hat{c}_{p_0} T_0} \left(1 + \sum_m^{n_I} \Delta_m Z_m \right)^2 \theta^2 \left(\frac{P_0}{P} \right)^2$$

$$- \sum_m^{n_I} \text{DHR}_m \left[1 - \frac{\dot{m} v_0{}^2}{A^2 F_0 R T_0} \left(1 + \sum_m^{n_I} \Delta_m Z_m \right) \theta \left(\frac{P_0}{P} \right)^2 \right] \frac{dZ_m}{d\tau}$$

$$- \left(\frac{\dot{m} v_0{}^2}{A^2 F_0 \hat{c}_{p_0} T_0} \right) \left(1 + \sum_m^{n_I} \Delta_m Z_m \right) \theta^2 \left(\frac{P_0}{P} \right)^2 \sum_m^{n_I} \Delta_m \frac{dZ_m}{d\tau}$$

and

$$(\text{Denominator}) = \text{CF}(Z_m, \theta) \left[1 - \frac{\dot{m} v_0{}^2}{A^2 F_0 R T_0} \left(1 + \sum_m^{n_I} \Delta_m Z_m \right) \theta \left(\frac{P_0}{P} \right)^2 \right]$$

$$+ \frac{\dot{m} v_0{}^2}{A^2 F_0 \hat{c}_{p_0} T_0} \left(1 + \sum_m^{n_I} \Delta_m Z_m \right)^2 \theta \left(\frac{P_0}{P} \right)^2$$

Equation 7.5.16 is the dimensionless, differential energy balance equation for cylindrical tubular flow reactors, relating the temperature, θ, to the extents of the independent reactions, Z_m's, and P/P_0 as functions of space time τ. To design a plug-flow reactor, we have to solve design equations (Eq. 7.1.1), the energy balance equation (Eq. 7.5.16), and the momentum balance (Eq. 7.5.12), simultaneously subject to specified initial conditions.

Note that Eq. 7.5.16 was derived from the energy balance equation (first law of thermodynamics), and it does not impose a limit on the value of θ. However, the second law of thermodynamics imposes a restriction on the conversion of thermal energy to kinetic energy. For compressible fluids in tubes with uniform cross-sectional area, the velocity cannot exceed the sound velocity; hence,

$$u \leq \left(\frac{1}{\text{MW}} \frac{\hat{c}_p R T}{\hat{c}_p - R} \right)^{0.5} \tag{7.5.17}$$

Using Eq. 7.5.14, the second law imposes the following restriction

$$\left(\frac{P}{P_0} \right) \geq \left[\left(\frac{v_0}{A} \right)^2 \left(\frac{\text{MW}_0}{R T_0} \right) \left(\frac{\text{CF}(Z_m, \theta) \hat{c}_{p_0}}{\text{CF}(Z_m, \theta) \hat{c}_{p_0} - R} \right) \left(1 + \sum_m^{n_I} \Delta_m Z_m \right) \theta \right]^{0.5} \tag{7.5.18}$$

For flow through a packed-bed reactor, the pressure drop is expressed in terms of the Ergun equation:

$$-\frac{dP}{dL} = 150\frac{(1-\varepsilon)^2}{\varepsilon^3}\frac{\mu u}{(\phi d_p)^2} + 1.75\frac{(1-\varepsilon)}{\varepsilon^3}\frac{\rho u^2}{\phi d_p} \qquad (7.5.19)$$

where d_p is the particle diameter, ϕ is the shape factor, and ε is the bed void fraction. Noting that $\rho u = G$, the mass velocity, is constant, and, from Eq. 7.1.3, $dL = (v_0 t_{cr}/A)\, d\tau$, Eq. 7.5.19 reduces to

$$-\frac{d(P/P_0)}{d\tau} = \frac{v_0^2 t_{cr}}{AP_0}\left(150\frac{(1-\varepsilon)^2}{\varepsilon^3}\frac{\mu}{(\phi d_p)^2} + 1.75\frac{(1-\varepsilon)G}{\varepsilon^3 \phi d_p}\right) \times \left(1 + \sum_m^{n_I} \Delta_m Z_m\right)\theta\left(\frac{P_0}{P}\right)$$

$$(7.5.20)$$

Equation 7.5.20 provides a relation for the change in the pressure along a packed-bed reactor. In most applications, the gas velocity in packed-bed reactors is much lower than the sound velocity. Hence, to formulate the reactor design, we should solve design equations (Eq. 7.1.1), energy balance equation (Eq. 7.1.16), and momentum balance equation (Eq. 7.5.20) simultaneously, subject to specified initial conditions. The solutions provide Z_m, θ, and P/P_0 as functions of dimensionless space time, τ.

Example 7.12 A heavy hydrocarbon feedstock is being cracked in a tubular reactor placed in a furnace that maintains the wall of the reactor at 980 K. The cracking is represented by the following first-order chemical reactions:

$$\begin{array}{lll}
\text{Reaction 1:} & A \longrightarrow B + C \\
\text{Reaction 2:} & B \longrightarrow 2D \\
\text{Reaction 3:} & C \longrightarrow F + G \\
\text{Reaction 4:} & G \longrightarrow 2F
\end{array}$$

The feed stream consists of 90% species A and 10% species I (mole percent), its temperature is 900 K, its pressure is 5 atm, and it is fed to the reactor at a rate of 276 L/s. The inside diameter of the reactor is 10 cm, and its surface can be assumed "smooth." Accounting for pressure drop along the reactor:

a. Derive the reaction and species curves for isothermal operation and determine the local and average heat-transfer number (HTN).

b. Determine the reaction curves, the temperature curve, the pressure curve, and the species curves when the average isothermal HTN is maintained in the reactor.

c. What should be the reactor length to maximize the production of product C?

d. What reactor length should be specified if the pressure drop effect in neglected?

Data: The species molecular weights (in g/mol) are

$$MW_A = 104 \qquad MW_B = 56 \qquad MW_C = 48 \qquad MW_D = 28$$

$$MW_F = 16 \qquad MW_G = 32 \qquad MW_I = 28$$

At 900 K the reaction rate constants (in s^{-1}) are

$$k_1 = 2.0 \qquad k_2 = 1.0 \qquad k_3 = 0.3 \qquad k_4 = 0.2$$

The activation energies of the chemical reactions rate (in kcal/mol) are

$$E_{a_1} = 24.0 \qquad E_{a_2} = 32.0 \qquad E_{a_3} = 45.0 \qquad E_{a_4} = 55.0$$

At 900 K, the heats of reaction of the chemical reactions (in kcal/mol) are

$$\Delta H_{R_1} = 40.0 \qquad \Delta H_{R_2} = 50.0 \qquad \Delta H_{R_3} = 60.0 \qquad \Delta H_{R_4} = 90.0$$

The specific molar heat capacities of the species (in cal/mol K) are

$$\hat{c}_{pA} = 70 \qquad \hat{c}_{pB} = 40 \qquad \hat{c}_{pC} = 50 \qquad \hat{c}_{pD} = 30$$

$$\hat{c}_{pF} = 25 \qquad \hat{c}_{pG} = 40 \qquad \hat{c}_{pI} = 12$$

The estimated average Reynolds number in the reactor is 10^6, and the friction factor is $f = 0.005$.

Solution   The four given reactions are independent, and there are no dependent reactions. The stoichiometric coefficients of the chemical reactions are

$$s_{A_1} = -1 \qquad s_{B_1} = 1 \qquad s_{C_1} = 1 \qquad s_{D_1} = 0 \qquad s_{F_1} = 0 \qquad s_{G_1} = 0$$

$$s_{I_1} = 0 \qquad \Delta_1 = 1$$

$$s_{A_2} = 0 \qquad s_{B_2} = -1 \qquad s_{C_2} = 0 \qquad s_{D_2} = 2 \qquad s_{F_2} = 0 \qquad s_{G_2} = 0$$

$$s_{I_2} = 0 \qquad \Delta_2 = 1$$

$$s_{A_3} = 0 \qquad s_{B_3} = 0 \qquad s_{C_3} = -1 \qquad s_{D_3} = 0 \qquad s_{F_3} = 1 \qquad s_{G_3} = 1$$

$$s_{I_3} = 0 \qquad \Delta_3 = 1$$

$$s_{A_4} = 0 \qquad s_{B_4} = 0 \qquad s_{C_4} = 0 \qquad s_{D_4} = 0 \qquad s_{F_4} = 2 \qquad s_{G_4} = -1$$

$$s_{I_4} = 0 \qquad \Delta_4 = 1$$

We select the feed stream as the reference stream, and the reference temperature is $T_0 = 900$ K. Since the operating temperature is high, we assume ideal-gas behavior; hence, the reference concentration is

$$C_0 = \frac{P_0}{RT_0} = 0.0677\,\text{mol/L}$$

The composition of the reference stream is $y_{A_0} = 0.9$ and $y_{I_0} = 0.1$. The average molecular of the reference stream is

$$\text{MW}_0 = \sum_j^{n_I} y_{j_0}\text{MW}_j = 96.4\,\text{g/mol}$$

The density of the reference stream is

$$\rho_0 = \frac{(F_{\text{tot}})_0 \text{MW}_0}{v_0} = 6.52\,\text{g/L}$$

Using 5.2.30, the specific molar heat capacity of the reference stream is

$$\hat{c}_{p0} = \sum_j^{n_I} y_{j_0}\hat{c}_{p_j} = 64.2\,\text{cal/mol K}$$

Using 5.2.23, the dimensionless heats of reaction are

$$\text{DHR}_1 = 0.692 \quad \text{DHR}_2 = 0.865 \quad \text{DHR}_3 = 1.038 \quad \text{DHR}_4 = 1.558$$

Using 3.3.5, the dimensionless activation energies of the chemical reactions are

$$\gamma_1 = \frac{E_{a_1}}{RT_0} = 13.42 \quad \gamma_2 = 17.89 \quad \gamma_3 = 25.16 \quad \gamma_4 = 30.76$$

Using Eq. 7.1.1, the reactor design equations are

$$\frac{dZ_1}{d\tau} = r_1\left(\frac{t_{\text{cr}}}{C_0}\right) \tag{a}$$

$$\frac{dZ_2}{d\tau} = r_2\left(\frac{t_{\text{cr}}}{C_0}\right) \tag{b}$$

$$\frac{dZ_3}{d\tau} = r_3\left(\frac{t_{\text{cr}}}{C_0}\right) \tag{c}$$

$$\frac{dZ_4}{d\tau} = r_4\left(\frac{t_{\text{cr}}}{C_0}\right) \tag{d}$$

Using Eq. 7.1.15, for gas-phase reactions, the local concentration of species j is

$$C_j = C_0 \frac{y_{j_0} + \sum_m^{n_I} (s_j)_m Z_m}{\left(1 + \sum_m^{n_I} \Delta_m Z_m\right)\theta} \left(\frac{P}{P_0}\right) \tag{e}$$

The rates of the chemical reactions are

$$r_1 = k_1(T_0)C_0 \frac{y_{A_0} - Z_1}{(1 + Z_1 + Z_2 + Z_3 + Z_4)\theta} \left(\frac{P}{P_0}\right) e^{\gamma_1(\theta-1)/\theta} \tag{f}$$

$$r_2 = k_2(T_0)C_0 \frac{Z_1 - Z_2}{(1 + Z_1 + Z_2 + Z_3 + Z_4)\theta} \left(\frac{P}{P_0}\right) e^{\gamma_2(\theta-1)/\theta} \tag{g}$$

$$r_3 = k_3(T_0)C_0 \frac{Z_1 - Z_3}{(1 + Z_1 + Z_2 + Z_3 + Z_4)\theta} \left(\frac{P}{P_0}\right) e^{\gamma_3(\theta-1)/\theta} \tag{h}$$

$$r_4 = k_4(T_0)C_0 \frac{Z_3 - Z_4}{(1 + Z_1 + Z_2 + Z_3 + Z_4)\theta} \left(\frac{P}{P_0}\right) e^{\gamma_4(\theta-1)/\theta} \tag{i}$$

We select the characteristic reaction time on the basis of Reaction 1:

$$t_{cr} = \frac{1}{k_1(T_0)} = 0.5 \, \text{s} \tag{j}$$

Substituting (e) through (j) into (a), (b), (c), and (d), the design equations reduce to

$$\frac{dZ_1}{d\tau} = \frac{y_{A_0} - Z_1}{(1 + Z_1 + Z_2 + Z_3 + Z_4)\theta} \left(\frac{P}{P_0}\right) e^{\gamma_1(\theta-1)/\theta} \tag{k}$$

$$\frac{dZ_2}{d\tau} = \frac{k_2(T_0)}{k_1(T_0)} \frac{Z_1 - Z_2}{(1 + Z_1 + Z_2 + Z_3 + Z_4)\theta} \left(\frac{P}{P_0}\right) e^{\gamma_2(\theta-1)/\theta} \tag{l}$$

$$\frac{dZ_3}{d\tau} = \frac{k_3(T_0)}{k_1(T_0)} \frac{Z_1 - Z_3}{(1 + Z_1 + Z_2 + Z_3 + Z_4)\theta} \left(\frac{P}{P_0}\right) e^{\gamma_3(\theta-1)/\theta} \tag{m}$$

$$\frac{dZ_4}{d\tau} = \frac{k_4(T_0)}{k_1(T_0)} \frac{Z_3 - Z_4}{(1 + Z_1 + Z_2 + Z_3 + Z_4)\theta} \left(\frac{P}{P_0}\right) e^{\gamma_4(\theta-1)/\theta} \tag{n}$$

Since θ appears in the design equations, we have to use the energy balance equation to express the changes in temperature. Assuming negligible kinetic energy and friction work, the energy balance equation is given by Eq. 7.1.16,

$$\frac{d\theta}{d\tau} =$$

$$\frac{1}{\text{CF}(z_m, \theta)} \left(\text{HTN}(\theta_F - \theta) - \text{DHR}_1 \frac{dZ_1}{d\tau} - \text{DHR}_2 \frac{dZ_2}{d\tau} - \text{DHR}_3 \frac{dZ_3}{d\tau} - \text{DHR}_4 \frac{dZ_4}{d\tau} \right)$$

$$\tag{o}$$

For gas-phase reactions, the local correction factor of the heat capacity of the reacting fluid is

$$CF(Z_m, \theta) = \frac{1}{(F_{tot})_0 \hat{c}_{p0}} (F_A \hat{c}_{pA} + F_B \hat{c}_{pB} + F_C \hat{c}_{pC} + F_D \hat{c}_{pD} + F_F \hat{c}_{pF} + F_G \hat{c}_{pG} + F_I \hat{c}_{pI})$$

(p)

where F_j is expressed by Eq. 2.7.8. Similarly, since P/P_0 appears in the design equations, we have to use the momentum balance equation to express the change in the pressure. Assuming negligible effect due to inertia changes, we use Eq. 7.5.4,

$$-\frac{d(P/P_0)}{d\tau} = \left(2f \frac{\rho_0 u_0^3 t_{cr}}{DP_0}\right)(1 + Z_1 + Z_2 + Z_3 + Z_4)\theta \left(\frac{P_0}{P}\right)$$

(q)

where u_0 is the velocity at the reactor inlet:

$$u_0 = \frac{4v_0}{\Pi D^2} = 37.7 \, \text{m/s}$$

To comply with the second law of thermodynamics, the local velocity in the reactor cannot exceed the sound velocity. For ideal gas, the sound velocity is

$$u_s = \left[\left(\frac{\hat{c}_p}{\hat{c}_p - R}\right)\frac{RT_0\theta}{MW}\right]^{0.5}$$

(r)

Combining (r) with Eq. 7.5.14, the local pressure should satisfy the following condition:

$$\frac{P}{P_0} \geq \frac{u_0(1 + Z_1 + Z_2 + Z_3 + Z_4)\theta}{\left[\left(\frac{\hat{c}_p}{\hat{c}_p - R}\right)\frac{RT_0\theta}{MW}\right]^{0.5}}$$

(s)

Note that both $\hat{c}_p$ and MW vary along the reactor as the reactions proceed, and they are given by

$$MW = \sum_{j}^{J} y_j MW_j$$

$$\hat{c}_p = \sum_{j}^{J} y_j \hat{c}_{p_j}(\theta)$$

where y_j is the local species molar faction given by

$$y_j = \frac{F_j}{F_{tot}} = \frac{y_{j_0} + \sum\limits_{m}^{n_I} (s_j)_m Z_m}{1 + \sum\limits_{m}^{n_I} \Delta_m Z_m}$$

To design the reactor, we have to solve (k), (l), (m), (n), simultaneously with (o) and (q), and verify that (s) is satisfied. We solve these equations numerically, subject to the initial conditions that at $\tau = 0$, $Z_1 = Z_2 = Z_3 = Z_4 = 0$, $\theta = 1$, and $P/P_0 = 1$.

a. For isothermal operation, $\theta = 1$, and we use (o) to determine the local HTN_{iso}:

$$HTN_{iso} = \frac{1}{\theta_F - 1}\left(DHR_1 \frac{dZ_1}{d\tau} + DHR_2 \frac{dZ_2}{d\tau} + DHR_3 \frac{dZ_3}{d\tau} + DHR_4 \frac{dZ_4}{d\tau} \right) \quad (t)$$

Figure E7.12.1 shows the reaction curves, Figure E7.12.2 shows the pressure curve, and Figure E7.12.3 shows the HTN curve for this case. The average isothermal HTN is 5.6.

b. Now that the HTN is known ($HTN_{ave} = 5.6$), we solve (k), (l), (m), and (n) simultaneously with (o) and (q). Figure E7.12.4 shows the actual reaction curves, Figure E7.12.5 shows the temperature curve, Figure E7.12.6 shows the species curves, and Figure E7.12.7 shows the pressure curve. Note that for the specified HTN, the temperature increases along the reactor except for a short section near the inlet. Also note that, at any point, the pressure is well above the limit imposed by the second law; hence, the assumption of negligible kinetic energy is justified. **c.** From Figure E7.12.6, the highest production of product C is achieved at $\tau = 2.73$, with $F_C/(F_{tot})_0 = 0.6038$, or $F_C = 12.1$ mol/s. The length of the reactor

$$L = \frac{4(\tau v_0 t_{cr})}{\pi D^2} = 51.4 \text{m}$$

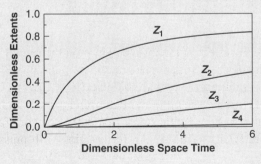

Figure E7.12.1 Reaction operating curves—isothermal operation.

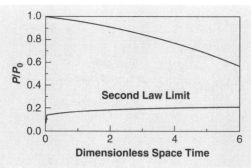

Figure E7.12.2   Pressure curve—isothermal operation.

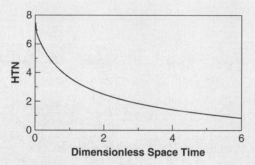

Figure E7.12.3   HTN curve—isothermal operation.

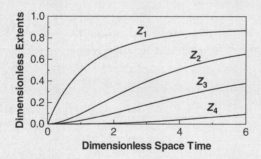

Figure E7.12.4   Reaction operating curves—actual operation.

d. To examine the effect of the pressure drop on the reactor performance, we carry out the design while neglecting the pressure drop. We solve (j), (k), (l), and (m) simultaneously with (o), and $d(P/P_0)/d\tau = 0$. From the curve of product C (not shown here), we determine that the highest production is achieved at $\tau = 2.29$, with $F_C/(F_{tot})_0 = 0.5917$, or $F_C = 11.86$ mol/s. From Figure E7.12.6 (the design formulation with pressure drop effect), at $\tau = 2.29$, $F_C/(F_{tot})_0 = 0.598$, or $F_C = 11.98$ mol/s. Hence, in this case, the error due to neglecting the effect of pressure drop is only 1%.

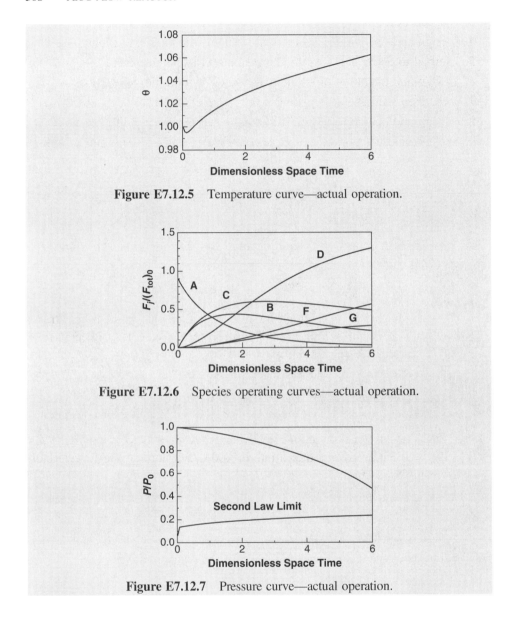

Figure E7.12.5 Temperature curve—actual operation.

Figure E7.12.6 Species operating curves—actual operation.

Figure E7.12.7 Pressure curve—actual operation.

7.6 SUMMARY

In this chapter, we analyzed the operation of plug-flow reactors. We covered the following topics:

1. The underlying assumptions of the plug-flow reactor model and when they are satisfied in practice
2. The design equations, the energy balance equation, and the auxiliary relations for species concentrations

3. The reaction operating curves and species operating curves
4. Design and operation of isothermal reactors with single reactions
5. Method to determine the rate expression and its parameters
6. Design and operation of isothermal reactors with multiple reactions
7. Described a procedure to estimate the range of the HTN
8. Design and operation of nonisothermal reactors with multiple reactions
9. Design and operation of gas-phase plug-flow reactors with multiple reactions
   where the pressure drop along the reactor is not negligible

PROBLEMS*

7.1₂ At 650°C, phosphine vapor decomposes according to the first-order, chemical reaction

$$4PH_3 \longrightarrow P_4(g) + 6H_2$$

Phosphine is fed into a plug-flow reactor at a rate of 10 mol/h in a stream which consists of 2/3 phosphine and 1/3 inert. The reactor is operated at 650°C and 11.4 atm. At 650°C, $k = 10\,hr^{-1}$.

a. Derive and solve the design equation.

b. Plot the reaction and species operating curves.

c. Determine the needed reactor volumes for 75% and 90% conversion.

7.2₂ An aqueous feed containing Reactants A and B ($C_{A_0} = 0.1$ mol/L, $C_{B_0} = 0.2$ mol/L) is fed at a rate of 400 L/min into a plug-flow reactor, where the reaction

$$A + B \longrightarrow R$$

takes place. The rate expression of the reaction is $r = 200C_A C_B$ mol/L min.

a. Derive the design equation and plot the reaction curve.

b. Derive and plot the species curves.

c. Determine the needed reactor volumes for 90%, 99%, and 99.9% conversion.

7.3₂ The homogeneous, gas-phase chemical reaction

$$A \longrightarrow 3R$$

*Subscript 1 indicates simple problems that require application of equations provided in the text. Subscript 2 indicates problems whose solutions require some more in-depth analysis and modifications of given equations. Subscript 3 indicates problems whose solutions require more comprehensive analysis and involve application of several concepts. Subscript 4 indicates problems that require the use of a mathematical software or the writing of a computer code to obtain numerical solutions.

follows second-order kinetics. For a feed rate of $4 \, m^3/h$ of Reactant A at 5 atm and 350°C, an experimental reactor (2.5 cm ID pipe 2 m long) gives 60% conversion. You are assigned to design a commercial plant to treat $320 \, m^3/h$ of feed consisting of 50% A, 50% inert at 35 atm, and 350°C to obtain 80% conversion. Assume plug-flow, negligible pressure drop, and ideal gas behavior.

a. How many 5-m length of 20 cm ID pipes are required?

b. Should they be placed in parallel or in series?

7.4$_2$ Formic acid is fed at a rate of 8 mol/h into a 10-L tubular reactor operated at 150°C and 1 atm. The acid decomposes according to the first-order, chemical reaction

$$HCOOH \longrightarrow H_2O + CO$$

At 150°C, $k = 2.46 \, min^{-1}$. Assume ideal-gas behavior.

a. Derive the design equation.

b. Plot the reaction curve and the species curves.

c. Determine the conversion of the formic acid.

7.5$_4$ Methanol is produced by the gas-phase reaction

$$CO + 2H_2 \longrightarrow CH_3OH$$

However, at the reactor operating conditions, the undesirable reaction

$$CO + 3H_2 \longrightarrow CH_4 + H_2O$$

is also taking place. Both reactions are second-order (each is first-order with respect to each reactant) and $k_2/k_1 = 1.2$. A synthesis gas stream is fed into a plug-flow reactor operated at 450°C and 5 atm. Plot the reaction and species curves for isothermal operation with a feed consisting of 50% CO and 50% H_2 (*mole basis*).

7.6$_4$ Below is a simplified kinetic model of the cracking of propane to produce ethylene

$$C_3H_8 \longrightarrow C_2H_4 + CH_4 \qquad\qquad r_1 = k_1 C_{C_3H_8}$$

$$C_3H_8 \rightleftarrows C_3H_6 + H_2 \qquad\qquad r_2 = k_2 C_{C_3H_8} \quad r_{-2} = \frac{k_2}{K_2} C_{C_3H_6} C_{H_2}$$

$$C_3H_8 + C_2H_4 \longrightarrow C_2H_6 + C_3H_6 \qquad r_3 = k_3 C_{C_3H_8} C_{C_2H_4}$$

$$2C_3H_6 \longrightarrow 3C_2H_4 \qquad\qquad r_4 = k_4 C_{C_3H_6}$$

$$C_3H_6 \rightleftarrows C_2H_2 + CH_4 \qquad\qquad r_5 = k_5 C_{C_3H_8} \quad r_{-5} = \frac{k_5}{K_5} C_{C_2H_2} C_{CH_4}$$

$$C_2H_4 + C_2H_2 \longrightarrow C_4H_6 \qquad\qquad r_6 = k_6 C_{C_2H_4} C_{C_2H_2}$$

At 800°C, $k_1 = 2.341\ s^{-1}$, $k_2 = 2.12\ s^{-1}$, $K_2 = 1000$, $k_3 = 23.63\ m^3/$ kmol s, $k_4 = 0.816\ s^{-1}$, $k_5 = 0.305\ s^{-1}$, $K_5 = 2000$, and $k_6 = 4.06 \times 10^3$ $m^3/$kmol s. Design an isothermal plug-flow reactor for cracking of propane to be operated at 2 atm and 800°C. Derive and plot the reaction and species curves. Assume propane is fed to the reactor.

7.7₄ The liquid-phase chemical reactions

$$\text{Reactions 1 \& 2:}\quad A + B \ \rightleftharpoons\ C$$
$$\text{Reaction 3:}\quad C + B \ \longrightarrow\ D$$

take place in a plug-flow reactor, operating isothermally at 90°C. A 200 L/min stream with $C_A = 2\ mol/L$, $C_B = 2\ mol/L$ is to be processed in the reactor. Based on the data below, derive and plot the reaction and species operating curves.

a. Derive and solve the design equations and plot the reaction and species curves.

b. Determine the reactor volume needed for 70% conversion of reactant B.

c. Determine the production rates of species C and D for $f_B = 0.7$.

d. If we want to maximize the production of C, what should be the reactor volume?

e. What is the maximum yield of C?

Data: The rate expressions of the chemical reactions are:

$$r_1 = k_1 C_A C_B \qquad r_2 = k_2 C_C \qquad r_3 = k_3 C_C C_B$$

At 90°C, $k_1 = 3\ L\ mol^{-1}\ min^{-1}$, $k_2 = 0.5\ min^{-1}$
$$k_3 = 1\ L\ mol^{-1}\ min^{-1}$$

7.8₄ The first-order gas-phase reaction

$$A \ \longrightarrow\ B + C$$

takes place in a plug-flow reactor. A stream consisting of 90% A and 10% I (% mole) is fed into a 200 L reactor at a rate of 50 L/s. The feed is at 731 K and 3 atm. Based on the data below, derive the reaction and species curves and the temperature curve for each of the operations below. Determine:

a. Determine the conversion of A when the reactor is operated isothermally.

b. Determine the heating load in (a).

c. Determine the local HTN and average isothermal HTN.

d. Determine the conversion of A and the outlet temperature when the reactor is operated adiabatically.

e. Determine the conversion of A and the outlet temperature when *HTN* is 25% of the value estimated in (c).

Data: At 731 K, $k = 0.2\,\text{s}^{-1}$, $E_a = 12,000\,\text{cal/mol}$
$$\Delta H_R = -10,000\,\text{cal/mol extent}$$

$$\hat{c}_{p_A} = 25\,\text{cal/mol K} \qquad \hat{c}_{p_B} = 15\,\text{cal/mol K}$$

$$\hat{c}_{p_C} = 18\,\text{cal/mol K} \qquad \hat{c}_{p_I} = 9\,\text{cal/mol K}$$

7.9$_4$ The elementary liquid-phase reactions

$$A + B \longrightarrow C$$

$$C + B \longrightarrow D$$

take place in a plug-flow reactor with a diameter of 10 cm. A solution (with $C_A = 2\,\text{mol/L}$, $C_B = 2\,\text{mol/L}$) at $80\,^\circ\text{C}$ is fed into the reactor at a rate of $200\,\text{L/min}$.

a. Determine the length of the reactor for maximum production of product C if the reactor is operated isothermally.
b. Determine the heating/cooling load in (a).
c. Determine the local and average isothermal HTN.
d. Determine the length of the reactor for maximum production of product C if the reactor is operated adiabatically.
e. Plot the temperature profile along the reactor in (d).
f. Determine the length of the reactor for maximum production of product C if the shell of the reactor is maintained at $80\,^\circ\text{C}$, and the HTN is half of the value estimated in (c).
g. Plot the temperature profile along the reactor in (f).

Data: At $80\,^\circ\text{C}$,
$$k_1 = 0.1\,\text{L mol}^{-1}\,\text{min}^{-1} \qquad\qquad E_{a_1} = 12,000\,\text{cal/mol}$$
$$k_2 = 0.2\,\text{L mol}^{-1}\,\text{min}^{-1} \qquad\qquad E_{a_2} = 16,000\,\text{cal/mol}$$
$$\Delta H_{R_1} = -15,000\,\text{cal/mol extent} \qquad \Delta H_{R_2} = -10,000\,\text{cal/mol extent}$$
$$\text{Density} = 900\,\text{g/L} \qquad\qquad \text{Heat capacity} = 0.8\,\text{cal/g}\,^\circ\text{C}$$

7.10$_4$ The irreversible gas-phase reactions (both are first-order)

$$A \longrightarrow 2V$$

$$V \longrightarrow 2W$$

Take place in a tubular reactor (plug-flow). A stream of species A is fed into a 10-cm ID reactor at a rate of $20\,\text{L/min}$. The feed is at 3 atm and 731 K ($C_A = 0.05\,\text{mol/L}$). Based on the data below,

a. Determine the length of an isothermal reactor to maximize the production of Product V. What is the maximum production rate of Product V?

b. Determine the conversion of Reactant A and the yield of Product V in (a).

c. Determine the heating load in (a), and estimate the average isothermal HTN.

d. Determine the length of an adiabatic reactor to maximize the production of Product V.

e. What is the maximum production rate of V in (d)?

f. Determine the length of a reactor whose wall temperature is 750 K if the *HTN* is 25% the value of the isothermal, that provides maximum production rate of Product V. What is the production rate of Product V?

Data: At 731 K, $k_1 = 2\,min^{-1}$, $k_2 = 0.5\,min^{-1}$

$$E_{a_1} = 8000\,cal/mol \qquad E_{a_2} = 12{,}000\,cal/mol$$

$$\Delta H_{R_1} = 3000\,cal/mol\ of\ V \qquad \Delta H_{R_2} = 4{,}500\,cal/mol\ of\ W$$

$$\hat{c}_{p_A} = 65\,cal/mol\,K \quad \hat{c}_{p_V} = 40\,cal/mol\,K \quad \hat{c}_{p_W} = 25\,cal/mol\,K$$

7.11₄ Cracking of naphtha cut to produce olefins is a common process in the petrochemical industry. The cracking reactions are represented by the simplified elementary gas-phase reactions:

$$
\begin{aligned}
C_{10}H_{22} &\longrightarrow C_4H_{10} + C_6H_{12}\\
C_4H_{10} &\longrightarrow C_3H_6 + CH_4\\
C_6H_{12} &\longrightarrow C_2H_4 + C_4H_8\\
C_4H_8 &\longrightarrow 2C_2H_4
\end{aligned}
$$

The reactions take place in a 90 m long 10 cm ID tubular reactor (plug flow) placed in a furnace chamber. The reactor wall is maintained at 800°C. A gas mixture at $T = 700°C$ and $P = 5$ atm, consisting of 90% naphtha ($C_{10}H_{22}$) and 10% I, is fed into the reactor at a rate of 100 L/s. Based on the data below:

a. Estimate the HTN for isothermal operation at $T = 700°C$.

b. Plot the reaction and species curves and the temperature profile for "actual" isothermal operation.

c. What should the reactor length be to optimize the production rate of butane?

d. Determine the production rate of ethylene and propylene in (c).

e. Determine the heating load of the reactor in (c).

f. Estimate the needed value of the heat-transfer coefficient.

Data: $r_1 = k_1 C_{C_{10}H_{22}}$ at $700°C$, $k_1 = 2.0\,s^{-1}$ $E_{a_1} = 25,000\,cal/mol$
$r_2 = k_2 C_{C_4H_{10}}$ at $700°C$, $k_2 = 0.5\,s^{-1}$ $E_{a_2} = 35,000\,cal/mol$

$r_3 = k_3 C_{C_6H_{12}}$ at $700°C$, $k_3 = 0.01\,s^{-1}$ $E_{a_3} = 40,000\,cal/mol$

$r_4 = k_4 C_{C_4H_8}$ at $700°C$, $k_4 = 0.001\,s^{-1}$ $E_{a_4} = 45,000\,cal/mol$

$\Delta H_{R_1}(T_0) = 20,000\,cal/mol\,extent$ $\Delta H_{R_2}(T_0) = 35,000\,cal/mol\,extent$

$\Delta H_{R_3}(T_0) = 45,000\,cal/mol\,extent$ $\Delta H_{R_4}(T_0) = 55,000\,cal/mol\,extent$

$\hat{c}_{pC_{10}H_{22}} = -280\,cal/mol°C$ $\hat{c}_{pC_6H_{12}} = -180\,cal/mol°C$ $\hat{c}_{pCH_4} = 20\,cal/mol°C$

$\hat{c}_{pC_4H_{10}} = 150\,cal/mol°C$ $\hat{c}_{pC_4H_8} = 140\,cal/mol°C$ $\hat{c}_{pI} = 10\,cal/mol°C$

$\hat{c}_{pC_3H_6} = 120\,cal/mol°C$ $\hat{c}_{pC_2H_4} = 30\,cal/mol°C$

7.12$_4$ The first-order gas-phase reaction

$$A \longrightarrow B + C$$

takes place in a plug-flow reactor. A stream consisting of species 90% A and 10% I (% mole) is fed into a 200-L reactor at a rate of $20\,L/s$. The feed is at 731 K and 3 atm. Based on the data below, calculate:

a. Determine the conversion of Reactant A when the reactor is operated isothermally.

b. Determine the heating/cooling load in (a).

c. Determine the local HTN and estimate the average HTN for isothermal operation ($T_F = 710$ K).

d. Determine the conversion of reactant A and the outlet temperature when the reactor is operated adiabatically.

e. Determine the conversion of reactant A if HTN is 25% of the isothermal HTN, and the temperature of the cooling fluid is 710 K.

Data: At 731 K, $k = 0.2\,s^{-1}$, $E_a = 12,000$ cal/mol
$$\Delta H_R = -10,000 \text{ cal/mol extent}$$

$$\hat{c}_{pA} = 25\,cal/mol°K \qquad \hat{c}_{pB} = 15\,cal/mol°K$$

$$\hat{c}_{pC} = 18\,cal/mol°K \qquad \hat{c}_{pI} = 9\,cal/mol°K$$

7.13$_2$ The second-order, gas-phase reaction

$$A \longrightarrow B + C$$

is carried out in a cascade of two tubular reactors connected in series. Reactant A is fed at a rate of $100\,mol/h$ into the first reactor, whose volume is 1000 L. The molar flow rate of reactant A at the exit of the first reactor is $60\,mol/h$, and its flow rate at the exit of the second reactor is 20 mol/h. The temperature throughout the system is 150°C, and the pressure is 2 atm. Determine the volume of the second reactor by:

a. Taking the inlet stream to the first reactor as the reference stream.

b. Taking the inlet stream to the second reactor as the reference stream.

7.14₄ The elementary liquid-phase reactions

$$A + B \longrightarrow C$$

$$C + B \longrightarrow D$$

$$D + B \longrightarrow E$$

take place in a 0.1-m ID tubular reactor. A solution ($C_A = C_B = 2$ mol/L) at 80°C is fed into the reactor at a rate of 100 L/min. Based on the data below, derive the reaction operating curves and the temperature curve for each of the operations below.

a. Determine the reactor length needed to maximize the production of Product C for isothermal operation at 80°C.

b. Determine the heating load on the reactor in (a).

c. Determine the local HTN, and estimate the average HTN for isothermal operation ($T_F = 65$°C).

d. Determine the reactor length needed to maximize production of product C for adiabatic operation. What is the outlet temperature?

e. Determine the reactor length needed to maximize the production of product C with HTN 25% of average isothermal HTN, and $T_F = 65$°C.

Data: At 80°C,

$k_1 = 0.1$ L mol^{-1} min^{-1} $E_{a_1} = 6000$ cal/mol
$k_2 = 0.2$ L mol^{-1}min^{-1} $E_{a_2} = 8000$ cal/mol
$k_3 = 0.3$ L mol^{-1}min^{-1} $E_{a_3} = 10{,}000$ cal/mol
$\Delta H_{R_1} = -15{,}000$ cal/mol extent $\Delta H_{R_2} = -10{,}000$ cal/mol extent
$\Delta H_{R_3} = -8000$ cal/mol extent

Density of the solution $= 1000$ g/L
Heat Capacity of the solution $= 1$ cal/g°C.

7.15₄ The elementary gas-phase reversible chemical reactions

Reactions 1 & 2: $A + B \rightleftarrows C$
Reactions 3 & 4: $C + B \rightleftarrows D$
Reactions 5 & 6: $D + B \rightleftarrows E$

take place in a tubular reactor. A stream consisting of 50% A and 50% B (mole %) at 600 K and 5 atm, is fed into the reactor at a rate of

1000 L/min. Based on the data below, derive the reaction operating curves and the temperature curve for each of the operations below.

a. Determine the reactor volume needed to maximize the production of product C for isothermal operation at 600 K.

b. Determine the heating load on the reactor in (a).

c. Determine the local HTN, and estimate the average HTN for isothermal operation ($T_F = 580$ K).

d. Determine the reactor volume needed to maximize production of Product C for adiabatic operation. What is the outlet temperature?

e. Determine the reactor length needed to maximize the production of Product C with HTN 25% of average isothermal HTN, and $T_F = 580$ K.

Data: At 600 K, $k_1 = 100$; $k_3 = 200$; $k_5 = 300$ L mol^{-1} min^{-1}

$$k_2 = 0.3; \quad k_4 = 0.5; \quad k_6 = 0.6 \text{ min}^{-1}$$

$$E_{a_1} = 6{,}000 \text{ cal/mol} \qquad E_{a_2} = 12{,}000 \text{ cal/mol}$$

$$E_{a_3} = 8{,}000 \text{ cal/mol} \qquad E_{a_4} = 13{,}000 \text{ cal/mol}$$

$$E_{a_5} = 10{,}000 \text{ cal/mol} \qquad E_{a_6} = 14{,}000 \text{ cal/mol}$$

$\Delta H_{R_1} = -6{,}000$ cal/mol extent $\Delta H_{R_3} = -5{,}000$ cal/mol extent

$\Delta H_{R_5} = -4{,}000$ cal/mol extent

$\hat{c}_{p_A} = 25$ cal/mol°K $\hat{c}_{p_B} = 10$ cal/mol°K $\hat{c}_{p_C} = 30$ cal/mol°K

$\hat{c}_{p_D} = 35$ cal/mol°K $\hat{c}_{p_E} = 40$ cal/mol°K

8

CONTINUOUS STIRRED-TANK REACTOR

The continuous stirred-tank reactor (CSTR) is a mathematical model that describes an important class of continuous reactors—continuous, steady, well-agitated tank reactors. The CSTR model is based on two assumptions:

- Steady-state operation
- Uniform conditions (concentrations and temperature) exist throughout the reactor volume (due to good mixing)

These conditions are readily achieved in small-scale agitated reactors. However, in large industrial reactors, it is not easy to achieve good mixing, and special care should be given to the design of the tank and agitator. Furthermore, even when

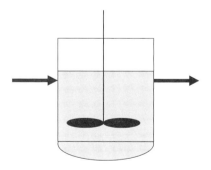

Figure 8.1 Schematic description liquid-phase CSTR.

Principles of Chemical Reactor Analysis and Design, Second Edition. By Uzi Mann
Copyright © 2009 John Wiley & Sons, Inc.

the same conditions exist in most sections of the reactor, the conditions near the reactor inlet and near the reactor wall are different from those in the remainder of the reactor. Since these zones usually represent a small portion of the reactor, the CSTR model provides a reasonable description of the well-agitated large reactors. Figure 8.1 shows schematically a liquid-phase CSTR.

8.1 DESIGN EQUATIONS AND AUXILIARY RELATIONS

The design equation of a CSTR was derived in Chapter 4. The design equation, written for the mth-independent reaction is

$$Z_{m_{\text{out}}} - Z_{m_{\text{in}}} = \left(r_{m_{\text{out}}} + \sum_{k}^{n_D} \alpha_{km} r_{k_{\text{out}}} \right) \tau \left(\frac{t_{\text{cr}}}{C_0} \right) \tag{8.1.1}$$

where Z_m is the dimensionless extent of the mth-independent reaction, defined by Eq. 2.7.2:

$$Z_m = \frac{\dot{X}_m}{(F_{\text{tot}})_0} \tag{8.1.2}$$

and τ is the dimensionless space time of the reactor defined by Eq. 4.4.8:

$$\tau = \frac{V_R}{v_0 t_{\text{cr}}} \tag{8.1.3}$$

where t_{cr} is a conveniently selected characteristic reaction time (defined by Eq. 3.5.1), and C_0 is a conveniently selected reference concentration defined by

$$C_0 = \frac{(F_{\text{tot}})_0}{v_0} \tag{8.1.4}$$

where $(F_{\text{tot}})_0$ and v_0 are, respectively, the total molar flow rate and the volumetric flow rate of the reference stream.

As discussed in Chapter 4, to describe the operation of a CSTR with multiple reactions, we have to write Eq. 8.1.1 for each independent chemical reaction. The solution of the design equations (the relationships between $Z_{m_{\text{out}}}$'s and τ) provide the reaction operating curves and describe the reactor operation. To solve the design equations, we have to express the rates of the chemical reactions that take place in the reactor in terms of Z_m's and τ. Below, we derive the auxiliary relations used in the design equations.

The volume-based rate expression of the ith chemical reaction (Eq. 3.3.1 and Eq. 3.3.6) is

$$r_i = k_i(T_0) e^{\gamma_i(\theta - 1)/\theta} h_i(C_j\text{'s}) \tag{8.1.5}$$

where $k_i(T_0)$ is the reaction rate constant at reference temperature T_0, γ_i is the dimensionless activation energy ($\gamma_i = E_{a_i}/RT_0$), and $h_i(C_j\text{'s})$ is a function of the species concentrations, given by the rate expression. For a CSTR, the concentration of species j is the same everywhere inside the reactor and is equal to the outlet concentration:

$$C_{j_{\text{out}}} = \frac{F_{j_{\text{out}}}}{v_{\text{out}}} \tag{8.1.6}$$

where $F_{j_{\text{out}}}$ and v_{out} are, respectively, the molar flow rate of species j and the volumetric flow rate at the reactor outlet. Using Eq. 2.7.7,

$$F_{j_{\text{out}}} = (F_{\text{tot}})_0 \left[\frac{(F_{\text{tot}})_{\text{in}}}{(F_{\text{tot}})_0} y_{j_{\text{in}}} + \sum_m^{n_I} (s_j)_m Z_{m_{\text{out}}} \right]$$

and the outlet concentration of species j is

$$C_{j_{\text{out}}} = \frac{(F_{\text{tot}})_0}{v_{\text{out}}} \left[\frac{(F_{\text{tot}})_{\text{in}}}{(F_{\text{tot}})_0} y_{j_{\text{in}}} + \sum_m^{n_I} (s_j)_m Z_{m_{\text{out}}} \right] \tag{8.1.7}$$

When the inlet stream is selected as the reference stream, Eq. 8.1.7 reduces to

$$C_{j_{\text{out}}} = \frac{(F_{\text{tot}})_0}{v_{\text{out}}} \left[y_{j_0} + \sum_m^{n_I} (s_j)_m Z_{m_{\text{out}}} \right] \tag{8.1.8}$$

For *liquid-phase* reactions, the density of the reacting fluid is assumed to be constant; hence, $v_{\text{out}} = v_0$, and Eq. 8.1.8 reduces to

$$C_{j_{\text{out}}} = C_0 \left[y_{j_0} + \sum_m^{n_I} (s_j)_m Z_{m_{\text{out}}} \right] \tag{8.1.9}$$

Equation (8.1.9) provides the species concentrations in terms of the extents of the independent reactions for liquid-phase reactions.

For *gas-phase* reactions, the volumetric flow rate depends on the total molar flow rate and the temperature and pressure in the reactor. Assuming ideal-gas behavior, the volumetric flow rate is

$$v_{\text{out}} = v_0 \left[\frac{(F_{\text{tot}})_{\text{out}}}{(F_{\text{tot}})_0} \right] \left(\frac{T_{\text{out}}}{T_0} \right) \left(\frac{P_0}{P_{\text{out}}} \right) \tag{8.1.10}$$

Using Eq. 2.7.9 to express the total molar flow rate in terms of the extents of the independent reactions,

$$F_{\text{tot}} = (F_{\text{tot}})_0 \left[\frac{(F_{\text{tot}})_{\text{in}}}{(F_{\text{tot}})_0} + \sum_m^{n_I} \Delta_m Z_m \right]$$

Equation 8.1.10 becomes

$$v_{\text{out}} = v_0 \left[\frac{(F_{\text{tot}})_{\text{in}}}{(F_{\text{tot}})_0} + \sum_{m}^{n_I} \Delta_m Z_{m_{\text{out}}} \right] \theta_{\text{out}} \left(\frac{P_0}{P_{\text{out}}} \right) \tag{8.1.11}$$

where $\theta_{\text{out}} = T_{\text{out}}/T_0$. Substituting Eq. 8.1.11 into Eq. 8.1.7, the outlet concentration is

$$C_{j_{\text{out}}} = C_0 \frac{[(F_{\text{tot}})_{\text{in}}/(F_{\text{tot}})_0] y_{j_{\text{in}}} + \sum_{m}^{n_I} (s_j)_m Z_{m_{\text{out}}}}{\left((F_{\text{tot}})_{\text{in}}/(F_{\text{tot}})_0 + \sum_{m}^{n_I} \Delta_m Z_{m_{\text{out}}} \right) \theta_{\text{out}}} \left(\frac{P_{\text{out}}}{P_0} \right) \tag{8.1.12}$$

When we select the inlet stream as the reference stream, Eq. 8.1.12 becomes

$$C_{j_{\text{out}}} = C_0 \frac{y_{j_0} + \sum_{m}^{n_I} (s_j)_m Z_{m_{\text{out}}}}{\left(1 + \sum_{m}^{n_I} \Delta_m Z_{m_{\text{out}}} \right) \theta_{\text{out}}} \left(\frac{P_{\text{out}}}{P_0} \right) \tag{8.1.13}$$

Equation 8.1.13 provides the species concentrations in terms of the extents of the independent reactions for gas-phase reactions in CSTRs.

Since the reactor temperature, θ_{out}, may be different than the inlet temperature, we have to solve the energy balance equation simultaneously with the design equations. For CSTRs with negligible viscous and shaft work, the energy balance equation, derived in Chapter 5, is

$$\text{HTN}\tau(\theta_F - \theta_{\text{out}}) = \sum_{m}^{n_I} \text{DHR}_m(Z_{m_{\text{out}}} - Z_{m_{\text{in}}}) + \text{CF}(Z, \theta)_{\text{out}}(\theta_{\text{out}} - 1) -$$

$$\text{CF}(Z, \theta)_{\text{in}}(\theta_{\text{in}} - 1) \tag{8.1.14}$$

where HTN is the dimensionless heat-transfer number defined by Eq. 5.2.22:

$$\text{HTN} = \frac{U t_{\text{cr}}}{C_0 \hat{c}_{p0}} \left(\frac{S}{V} \right) \tag{8.1.15}$$

DHR_m is the dimensionless heat of reaction of the mth-independent chemical reaction, defined by Eq. 5.2.23:

$$\text{DHR}_m = \frac{\Delta H_{R_m}(T_0)}{T_0 \hat{c}_{p0}} \tag{8.1.16}$$

and $\text{CF}(Z_m, \theta)$ is the correction factor of the heat capacity of the reacting fluid, defined by Eq. 5.2.54. The term on the left of Eq. 8.1.14 is the dimensionless heat-transfer rate:

$$\frac{\dot{Q}}{(F_{\text{tot}})_0 \hat{c}_{p_0} T_0} = \text{HTN}\tau(\theta_F - \theta) \tag{8.1.17}$$

The first term on the right of Eq. 8.1.14 represents the heat generated (or consumed) by the chemical reactions in the reactor.

The specific molar specific heat capacity of the reference state, $\hat{c}_{p0}$, is determined differently for gas-phase and liquid-phase reactions. For *gas-phase* reactions, it is defined by Eq. 5.2.58,

$$\hat{c}_{p0} \equiv \sum_{j}^{J} y_{j_0} \hat{c}_{p_j}(T_0) \tag{8.1.18}$$

and for *liquid-phase* reactions, it is defined by Eq. 5.2.60,

$$\hat{c}_{p0} \equiv \frac{\dot{m}\bar{c}_p}{(F_{\text{tot}})_0} \tag{8.1.19}$$

To solve Eq. 8.1.14, we have to specify the value of HTN. However, its value depends on the heat transfer coefficient, U, which in turn depends on the flow conditions in the reactor, the properties of the fluid, and the heat-transfer area per unit volume (S/V). These parameters are not known a priori. Therefore, we develop a procedure to estimate the range of HTN. For isothermal operation $(\theta_{\text{out}} = \theta_{\text{in}})$, we can determine the HTN of a given reactor from Eq. 8.1.14 (taking the reactor temperature as the reference temperature, $\theta_{\text{in}} = 1$),

$$\text{HTN}_{\text{iso}}(\tau) = \frac{1}{(\theta_F - 1)\tau} \sum_{m}^{n_I} \text{DHR}_m (Z_{m_{\text{out}}} - Z_{m_{\text{in}}}) \tag{8.1.20}$$

Note that the value of HTN varies with the volume of the CSTR. We define an *average* isothermal HTN over a range of reactor volumes by

$$\text{HTN}_{\text{ave}} \equiv \frac{1}{\tau_{\text{tot}}} \int_{0}^{\tau_{\text{tot}}} \text{HTN}(\tau)\, d\tau \tag{8.1.21}$$

where τ_{tot} is the total dimensionless space time of the largest reactor. Recall that for adiabatic operation HTN $= 0$. In most cases, the heat-transfer number would be

$$0 < \text{HTN} \leq \text{HTN}_{\text{ave}} \tag{8.1.22}$$

Equation 8.1.22 provides only an estimate on the range of the value of HTN to be used in the design. We select a specific value after examining the reactor performance with different values of HTN. When multiple reactions take place, it is important to examine the reactor operation for different values of HTN, since it is difficult to predict the effect of the heat transfer on the relative rates of the individual reactions. Once a specific reactor vessel has been designed, it is necessary to verify that the reactor configuration and the flow conditions in the reactor actually provide the specified value of HTN.

For convenience, Tables A.3a and A.3b in Appendix A provide the design equation and the auxiliary relations used in the design of CSTRs. Table A.4 provides the energy balance equation.

In the remainder of the chapter, we discuss how to apply the design equations and the energy balance equations to determine various quantities related to the operations of CSTRs. In Section 8.2 we examine isothermal operations with *single* reactions to illustrate how the rate expressions are incorporated into the design equation and how rate expressions are determined. In Section 8.3, we expand the analysis to isothermal operations with *multiple* reactions. In Section 8.4, we consider nonisothermal operations with multiple reactions.

8.2 ISOTHERMAL OPERATIONS WITH SINGLE REACTIONS

We start the analysis of CSTRs by considering isothermal operations with single chemical reactions. Isothermal CSTRs are defined as those where $\theta_{out} = \theta_{in}$. Since we do not have to determine the reactor temperature, we have to solve only the design equations. The energy balance equation provides the heating (or cooling) load necessary to maintain the isothermal conditions. Also, for isothermal operations, the individual reaction rates depend only on the species concentrations, and, when the reactor temperature is taken as the reference temperature, $T = T_0$, and Eq. 8.1.5 reduces to

$$r_i = k_i(T_0)h_i(C_j\text{'s}) \tag{8.2.1}$$

When a single chemical reaction takes place in a CSTR, there is only one independent reaction and no dependent reactions, and Eq. 8.1.1 reduces to

$$Z_{out} - Z_{in} = r_{out}\tau\left(\frac{t_{cr}}{C_0}\right) \tag{8.2.2}$$

where Z_{out} and Z_{in} are the dimensionless extents of the reaction at the reactor outlet and inlet, respectively, and r_{out} is the reaction rate. We can rearrange Eq. 8.2.2 as

$$\tau = \left(\frac{C_0}{t_{cr}}\right)\frac{Z_{out} - Z_{in}}{r_{out}} \tag{8.2.3}$$

Note that the values of Z_{in} and Z_{out} depend on the selection of the reference stream. Also note that if we use the definition of the characteristic reaction time (Eq. 3.5.1), Eq. 8.2.3 reduces to

$$\tau = \left(\frac{r_0}{r_{out}}\right)(Z_{out} - Z_{in}) \tag{8.2.4}$$

The solution of the design equation, Z_{out} versus τ, provides the dimensionless reaction operating curve. It describes the progress of the chemical reaction as a function of the reactor volume. Also, once Z_{out} is known, we can apply Eq. 2.7.8 to obtain the species operating curves, indicating the species molar flow rates as a function of the reactor volume. Note that for given inlet conditions, Eq. 8.2.2 has three variables: the dimensionless space time, τ, the reaction extent at the reactor outlet, Z_{out}, and the reaction rate, r_{out}. We use the design equation to determine any one of those variables when the other two are provided. A typical design problem is to determine the reactor volume needed to obtain a specified extent (or conversion) for a given feed rate and reaction rate. The second application is to determine the extent (or conversion) at the reactor outlet for a given reactor volume and reaction rate. The third application is to determine the reaction rate when the extent (or conversion) at the reactor outlet is provided for a given feed rate.

If one prefers to express the design equation in terms of the reactor volume, V_R, rather than the dimensionless space time τ, substituting Eq. 8.1.3, into Eq. 8.2.2:

$$V_R = (F_{tot})_0 \frac{Z_{out} - Z_{in}}{r_{out}} \tag{8.2.5}$$

For CSTRs with single chemical reactions, the common practice has been to express the design equation in terms of the conversion of the limiting reactant, f_A. Using Eq. 2.6.5,

$$Z = -\frac{y_{A_0}}{s_A} f_A$$

and Eq. 8.2.5 becomes

$$V_R = F_{A_0} \frac{f_{A_{out}} - f_{A_{in}}}{(-r_A)_{out}} \tag{8.2.6}$$

where $(-r_A)_{out} = -s_A r_{out}$, is the depletion rate of reactant A.

To solve Eq. 8.2.2, we have to express the reaction rate, r_{out}, in terms of the dimensionless extent, Z_{out}. To do so, we express the species concentrations in terms of Z_{out}. For single liquid-phase reactions, Eq. 8.1.9 reduces to

$$C_{j_{out}} = C_0(y_{j_0} + s_j Z_{out}) \tag{8.2.7}$$

For single gas-phase reactions, Eq. 8.1.12 reduces to

$$C_{j_{out}} = C_0 \frac{y_{j_0} + s_j Z_{out}}{1 + \Delta Z_{out}} \tag{8.2.8}$$

Below, we analyze the operation of isothermal CSTRs with single reactions for different types of chemical reactions. For convenience, we divide the analysis into two sections: reactor design and determination of the rate expression. In the former, we determine the size of the reactor or the production rate of a given reactor for a known reaction rate, and, in the latter, we determine the rate expression from experimental reactor operating data.

8.2.1 Design of a Single CSTR

Examining Eq. 8.2.4, we see that if we know how the reaction rate, r, varies with Z, we can plot r_0/r versus Z, as shown in Figure 8.2. The area of the rectangle between Z_{in} and Z_{out} is equal to the dimensionless space time, τ. The required reactor volume, V_R, is readily determined from the dimensionless space time by $V_R = v_0 t_{cr}\tau$. Recall that for a plug-flow reactor (Fig. 7.2) the dimensionless space time, τ, is provided by the area under the curve. Hence, for common rate expressions, the volume of a CSTR needed to achieve a certain extent is larger than that of a plug-flow reactor. The reason for the different performance is the concentration profiles of the reactants in each reactor. In a plug-flow reactor, the reactant concentrations decrease gradually along the reactor, whereas in a CSTR, the outlet low concentrations exist everywhere in the reactor. For common kinetics, the reaction rate is faster at higher concentrations. Hence, a plug-flow reactor with the same volume as a CSTR (same space time) provides a higher conversion (or extent) than the CSTR. In fact, a plug-flow reactor represents the "best" reactor configuration, whereas a CSTR represents the "worse" configuration from a volume utilization perspective. Actual reactors (neither pluglike nor well mixed) usually perform in between these two extremes.

We start the analysis of isothermal CSTRs with single chemical reactions and consider a first-order gas-phase chemical reaction:

$$A \longrightarrow \text{Products}$$

and a stream with known composition to be fed to the reactor ($y_{j_{in}}$'s, are specified). For this case, $s_A = -1$, Δ depends on the reaction's stoichiometry, and the rate expression is $r = kC_A$. Using Eq. 8.2.8, the reaction rate, expressed in terms of extent, is

$$r = kC_0 \frac{y_{A_0} - Z}{1 + \Delta\, Z} \tag{8.2.9}$$

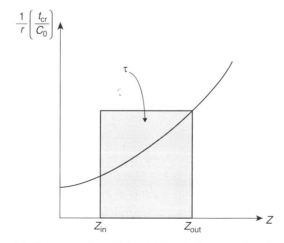

Figure 8.2 Graphical presentation of the CSTR design equation for single reactions.

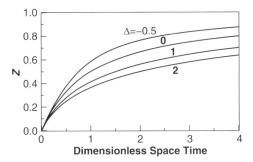

Figure 8.3 Reaction operating curves for a single first-order reaction of the form A →
products; effect of expansion factor.

Substituting Eq. 8.2.9 into Eq. 8.2.2,

$$Z_{out} - Z_{in} = kC_0\tau \frac{y_{A_0} - Z_{out}}{1 + \Delta Z_{out}} \left(\frac{t_{cr}}{C_0}\right) \tag{8.2.10}$$

We select the inlet stream as the reference stream, $Z_{in} = 0$, when only reactant A is
fed into the reactor, $y_{A_0} = 1$. Using Eq. 3.5.4, the characteristic reaction time is

$$t_{cr} = \frac{1}{k}$$

and the design equation reduces to

$$Z_{out} = \tau \frac{1 - Z_{out}}{1 + \Delta Z_{out}} \tag{8.2.11}$$

The solution of Eq. 8.2.11 for different values of Δ is shown in Figure 8.3. Note that for
larger Δ, a larger reactor volume is required to achieve a given level of extent.

The design of CSTRs and the values of calculated quantities (reactor volume,
species production rates, etc.) do not depend on the specific reference stream
selected. Example 8.2 illustrates the use of a fictitious stream as a reference
stream and the behavior of a CSTR when one product is condensable. Example
8.3 illustrates the design of a CSTR with a chemical reaction whose rate expression
is not a power function of the concentrations.

Example 8.1 The first-order, gas-phase, reaction

$$A \longrightarrow B + C$$

takes place in an isothermal CSTR. The reaction is first order, and its rate constant at the operating temperature is $k = 0.1 \text{ s}^{-1}$. A gaseous stream of reactant A
($C_{A_0} = 0.04 \text{ mol/L}$) is fed into the reactor at a rate of 10 mol/min.

a. Determine the volume of the reactor needed to achieve 80% conversion.

b. Determine the volume of the reactor needed to achieve 80% conversion if, by
mistake, we use $\Delta = 0$.

c. Determine the actual conversion obtained in a CSTR with the volume calculated in (b).

d. Determine the feed flow rate to a CSTR with the volume calculated in (b) if we want to maintain 80%.

Solution   The stoichiometric coefficients are

$$s_A = -1 \qquad s_B = 1 \qquad s_C = 1 \qquad \Delta = 1$$

We select the inlet stream as the reference stream $Z_{in} = 0$ and $(F_{tot})_0 = F_{A_{in}}$, and, since only reactant A is fed into the reactor, $y_{A_0} = 1$, $y_{B_0} = y_{C_0} = 0$, $C_0 = C_{A_{in}}$. The volumetric feed rate is

$$v_0 = \frac{(F_{tot})_0}{C_0} = 250 \text{ L/min}$$

Using Eq. 2.6.5 to relate the extent to the conversion,

$$Z_{out} = -\frac{y_{A_0}}{s_A} f_{A_{out}} = 0.80$$

a. For a CSTR with a single chemical reaction, the design equation Eq. 8.2.2 is

$$Z_{out} - Z_{in} = r\tau \left(\frac{t_{cr}}{C_0}\right) \qquad (a)$$

We use Eq. 8.2.8 to express C_A, and the reaction rate is

$$r = kC_0 \frac{1 - Z}{1 + \Delta Z} \qquad (b)$$

Using Eq. 3.5.4, for a first-order reaction, the characteristic reaction time is

$$t_{cr} = \frac{1}{k} = 10 \text{ s} = 0.167 \text{ min} \qquad (c)$$

Substituting (b) and (c) into (a), the design equation is

$$Z_{out} = \tau \frac{1 - Z_{out}}{1 + Z_{out}} \qquad (d)$$

We solve (d) for different values of τ to obtain the reaction curve, shown in Figure E8.1.1. Note that each τ represents a tank with a different volume. Once we have Z_{out}, we use Eq. 2.7.8 to obtain the species curves:

$$\frac{F_{A_{out}}}{(N_{tot})_0} = 1 - Z_{out}$$

$$\frac{F_{B_{out}}}{(N_{tot})_0} = \frac{F_{C_{out}}}{(N_{tot})_0} = Z_{out}$$

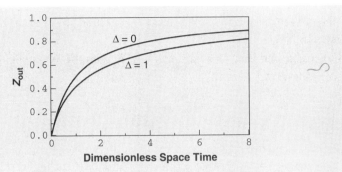

Figure E8.1.1   Reaction operating curve.

Figure E8.1.2 shows the species curves. From the reaction curve, for $Z_{out} = 0.8$, $\tau = 7.20$. Using Eq. 8.1.3 and (c), the needed reactor volume is

$$V_R = \tau v_0 t_{cr} = 300 \text{ L} \tag{e}$$

b. If we do not account for the change in the volumetric flow rate (and use $\Delta = 0$), the design equation is

$$Z_{out} = \tau(1 - Z_{out}) \tag{f}$$

and, for $Z_{out} = 0.8$, the solution is $\tau = 4.0$. Hence, the calculated reactor volume is

$$V_R = \tau v_0 t_{cr} = 166.7 \text{ L} \tag{g}$$

Note that if we use a wrong concentration expression, we specify a reactor volume that is only 55% of the required volume.

c. To calculate the actual outlet conversion on a CSTR with volume of 166.7 L, we solve (d) for $\tau = 4.0$, and the solution is $Z_{out} = f_{a_{out}} = 0.701$. Hence, we obtain only 70.1% conversion instead of the desired 80%.

d. To attain a conversion of 0.80 on the 166.7-L reactor, the feed flow rate should be reduced. We know that to obtain 80% conversion, τ should be

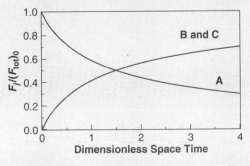

Figure E8.1.2   Species operating curves.

7.2. We substitute in (e) $V_R = 166.7$ L, $\tau = 7.20$; hence,

$$v_0 = \frac{V_R}{\tau t_{cr}} = 138.9 \, \text{L/min} \tag{h}$$

and the molar feed rate is

$$(F_{\text{tot}})_0 = v_0 C_0 = 5.55 \, \text{mol/min}$$

Hence, if we use the wrong expression for the species concentration and want to maintain the specified conversion, we can process only 5.55 mol/min instead of 10 mol/min.

Example 8.2 The elementary, gas-phase reaction

$$A + B \longrightarrow C$$

is carried out in an isothermal-isobaric CSTR operated at 2 atm and 170°C. At this temperature, $k = 90 \, \text{L mol}^{-1} \text{min}^{-1}$, and the vapor pressure of the product, C, is 0.3 atm. The reactor is fed with two gas streams: the first one consists of 80% A, 10% B, 10% inert (I), and is at 2.5 atm and 150°C; the second consists of 80% B, 20% I, and is at 3 atm and 180°C. The first stream is fed at a rate of 100 mol/min and the second at a rate of 120 mol/min.

a. Determine the conversion of reactant A when product C begins to condense.

b. What is the reactor volume if C just starts to condense?

c. What reactor volume is needed for 85% conversion of A?

d. Plot the reaction and species operating curves.

Solution At low conversion of reactant A, all the species are gaseous, and the reaction is

$$A(g) + B(g) \longrightarrow C(g) \tag{a}$$

and the stoichiometric coefficients are

$$s_A = -1 \qquad s_B = -1 \qquad s_C = 1 \qquad \Delta = \Delta_{\text{gas}} = -1$$

We select a fictitious reference stream formed by combining the two feed streams and select 2 atm and 170°C (the reactor operating conditions) as reference pressure and temperature. Hence,

$$F_{A_0} = 0.8F_1 = (0.8)(100) = 80 \, \text{mol/min}$$

$$F_{B_0} = 0.1F_1 + 0.8F_2 = (0.1)(100) + (0.8)(120) = 106 \, \text{mol/min}$$

$$F_{I_0} = 0.1F_1 + 0.2F_2 = (0.1)(100) + (0.2)(120) = 34 \, \text{mol/min}$$

$$(F_{\text{tot}})_0 = 220 \, \text{mol/min}$$

The composition of the reference stream is $y_{A_0} = 0.364$, $y_{B_0} = 0.482$, $y_{I_0} = 0.154$, and $Z_{\text{in}} = 0$. Assuming ideal-gas behavior, the reference concentration is

$$C_0 = \frac{P_0}{RT_0} = 0.055 \, \text{mol/L}$$

The volumetric flow rate of the reference stream is

$$v_0 = \frac{(F_{\text{tot}})_0}{C_0} = 4000 \, \text{L/min}$$

a. Product C starts to condense when its partial pressure in the reactor is 0.3 atm,

$$P_C = y_C P = \left(\frac{F_C}{F_{\text{tot}}}\right) P$$

Using Eqs. 2.7.8 and 2.7.10,

$$P_C = \left(\frac{s_C Z_{\text{out}}}{1 + \Delta_{\text{gas}} Z_{\text{out}}}\right) P = \frac{Z_{\text{out}}}{1 - Z_{\text{out}}} (2 \, \text{atm}) = 0.3 \, \text{atm} \qquad \text{(b)}$$

Solving (b), $Z_{\text{out}} = 0.130$, and using Eq. 2.6.5, the conversion is

$$f_{A_{\text{out}}} = -\frac{s_A}{y_{A_0}} Z_{\text{out}} = -\left(\frac{-1}{0.364}\right) 0.130 = 0.358$$

b. Using Eq. 8.2.2, with $Z_{\text{in}} = 0$,

$$Z_{\text{out}} = r_{\text{out}} \tau \left(\frac{t_{\text{cr}}}{C_0}\right) \qquad Z_{\text{out}} \leq 0.13 \qquad \text{(c)}$$

Since the reaction is elementary, the rate expression is $r = kC_A C_B$. Using Eq. 8.2.8 to express the species concentrations, the reaction rate is

$$r_{\text{out}} = kC_0^2 \frac{(y_{A0} - Z_{\text{out}})(y_{B0} - Z_{\text{out}})}{(1 + \Delta_{\text{gas}} Z_{\text{out}})^2} \qquad \text{(d)}$$

Using Eq. 3.5.4, for second-order reactions, the characteristic reaction time is

$$t_{\text{cr}} = \frac{1}{kC_0} = 0.202 \, \text{min} \qquad \text{(e)}$$

Substituting (d) and (e) into (c), the design equation becomes

$$\tau = \frac{Z_{\text{out}}(1 - Z_{\text{out}})^2}{(0.364 - Z_{\text{out}})(0.482 - Z_{\text{out}})} \qquad Z_{\text{out}} \leq 0.13 \qquad \text{(f)}$$

Solving (f) for $Z_{out} = 0.130$, $\tau = 1.195$. Using Eq. 8.1.3 and (e), the reactor volume is

$$V_R = v_0 t_{cr} \tau = 965.2\,L$$

c. If the reactor volume is larger than 965.2 L, the extent is larger than 0.130, and a portion of product C is formed by reaction (a) and a portion by the following reaction:

$$A(g) + B(g) \longrightarrow C(liquid) \tag{g}$$

Assuming that the volume of product C in the liquid phase is negligible, for this reaction, $\Delta_{gas} = -2$. Using Eq. 2.7.10, the total gas-phase molar flow rate at the reactor outlet is now

$$(F_{tot})_{gas} = (F_{tot})_0[1 + (-1)0.130 + (-2)(Z - 0.130)] = 1.13 - 2Z \tag{h}$$

Using Eq. 8.1.11, the concentration of reactant A is now

$$C_A = C_0 \frac{y_{A_0} + s_j Z}{1.13 - 2Z} \tag{i}$$

Substituting (i) into the rate expression (d) and the latter into (c), for $Z > 0.130$ (or $\tau > 1.195$), the design equation is

$$Z_{out} = \tau \frac{(y_{A_0} - Z_{out})(y_{B_0} - Z_{out})}{(1.13 - 2Z)^2} \qquad \tau > 1.195 \tag{j}$$

Using Eq. 2.6.5, for $f_A = 0.85$,

$$Z_{out} = -\frac{y_{A_0}}{s_A} f_{A_{out}} = -\left(\frac{0.364}{-1}\right) 0.85 = 0.3094 \tag{k}$$

Substituting $Z_{out} = 0.3094$ into (j), we obtain

$$\tau = \frac{Z_{out}(1.13 - 2Z)^2}{(0.364 - Z_{out})(0.482 - Z_{out})} = 8.58 \tag{l}$$

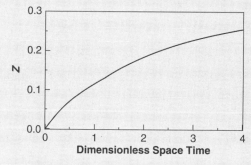

Figure E8.2.1   Reaction operating curve.

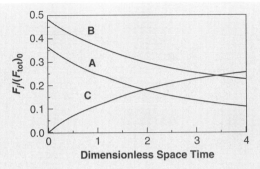

Figure E8.2.2 Species operating curves.

Using Eq. 8.1.3 and (e), the volume of the reactor is

$$V_R = v_0 t_{cr} \tau = 6,932 \text{ L}$$

d. The design of a CSTR is described by two design equations: (f) for $0 < Z_{out} < 0.130$ and (j) for $Z_{out} > 0.130$. Figure E8.2.1 shows the reaction curves, and Figure E8.2.2 shows the species curves.

Example 8.3 A biological waste, A, is decomposed in aqueous solution by an enzymatic reaction:

$$A \longrightarrow B + C$$

The rate expression of the reaction is

$$r = \frac{kC_A}{K_m + C_A}$$

An aqueous solution with $C_A = 2 \text{ mol/L}$ is fed into a CSTR at a rate of 200 L/min. For the enzyme type and concentration, $k = 0.1 \text{ mol/L min}$ and $K_m = 4 \text{ mol/L}$.
a. Derive and plot the reaction and species operating curves.
b. Determine the needed reactor volume to achieve 80% conversion.

Solution The stoichiometric coefficients of the reaction are

$$s_A = -1 \qquad s_B = 1 \qquad s_C = 1 \qquad \Delta = 1$$

We select the inlet stream as the reference stream; hence, $Z_{in} = 0$. Since reactant A is the only species fed to the reactor, $C_0 = C_{A_0}$, $y_{A_0} = 1$, $y_{B_0} = 1$, and $y_{C_0} = 1$.

Using Eq. 8.2.2, the design equation is

$$Z_{\text{out}} = \tau r_{\text{out}} \left(\frac{t_{\text{cr}}}{C_0} \right) \tag{a}$$

Using Eq. 8.2.7,

$$C_A = C_0 (1 - Z) \tag{b}$$

Substituting (b) in the rate expression and the latter in (a), the design equation is

$$Z_{\text{out}} = k \frac{1 - Z_{\text{out}}}{(K_m/C_0) + (1 - Z_{\text{out}})} \tau \left(\frac{t_{\text{cr}}}{C_0} \right) \tag{c}$$

Applying the procedure described in Section 3.5, the characteristic reaction time is

$$t_{\text{cr}} = \frac{C_0}{k} = 20 \, \text{min} \tag{d}$$

a. Substituting (d) into (c), for $K_m/C_0 = 2$, the design equation becomes

$$\tau = \frac{Z_{\text{out}}(3 - Z_{\text{out}})}{1 - Z_{\text{out}}} \tag{e}$$

Figure E8.3.1 shows the reaction curve. Using Eq. 2.7.8, the species curves are

$$\frac{F_{A_{\text{out}}}}{(F_{\text{tot}})_0} = 1 - Z_{\text{out}}$$

$$\frac{F_{B_{\text{out}}}}{(F_{\text{tot}})_0} = \frac{F_{C_{\text{out}}}}{(F_{\text{tot}})_0} = Z_{\text{out}}$$

Figure E8.3.2 shows the species curves.

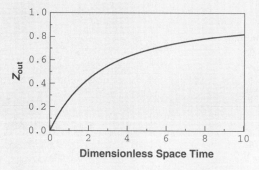

Figure E8.3.1 Reaction operating curve.

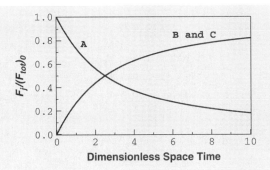

Figure E8.3.2   Species operating curves.

b. Using Eq. 2.6.5, for $f_{A_{out}} = 0.8$

$$Z_{out} = -\frac{y_A(0)}{s_A} f_{A_{out}} = -\frac{1}{-1}(0.80) = 0.80$$

From the reaction curve (or tabulated data), $Z_{out} = 0.80$ is reached at $\tau = 8.8$. Using Eq. 8.1.3 and (d), the needed reactor volume is

$$V_R = \frac{\tau v_0 C_0}{k} = 35{,}200\,\text{L}$$

8.2.2   Determination of the Reaction Rate Expression

Since the design equation contains the reaction rate at the outlet conditions, r_{out}, we can easily determine the rate expression from data obtained on a CSTR. Selecting the inlet stream as the reference stream, $Z_{in} = 0$, we write Eq. 8.2.2 as

$$r_{out} = \frac{1}{\tau}\left(\frac{C_0}{t_{cr}}\right) Z_{out} \tag{8.2.12}$$

Since t_{cr} is not known, we use Eq. 8.1.3 to express the design equation in terms of the reactor volume:

$$r_{out} = \frac{(F_{tot})_0}{V_R} Z_{out} \tag{8.2.13}$$

Hence, for known $(F_{tot})_0$ and V_R and for experimentally determined Z_{out}, we can calculate r_{out}. For nth-order reactions,

$$r = kC_A{}^n$$

Substituting in Eq. 8.2.13,

$$kC_{A_{out}}{}^n = \frac{(F_{tot})_0}{V_R} Z_{out}$$

To determine the reaction order, we take the logarithm of both sides and obtain

$$n \ln(C_{A_{\text{out}}}) = \ln\left(\frac{C_0 v_0}{kV_R} Z_{\text{out}}\right) \tag{8.2.14}$$

Modifying the term on the right-hand side, Eq. 8.2.14 reduces to

$$\ln(v_0 Z_{\text{out}}) = n \ln(C_{A_{\text{out}}}) + \ln\left(\frac{kV_R}{C_0}\right) \tag{8.2.15}$$

Thus, by operating a CSTR isothermally and measuring $C_{A_{\text{out}}}$ for different values of v_0, we can plot $\ln(v_0 Z_{\text{out}})$ versus $\ln(C_{A_{\text{out}}})$ and determine the reaction order from the slope of the line. Once the order is known, we can determine the value of the reaction rate constant from the design equation. We repeat this procedure at different reactor temperatures to determine the activation energy.

To apply Eq. 8.2.15, we have to relate Z_{out} to a measurable quantity. Usually, the species concentrations are measured at the reactor outlet and we use either Eq. 8.2.7 or Eq. 8.2.8. This is illustrated in Example 8.4.

Example 8.4   The gas-phase reaction

$$2\,A \longrightarrow R$$

is investigated on a CSTR. A stream of species A at 3 atm and 30°C (120 mmol/ L) is fed into a 1-L CSTR at different flow rates, and the exit concentration of reactant A is measured. From the data below, determine the order of the reaction and the value of the rate constant, k.

v_0 (L/min)	0.035	0.18	0.45	1.7
$C_{A_{\text{out}}}$ (mmol/L)	30	60	80	105

Solution   The stoichiometric coefficients of the chemical reaction are

$$s_A = -2 \qquad s_R = 1 \qquad \Delta = -1$$

The rate expression is

$$r = kC_A^{\alpha} \tag{a}$$

and we want to determine α and k. We select the inlet stream as the reference stream; hence, $Z_{\text{in}} = 0$, $y_{A_0} = 1$, and $y_{B_0} = 0$. To relate the extent to the exit concentration, we use Eq. 8.2.8:

$$C_{A_{\text{out}}} = C_0 \frac{1 - 2Z_{\text{out}}}{1 + \Delta Z_{\text{out}}} \tag{b}$$

Substituting $\Delta = -1$ into (b), and rearranging,

$$Z_{\text{out}} = \frac{C_0 - C_{A_{\text{out}}}}{2C_0 - C_{A_{\text{out}}}} \tag{c}$$

We calculate Z_{out} and then $(v_0 Z_{\text{out}})$ for each run. The table below shows the calculated values.

v_0 (L/min)	0.035	0.180	0.430	1.65
$C_{A_{\text{out}}}$ (mmol/L)	30	60	80	105
Z_{out}	0.429	0.333	0.250	0.111
$v_0 Z_{\text{out}}$ (L/min)	0.015	0.060	0.108	0.183

Using Eq. 8.2.18,

$$\ln(v_0 Z_{\text{out}}) = \alpha \ln C_{A_{\text{out}}} + \ln\left(\frac{kV_R}{C_0}\right) \tag{d}$$

We plot $\ln(v_0 Z_{\text{out}})$ versus $\ln(C_{A_{\text{out}}})$, shown in Figure E8.4.1, and the slope is α. From the figure,

$$\alpha = \frac{\ln(0.17) - \ln(0.0155)}{\ln(100) - \ln(30)} = 2.06 \approx 2$$

Now that we know that $\alpha = 2$, the characteristic reaction time is

$$t_{\text{cr}} = \frac{1}{kC_0} \tag{e}$$

Using Eq. 8.2.3 and (e), we calculate the dimensionless space time for each run:

$$\tau = Z_{\text{out}}\left(\frac{1 - Z_{\text{out}}}{1 - 2Z_{\text{out}}}\right)^2 \tag{f}$$

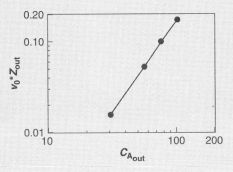

Figure E8.4.1 Determination of reaction order.

Using Eq. 8.1.3 and (e), (f) reduces to

$$kC_0 = \frac{v_0 Z_{\text{out}}}{V_R} \left(\frac{C_0}{C_{A_{\text{out}}}} \right)^2 \tag{g}$$

The average value of k for the four runs is 0.002 L/mmol min.

8.2.3   Cascade of CSTRs Connected in Series

Consider several CSTRs connected such that the effluent from one reactor is fed into the next reactor, as shown schematically in Figure 8.4. We select the inlet stream to the cascade as the reference stream and write Eq. 8.2.3, for *each* reactor. We can present the design equation graphically by plotting (r_0/r) versus Z, as shown schematically in Figure 8.5. Here, the dimensionless space time of each reactor is represented by a rectangle. Note that for a given outlet extent, the total volume of the cascade is smaller than the volume of a single CSTR. Also note that when a cascade of numerous small CSTRs is used, the rectangular areas approach the area under the curve. Hence, the total volume of the cascade converges to the volume of a plug-flow reactor.

Consider a liquid-phase, *first-order* reaction of the form A $\rightarrow$ P + R in an isothermal cascade of CSTRs where only reactant A is fed to the system. Taking the inlet stream to the cascade as the reference stream and since only reactant A is fed, $y_{A_0} = 1$. Using Eq. 8.2.3, the design equation for the nth CSTR is

$$\tau_n = \frac{Z_{n_{\text{out}}} - Z_{n_{\text{in}}}}{1 - Z_{n_{\text{out}}}} \tag{8.2.16}$$

where τ_n is the dimensionless space time of the nth reactor in the cascade, $\tau_n = V_{R_n}/(v_0 t_{\text{cr}})$. For the first reactor in the cascade,

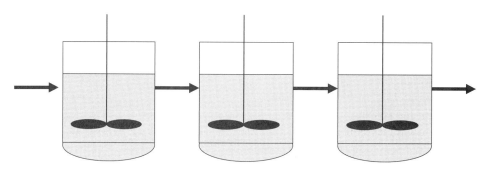

Figure 8.4   Cascade of CSTRs.

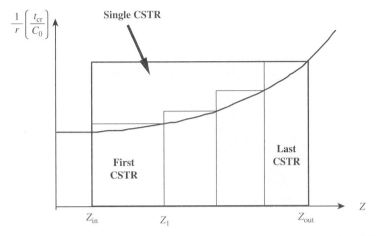

Figure 8.5 Graphical presentation of the design equation for a cascade of CSTRs.

$$\tau_1 = \frac{Z_1}{1 - Z_1} \quad \text{or} \quad Z_1 = \frac{\tau_1}{1 - \tau_1} \qquad (8.2.17)$$

For the second reactor in the cascade,

$$\tau_2 = \frac{Z_2 - Z_1}{1 - Z_2} \quad \text{or} \quad Z_2 = \frac{\tau_1 + \tau_2(1 + \tau_1)}{(1 - \tau_1)(1 - \tau_2)} \qquad (8.2.18)$$

Similarly, for the nth reactor in the cascade,

$$Z_n = \frac{\tau_1 + \tau_2(1 + \tau_1) + \cdots + \tau_n(1 + \tau_1) \cdots (1 + \tau_{n-1})}{(1 - \tau_1)(1 - \tau_2) \cdots (1 - \tau_n)} \qquad (8.2.19)$$

For the special case that each reactor has the same volume, $\tau_1 = \tau_2 = \cdots = \tau_n = \tau_{\text{tot}}/n$, Eq. 8.2.19 reduces to

$$Z_n = 1 - \frac{1}{[1 + (\tau_{\text{tot}}/n)]^n} \qquad (8.2.20)$$

In practice, the cascade usually consists of either equal-size reactors or a set of specified numbers of CSTRs in which the volume of each is selected such that total volume of the cascade, for a given outlet conversion, is the smallest. A cascade of equal-size CSTRs is represented in Figure 8.4 by rectangles whose areas are the same, and a cascade with the total smallest volume is represented by a given number of rectangles whose combined area is the smallest. Below, we consider the performance of a cascade of CSTRs when the reaction rate is given.

Example 8.5   The first-order, gas-phase reaction

$$A \longrightarrow B + C$$

is carried out in a cascade of CSTRs. A stream of reactant A is fed into the cascade at a rate of 10 mol/min. The feed concentration is $C_{A_0} = 0.04 \, \text{mol/L}$, and, at the reactor temperature, $k = 0.1 \, \text{s}^{-1}$. For outlet conversion of 80%, determine:

a. The volume of each reactor in a series of two equal-size CSTRs.
b. The volume of each reactor in a series of two optimized CSTRs.
c. The volume of each reactor in a series of three equal-size CSTRs.
d. The volume of each reactor in a series of three optimized CSTRs.
e. Compare the results in (a) and (b) to the values obtained in Example 7.1 and Example 8.1.

Solution   The stoichiometric coefficients of the reaction are

$$s_A = -1 \qquad s_B = 1 \qquad s_C = 1 \qquad \Delta = 1$$

We select the inlet stream to the cascade as the reference stream, and since only reactant A is fed, $C_0 = C_{A_0}$, $y_{A_0} = 1$, and $y_{B_0} = 0$. The volumetric flow rate of the reference stream is

$$v_0 = \frac{(F_{\text{tot}})_0}{C_0} = 250 \, \text{L/min}$$

Using Eq. 8.2.8 to express the species concentrations, the reaction rate is

$$r_{\text{out}} = kC_0 \frac{1 - Z_{\text{out}}}{1 + \Delta Z_{\text{out}}} \tag{a}$$

Using Eq. 3.5.4, the characteristic reaction time is

$$t_{\text{cr}} = \frac{1}{k} = 10 \, \text{s} = 0.167 \, \text{min} \tag{b}$$

Substituting (a) and (b) into Eq. 8.2.3, the design equation of each reactor is

$$\tau = \frac{(Z_{\text{out}} - Z_{\text{in}})(1 + Z_{\text{out}})}{1 - Z_{\text{out}}} \tag{c}$$

In a cascade of two CSTRs, for the first reactor,

$$\tau_1 = \frac{Z_1(1 + Z_1)}{1 - Z_1} \tag{d}$$

For the second reactor, $Z_{in} = Z_1$, $Z_{\text{out}} = Z_2$, and (c) reduces to

$$\tau_2 = \frac{(Z_2 - Z_1)(1 + Z_2)}{1 - Z_2} \tag{e}$$

a. For a cascade of equal-size reactors, $\tau_1 = \tau_2$, and we combine (d) and (e):

$$\frac{(Z_1)(1 + Z_1)}{1 - Z_1} = \frac{(Z_2 - Z_1)(1 + Z_2)}{1 - Z_2} \tag{f}$$

Substituting $Z_2 = 0.8$, we solve (f) and obtain $Z_1 = 0.569$. Now that we know Z_1 and Z_2, we can use either (d) or (e) to determine τ and obtain $\tau_1 = \tau_2 = 2.075$. We use Eq. 8.1.3 and (b) to calculate the volume of each CSTR,

$$V_R = \tau v_0 t_{cr} = 86.46 \text{ L} \tag{g}$$

The total volume of the cascade is $V_{R_1} + V_{R_2} = 172.92$ L.

b. To determine Z_1 for an optimized cascade of two CSTRs, we take the derivative of $\tau_1 + \tau_2$ with respect to Z_1 and equate it to zero:

$$\frac{d(\tau_1 + \tau_2)}{dZ_1} = 0 \tag{h}$$

Substituting (d) and (e), (h) reduces to

$$(1 + 2Z_1)(1 - Z_1) + Z_1(1 + Z_1) - \left(\frac{1 + Z_2}{1 - Z_2}\right)(1 - Z_1)^2 = 0 \tag{i}$$

Substituting $Z_2 = 0.8$, we solve (i) and obtain $Z_1 = 0.5528$. Now that we know the values of Z_1 and Z_2, we use (d) and (e) to calculate the τ of each reactor. We find that $\tau_1 = 1.919$ and $\tau_2 = 2.225$. Using Eq. 8.3.1 and (b), the volume of the two reactors are $V_{R_1} = 79.96$ L and $V_{R_2} = 92.71$ L, and the total volume of the cascade is 172.67 L.

c. In a cascade of three equal-size CSTRs, we write the design equation for each reactor. For the first reactor,

$$\tau_1 = \frac{Z_1(1 + Z_1)}{1 - Z_1} \tag{j}$$

For the second reactor,

$$\tau_2 = \frac{(Z_2 - Z_1)(1 + Z_2)}{1 - Z_2} \tag{k}$$

For the third reactor,

$$\tau_3 = \frac{(Z_3 - Z_2)(1 + Z_3)}{1 - Z_3} \tag{l}$$

For a cascade of equal-size reactors, $\tau_1 = \tau_2$ and $\tau_2 = \tau_3$. We combine (j), (k), and (l) and obtain two equations with two unknowns, Z_1 and Z_2:

$$\frac{Z_1(1 + Z_1)}{1 - Z_1} = \frac{(Z_3 - Z_2)(1 + Z_3)}{1 - Z_3} \tag{m}$$

$$\frac{(Z_2 - Z_1)(1 + Z_2)}{1 - Z_2} = \frac{(Z_3 - Z_2)(1 + Z_3)}{1 - Z_3} \tag{n}$$

Substituting $Z_3 = 0.8$, we solve (m) and (n) numerically and obtain $Z_1 = 0.4445$ and $Z_2 = 0.6716$. Now that we know the values of Z_1 and Z_2, we use (j), (k), or (l) to calculate the dimensionless space time of each reactor. We obtain $\tau_1 = \tau_2 = \tau_3 = 1.156$. To calculate the volume of each reactor, we use Eq. 8.1.3 and (b):

$$V_R = \frac{\tau v_0}{k} = 48.15 \text{ L}$$

The total volume of the cascade is 144.45 L.

d. To determine Z_1 and Z_2 for an optimized cascade of three CSTRs, we take the derivative of $(\tau_1 + \tau_2 + \tau_3)$ with respect to Z_1 and Z_2 and equate them to zero:

$$\frac{\partial(\tau_1 + \tau_2 + \tau_3)}{\partial Z_1} = 0 \qquad \frac{\partial(\tau_1 + \tau_2 + \tau_3)}{\partial Z_2} = 0 \tag{o}$$

Substituting (j), (k), and (l) and taking the derivatives, the two equations in (o) become

$$(1 + 2Z_1)(1 - Z_1) + Z_1(1 + Z_1) - \left(\frac{1 + Z_2}{1 - Z_2}\right)(1 - Z_1)^2 = 0 \tag{p}$$

$$(1 + 2Z_2 - Z_1)(1 - Z_2) + (Z_2 - Z_1)(1 + Z_2) - \left(\frac{1 + Z_3}{1 - Z_3}\right)(1 - Z_2)^2 = 0 \tag{q}$$

We substitute $Z_3 = 0.8$, solve (p) and (q) numerically, and obtain $Z_1 = 0.4152$ and $Z_2 = 0.6580$. Now that we know the values of Z_1 and Z_2, we use (j), (k), and (l) to calculate the dimensionless space time of each reactor. We find that $\tau_1 = 1.005$, $\tau_2 = 1.177$, and $\tau_3 = 1.278$. Using (g), the volumes of the three reactors are $V_{R_1} = 41.88$ L, $V_{R_2} = 49.04$ L, and $V_{R_3} = 53.25$ L. The total volume of the cascade is 144.17 L.

Figure E8.5.1 shows the reaction curve for a cascade of equal-size tanks, and compare them to those of a plug-flow reactor and a single CSTR.

e. The total volume of the cascade for the different cases is summarized below:

A single CSTR (from Example 8.1)	300.00 L
A cascade of two equal-size CSTRs	172.92 L
An optimized cascade of two CSTRs	172.67 L
A cascade of three equal-size CSTRs	144.45 L
An optimized cascade of three CSTRs	144.17 L
A plug-flow reactor (from Example 7.1)	100.08 L

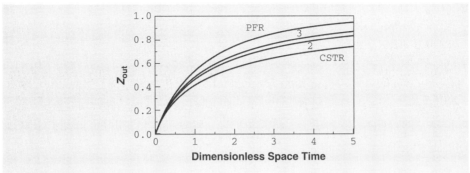

Figure E8.5.1   Comparison of reaction curves for cascade of equal-size tanks.

8.3 ISOTHERMAL OPERATIONS WITH MULTIPLE REACTIONS

When more than one chemical reaction takes place in the reactor, we have to determine how many independent reactions there are (and how many design equations are needed) and select a set of independent reactions. Next, we have to identify all the reactions that actually take place (including dependent reactions) and express their rates. We write Eq. 8.1.1 for *each* independent chemical reaction. To solve the design equations (obtain relationships between $Z_{m_{\text{out}}}$'s and τ), we express the rates of the individual chemical reactions in terms of the $Z_{m_{\text{out}}}$'s and τ. Since the temperature is constant, the energy balance equation is used to determine the heating load. The procedure for designing isothermal CSTRs with multiple reactions goes as follows:

1. Identify all the chemical reactions that take place in the reactor and define the stoichiometric coefficients of each species in each reaction.
2. Determine the number of independent chemical reactions.
3. Select a set of independent reactions among the reactions whose rate expressions are given.
4. For each *dependent* reaction, determine its α_{km} multipliers with each of the independent reactions, using Eq. 2.4.9.
5. Select a reference stream [determine $(F_{\text{tot}})_0$, C_0, v_0] and the reference species compositions, y_{j_0}'s.
6. Write Eq. 8.1.1 for each independent chemical reaction.
7. Select a leading (or desirable) reaction and determine the expression form and value of its characteristic reaction time, t_{cr}.
8. Express the reaction rates in terms of the dimensionless extents of the independent reactions, $Z_{m_{\text{out}}}$'s.
9. Specify the inlet conditions ($Z_{m_{\text{in}}}$'s).
10. Solve the design equations (determine $Z_{m_{\text{out}}}$'s as functions of τ) and obtain the reaction operating curves.

11. Determine the species operating curves, using Eq. 2.7.8.
12. Determine the reactor volume based on the most desirable value of τ, using Eq. 8.1.3.

Below, we describe the design formulation of isothermal CSTRs with multiple reactions for various types of chemical reactions (reversible, series, parallel, etc.). In most cases, we solve the equations numerically using mathematical software. In some simple cases, we obtain analytical solutions.

Example 8.6   The reversible, gas-phase chemical reaction

$$A \rightleftharpoons 2B$$

takes place in a CSTR operated at 2 atm and 120°C. The forward reaction is first order, and the backward reaction is second order. We want to process a 100-mol/min stream of pure A and achieve a level of 80% of the equilibrium conversion. At the operating conditions (120°C), $k_1 = 0.1\,\mathrm{min}^{-1}$ and $k_2 = 0.322\,\mathrm{L/mol\,min}^{-1}$.

a. Derive the design equation and plot the dimensionless reaction operating curve.

b. What is the equilibrium composition at 120°C in a CSTR for $k_2 C_0/k_1 = 0.5$?

c. What is the required reactor volume needed to reach 80% of the equilibrium conversion?

Solution   We treat a reversible reaction as two separate reactions; a forward reaction and a reverse reaction. But there is only one independent reaction. We select the forward reaction (Reaction 1) as the independent reaction and the reverse reaction as the dependent reaction. Hence, the index of the independent reaction is $m = 1$, the index of the dependent reaction is $k = 2$, and, $\alpha_{21} = -1$. The stoichiometric coefficients of the independent reaction are

$$s_{A_1} = -1 \qquad s_{B_1} = 2 \qquad \Delta_1 = 1$$

We select the inlet stream as the reference stream; hence $Z_{1_{in}} = 0$, and since only reactant A is fed into the reactor, $y_{A_0} = 1$ and $y_{B_0} = 0$. The reference concentration is

$$C_0 = \frac{P_0}{RT_0} = 0.0602\,\mathrm{mol/L} \tag{a}$$

Using Eq. 8.1.4, the volumetric flow rate of the reference stream is

$$v_0 = \frac{(F_{tot})_0}{C_0} = 1612\,\mathrm{L/min} \tag{b}$$

We write Eq. 8.1.1 for the independent reaction:

$$Z_{1_{out}} - Z_{1_{in}} = (r_{1_{out}} - r_{2_{out}})\tau\left(\frac{t_{cr}}{C_0}\right) \tag{c}$$

We use Eq. 8.1.13 to express the concentrations of the two species, and the rates of the two reactions are

$$r_1 = k_1 C_0 \frac{y_{A0} - Z_1}{1 + Z_1} \qquad r_2 = k_2 C_0^{\,2}\left(\frac{y_{b0} + 2Z_1}{1 + Z_1}\right)^2 \tag{d}$$

We select Reaction 1 as the leading reaction and using Eq. 3.5.4, define the characteristic reaction time as

$$t_{cr} = \frac{1}{k_1} = 10\,\text{min} \tag{e}$$

Substituting (d) and (e) into (c), the design equation reduces to

$$Z_{1_{out}} - \left[\frac{1 - Z_{1_{out}}}{1 + Z_{1_{out}}} - \left(\frac{k_2 C_0}{k_1}\right)\left(\frac{2Z_{1_{out}}}{1 + Z_{1_{out}}}\right)^2\right]\tau = 0 \tag{f}$$

We solve (f) numerically for different values of τ using a mathematical software. Figure E8.6.1 shows the reaction curve for various values of $k_2 C_0 / k_1$. Note that the curve for $k_2 C_0 / k_1 = 0$ represents the solution of the irreversible reaction. Once we have the reaction operating curve, we use Eq. 2.7.8 to obtain the species curves

$$\frac{F_{A_{out}}}{(F_{tot})_0} = y_{A_0} - Z_{out}$$

$$\frac{F_{B_{out}}}{(F_{tot})_0} = y_{B_0} + 2Z_{out}$$

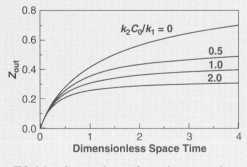

Figure E8.6.1 Comparison of reaction operating curves.

b. At equilibrium, $r_1 = r_2$, and we obtain

$$Z_{1_{eq}} = \left(1 + 4\frac{k_2 C_0}{k_1}\right)^{-0.5} \tag{g}$$

For $k_2 C_0 / k_1 = 0.5$, $Z_{1_{eq}} = 0.577$, and the mole fractions of the species are

$$y_{A_{eq}} = \frac{y_{A_0} - Z_{1_{eq}}}{1 + \Delta_1 Z_{1_{eq}}} = 0.268$$

$$y_{B_{eq}} = \frac{y_{B_0} + 2Z_{1_{eq}}}{1 + \Delta_1 Z_{1_{eq}}} = 0.732 \tag{h}$$

c. The extent for 80% of the equilibrium extent is 0.4616. From the operating curve for $k_2 C_0 / k_1 = 0.5$, an extent of 0.4616 is reached at $\tau = 2.74$. Using Eq. 8.1.3 and (e), the required reactor volume is

$$V_R = v_0 t_{cr} \tau = 2740 \text{ L}$$

Example 8.7   An organic solution containing reactant A ($C_A = 2.0\,\text{mol/L}$) is fed into a 1200-L CSTR, where the following chemical reactions take place:

$$\text{Reaction 1:} \quad 2A \longrightarrow B$$
$$\text{Reaction 2:} \quad B \longrightarrow C + 2D$$

Reaction 1 is a second-order reaction, and Reaction 2 is first order. Product B is the desired product. The feed rate is 100 L/min, and, at the operating temperature, $k_1 = 10\,\text{L mol}^{-1}\,\text{h}^{-1}$ and $k_2 = 4\,\text{h}^{-1}$.

a. Derive the design equations and plot the reaction and species operating curves.

b. Determine the conversion of reactant A, the yield of product B, and the production rates of products B and C for the given feed flow rate.

c. Determine the feed flow rate that provides the highest production rate of product B.

d. Repeat (b) for the optimal feed rate.

Solution   The stoichiometric coefficients of the chemical reactions are

$$
\begin{array}{lllll}
s_{A_1} = -2 & s_{B_1} = 1 & s_{C_1} = 0 & s_{D_1} = 0 & \Delta_1 = -1 \\
s_{A_2} = 0 & s_{B_2} = -1 & s_{C_2} = 1 & s_{D_2} = 2 & \Delta_2 = 2
\end{array}
$$

The two reactions are independent, and there are no dependent reactions. We select the inlet steam as the reference stream; hence, $Z_{1_{in}} = Z_{2_{in}} = 0$. Since only reactant A is fed into the reactor, $C_0 = C_{A_{in}} = 2\,\text{mol/L}$, $y_{A_0} = 1$, and

$y_{B_0} = y_{C_0} = y_{D_0} = 0$. The molar flow rate of the reference stream is

$$(F_{tot})_0 = v_0 C_0 = 200 \, \text{mol/min}$$

a. We write Eq. 8.1.1 for each independent reaction:

$$Z_{1_{out}} = r_{1_{out}} \tau \left(\frac{t_{cr}}{C_0} \right) \tag{a}$$

$$Z_{2_{out}} = r_{2_{out}} \tau \left(\frac{t_{cr}}{C_0} \right) \tag{b}$$

We use Eq. 8.1.9 to express the species concentrations, and the reaction rates are

$$r_{1_{out}} = k_1 C_0^2 (1 - 2Z_{1_{out}})^2 \tag{c}$$

$$r_{2_{out}} = k_2 C_0 (Z_{1_{out}} - Z_{2_{out}}) \tag{d}$$

We select Reaction 1 as the leading reaction and using Eq. 3.5.4, define the characteristic reaction time by

$$t_{cr} = \frac{1}{k_1 C_0} = 3 \, \text{min} \tag{e}$$

Substituting (c), (d), and (e) into (a) and (b), the design equations become

$$Z_{1_{out}} - (1 - 2Z_{1_{out}})^2 \tau = 0 \tag{f}$$

$$Z_{2_{out}} - \left(\frac{k_2}{k_1 C_0} \right) (Z_{1_{out}} - Z_{2_{out}}) \tau = 0 \tag{g}$$

Once we solve (f) and (g), we can determine the species operating curves using Eq. 2.7.8:

$$\frac{F_{A_{out}}}{(F_{tot})_0} = 1 - 2Z_{1_{out}} \tag{h}$$

$$\frac{F_{B_{out}}}{(F_{tot})_0} = Z_{1_{out}} - Z_{2_{out}} \tag{i}$$

$$\frac{F_{C_{out}}}{(F_{tot})_0} = Z_{2_{out}} \tag{j}$$

$$\frac{F_{D_{out}}}{(F_{tot})_0} = 2Z_{2_{out}} \tag{k}$$

In this case we can solve (f) and (g) analytically,

$$Z_{1_{out}} = \frac{(1 + 4\tau) - \sqrt{1 + 8\tau}}{8\tau} \tag{l}$$

$$Z_{2_{out}} = \left(\frac{0.2\tau}{1 + 0.2\tau} \right) \frac{(1 + 4\tau) - \sqrt{1 + 8\tau}}{8\tau} \qquad (m)$$

Figure E8.7.1 shows the two reaction curves. We substitute (l) and (m) into (h) through (k) and obtain the species curves shown in Figure E8.7.2.

b. Using Eq. 8.1.3 and (e), for the given feed volumetric rate,

$$\tau = \frac{V_R}{v_0 t_{cr}} = 4$$

Substituting $\tau = 4$ in (f) and (g), $Z_{1_{out}} = 0.3517$ and $Z_{2_{out}} = 0.1563$. Using (h) through (k), the species flow rates at the reactor outlet are $F_{B_{out}} = 39.1 \, mol/min$, $F_{B_{out}} = 39.1 \, mol/min$, $F_{C_{out}} = 31.3 \, mol/min$, and $F_{D_{out}} = 62.5 \, mol/min$.

The conversion of reactant A is

$$f_{A_{out}} = \frac{F_{A_0} - F_{A_{out}}}{F_{A_0}} = 0.704$$

Using Eq. 2.6.14, the yield of product B is

$$\eta_{B_{out}} = -\left(\frac{-2}{1} \right) \frac{F_{B_{out}} - F_{B_{in}}}{F_{A_0}} = 0.391$$

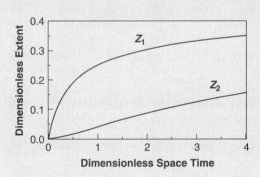

Figure E8.7.1   Reaction operating curves.

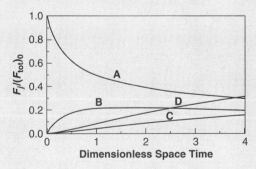

Figure E8.7.2   Species operating curves.

c. From the curve of product B, the highest $F_{B_{out}}$ is reached at $\tau = 1.75$, and using (f) and (g), $Z_{1_{out}} = 0.295$ and $Z_{2_{out}} = 0.0764$. Using Eq. 8.1.3 and (e), for $\tau = 1.75$, the volumetric feed flow rate is

$$v_0 = \frac{V_R}{\tau t_{cr}} = 228.6 \, \text{L/min}$$

The optimal feed molar flow rate is

$$(F_{tot})_0 = v_0 C_0 = 457.1 \, \text{mol/min}$$

d. Using (h) through (k), the species flow rates at the reactor outlet are $F_{A_{out}} = 187.6 \, \text{mol/min}$, $F_{B_{out}} = 99.3 \, \text{mol/min}$, $F_{C_{out}} = 34.9 \, \text{mol/min}$, $F_{D_{out}} = 69.8 \, \text{mol/min}$. The conversion of reactant A is

$$f_{A_{out}} = \frac{F_{A_{in}} - F_{A_{out}}}{F_{A_{in}}} = 0.590$$

The yield of product B is

$$\eta_{B_{out}} = -\left(\frac{-2}{1}\right) \frac{F_{B_{out}} - F_{B_{in}}}{F_{A_{in}}} = 0.435$$

The following table provides a comparison between the given and the optimal operations:

	Given Operation	Optimal Operation
Reactor volume (L)	1200	1200
Feed flow rate (L/min)	100	229.9
Conversion of reactant A	0.704	0.590
Yield of product B	0.391	0.435
Flow rate of product B (mol/min)	39.1	99.3
Flow rate of product C (mol/min)	31.3	34.9
Flow rate of product D (mol/min)	62.5	69.8

Example 8.8 Product B is produced in a CSTR where the following gas-phase, first-order chemical reactions take place:

$$\text{Reaction 1:} \quad A \longrightarrow 2B$$
$$\text{Reaction 2:} \quad B \longrightarrow C + D$$

A stream of reactant A ($C_{A_{in}} = 0.04 \, \text{mol/L}$) is fed into a 200-L CSTR at a rate of $100 \, \text{L/min}$. At the reactor operating temperature, $k_1 = 2 \, \text{min}^{-1}$ and $k_2 = 1 \, \text{min}^{-1}$.

a. Derive the design equations and plot the reaction and species curves.

b. Determine the conversion of reactant A, the yield of product B, and the production rate of products B and C for the given feed rate.

c. Determine the reactor volume that provides the highest production rate of product B.

d. Determine the conversion of reactant A and the yield of product B for the optimal reactor.

Solution   This is an example of series (consecutive) chemical reactions. The stoichiometric coefficients of the two reactions are

$$s_{A_1} = -1 \qquad s_{B_1} = 2 \qquad s_{C_1} = 0 \qquad s_{D_1} = 0 \qquad \Delta_1 = 1$$
$$s_{A_2} = 0 \qquad s_{B_2} = -1 \qquad s_{C_2} = 1 \qquad s_{D_2} = 1 \qquad \Delta_2 = 1$$

Since each reaction has a species that does not participate in the other, the two reactions are independent, and there is no dependent reaction. We select the inlet stream as the reference stream; hence, $Z_{1_{in}} = Z_{2_{in}} = 0$. Since only reactant A is fed into the reactor, $C_0 = C_{A_{in}} = 0.04 \, \text{mol/L}$, $y_{A_0} = 1$, and $y_{B_0} = y_{C_0} = y_{D_0} = 0$. The molar flow rate of the reference stream is

$$(F_{tot})_0 = v_0 C_0 = 4 \, \text{mol/min}$$

We write Eq. 8.1.1 for each independent reaction,

$$Z_{1_{out}} = r_{1_{out}} \tau \left(\frac{t_{cr}}{C_0} \right) \tag{a}$$

$$Z_{2_{out}} = r_{2_{out}} \tau \left(\frac{t_{cr}}{C_0} \right) \tag{b}$$

a. We use Eq. 8.1.13 to express the species concentrations, and the reaction rates are

$$r_{1_{out}} = k_1 C_0 \frac{1 - Z_{1_{out}}}{1 + Z_{1_{out}} + Z_{2_{out}}} \tag{c}$$

$$r_{2_{out}} = k_2 C_0 \frac{2 Z_{1_{out}} - Z_{2_{out}}}{1 + Z_{1_{out}} + Z_{2_{out}}} \tag{d}$$

We define the characteristic reaction time on the basis of Reaction 1; hence,

$$t_{cr} = \frac{1}{k_1} = 0.5 \, \text{min} \tag{e}$$

Substituting (c), (d), and (e) into (a) and (b), the design equations reduce to

$$Z_{1_{out}} - \left(\frac{1 - Z_{1_{out}}}{1 + Z_{1_{out}} + Z_{2_{out}}} \right) \tau = 0 \tag{f}$$

$$Z_{2_{out}} - \left(\frac{k_2}{k_1}\right)\left(\frac{2Z_{1_{out}} - Z_{2_{out}}}{1 + Z_{1_{out}} + Z_{2_{out}}}\right)\tau = 0 \qquad \text{(g)}$$

We solve (f) and (g) numerically for different values of τ. The reaction curves are shown in Figure E8.8.1. Once we have Z_1 and Z_2 as a function of τ, we use Eq. 2.7.8 to determine the species curves:

$$\frac{F_{A_{out}}}{(F_{tot})_0} = 1 - Z_{1_{out}} \qquad \text{(h)}$$

$$\frac{F_{B_{out}}}{(F_{tot})_0} = 2Z_{1_{out}} - Z_{2_{out}} \qquad \text{(i)}$$

$$\frac{F_{C_{out}}}{(F_{tot})_0} = \frac{F_{D_{out}}}{(F_{tot})_0} = Z_{2_{out}} \qquad \text{(j)}$$

Figure E8.8.2 shows the species curves.

b. For the given feed flow rate, using Eq. 8.1.3 and (e), the dimensionless space time of the current operation is

$$\tau = \frac{V_R}{v_0 t_{cr}} = 4$$

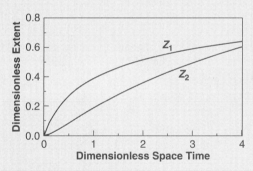

Figure E8.8.1 Reaction operating curves.

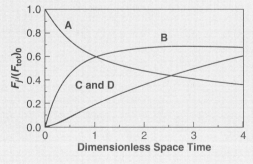

Figure E8.8.2 Species operating curves.

At $\tau = 4$, the solutions of (f) and (g) are $Z_{1_{out}} = 0.6406$ and $Z_{2_{out}} = 0.6037$. Using (h) through (j), the species molar flow rates at the reactor outlet are $F_{A_{out}} = 1.438\,\text{mol/min}$, $F_{B_{out}} = 2.71\,\text{mol/min}$, and $F_{C_{out}} = F_{D_{out}} = 2.42\,\text{mol/min}$. The conversion of reactant A is

$$f_{A_{out}} \equiv \frac{F_{A_{in}} - F_{A_{out}}}{F_{A_{in}}} = 0.641$$

Using Eq. 2.6.14, the yield of product B is

$$\eta_{B_{out}} = -\left(\frac{-1}{2}\right) \frac{F_{B_{out}} - F_{B_{in}}}{F_{A_{in}}} = 0.339$$

c. From the curve of product B, the highest $F_{B_{out}}$ is reached at $\tau = 3.0$. Using Eq. 8.1.3 and (e), the optimal reactor volume is

$$V_R = v_0 t_{cr} \tau = (100\,\text{L/min})(0.5\,\text{min})(3.0) = 150\,\text{L}$$

d. At $\tau = 3$, the solutions of (f) and (g) are $Z_{1_{out}} = 0.5901$ and $Z_{2_{out}} = 0.4939$. Using (h) through (j), the species molar flow rates at the reactor outlet are $F_{A_{out}} = 1.64\,\text{mol/min}$, $F_{B_{out}} = 2.75\,\text{mol/min}$, and $F_{C_{out}} = F_{D_{out}} = 1.98\,\text{mol/min}$. The conversion of reactant A is

$$f_{A_{out}} = \frac{F_{A_{in}} - F_{A_{out}}}{F_{A_{in}}} = 0.59$$

The yield of product B is

$$\eta_{B_{out}} = -\left(\frac{-1}{2}\right) \frac{F_{B_{out}} - F_{B_{in}}}{F_{A_{in}}} = 0.343$$

The following table provides a comparison between the given and the optimal reactors:

	Given Reactor	Optimal Reactor
Reactor volume (L)	200	150
Volumetric feed rate (L/min)	100	100
Conversion of reactant A	0.641	0.590
Flow rate of product B (mol/min)	2.71	2.75
Flow rate of product C (mol/min)	2.41	1.98

We see that the volume of the optimal reactor is 25% smaller than the given reactor. While processing the same feed rate, the conversion of reactant A is slightly lower, the production rate of product B is slightly higher, and the production rate of products C and D is reduced by about 20%.

Example 8.9 A chemical plant generates a stream of reactant A and a stream of reactant B that presently are being discarded. The engineering department suggested to use an available 2000-L CSTR to produce valuable product C, by combining these streams. The reactor operates at 184°C and 1.5 atm. At these conditions, the following gas-phase reactions take place:

$$\text{Reactions 1 \& 2:} \quad A + B \rightleftarrows C$$
$$\text{Reaction 3:} \quad C + B \longrightarrow D$$

The reaction rates are $r_1 = k_1 C_A C_B$; $r_2 = k_2 C_C$; and $r_3 = k_3 C_A C_B$. The available pumping equipment in the plant can provide a maximum feed rate of 800 L/min.
a. Determine the proportion of reactants A and B in the feed to maximize the production of product C.
b. Determine the rates reactant A and reactant B are fed to the reactor.
c. Determine the flow rates of all species at the reactor outlet (at optimal feed composition).
d. Determine the conversions of reactants A and B and the yield of product C.

Data: At 184 °C, $k_1 = 20\,\text{L/mol min}^{-1}$, $k_2 = 0.16\,\text{min}^{-1}$, $k_3 = \text{L/mol min}^{-1}$

Solution We have here two independent reactions and one dependent reaction. We select Reactions 1 and 3 as the independent reactions and Reaction 2 as the dependent reaction; hence, $m = 1, 3$, and $k = 2$. We use Eq. 2.4.9 and obtain $\alpha_{21} = -1, \alpha_{23} = 0$. The stoichiometric coefficients of the two independent reactions are

$$\begin{array}{ccccc} s_{A_1} = -1 & s_{B_1} = -1 & s_{C_1} = 1 & s_{D_1} = 0 & \Delta_1 = -1 \\ s_{A_3} = 0 & s_{B_3} = -1 & s_{C_3} = -1 & s_{D_3} = 1 & \Delta_3 = -1 \end{array}$$

We select the inlet stream as the reference stream and denote y_{A_0} and y_{B_0} to be the mole fractions of the two reactants in the feed stream ($y_{C_0} = y_{D_0} = 0$). The concentration and molar flow rate of the reference stream are

$$C_0 = \frac{P}{RT} = 0.04\,\text{mol/L}$$

$$(F_{\text{tot}})_0 = v_0 C_0 = 32\,\text{mol/min}$$

We select Reaction 1 as the leading reaction, and using Eq. 3.5.4, the characteristic reaction time is

$$t_{\text{cr}} = \frac{1}{k_1 C_0} \tag{a}$$

For the given reactor,

$$\tau = \frac{V_R}{v_0} k_1 C_0 = 2 \tag{b}$$

We write Eq. 8.1.1, for each independent reaction,

$$Z_{1_{out}} = (r_{1_{out}} - r_{2_{out}})\tau \left(\frac{t_{cr}}{C_0}\right) \tag{c}$$

$$Z_{3_{out}} = r_{3_{out}}\tau \left(\frac{t_{cr}}{C_0}\right) \tag{d}$$

Using Eq. 8.1.13 to express the species concentrations, the rates of the reactions are

$$r_{1_{out}} = k_1 C_0{}^2 \frac{(y_{A_0} - Z_{1_{out}})(y_{B_0} - Z_{1_{out}} - Z_{3_{out}})}{(1 - Z_{1_{out}} - Z_{3_{out}})^2} \tag{e}$$

$$r_{2_{out}} = k_2 C_0 \frac{Z_{1_{out}} - Z_{3_{out}}}{1 - Z_{1_{out}} - Z_{3_{out}}} \tag{f}$$

$$r_{3_{out}} = k_3 C_0{}^2 \frac{(Z_{1_{out}} - Z_{3_{out}})(y_{B_0} - Z_{1_{out}} - Z_{3_{out}})}{(1 - Z_{1_{out}} - Z_{3_{out}})^2} \tag{g}$$

Substituting (e), (f), (g), and (a) into (c) and (d), the design equations become

$$Z_{1_{out}} = \left[\frac{(y_{A_0} - Z_{1_{out}})(y_{B_0} - Z_{1_{out}} - Z_{3_{out}})}{(1 - Z_{1_{out}} - Z_{3_{out}})^2} - \left(\frac{k_2}{k_1 C_0}\right)\frac{Z_{1_{out}} - Z_{3_{out}}}{1 - Z_{1_{out}} - Z_{3_{out}}}\right]\tau \tag{h}$$

$$Z_{3_{out}} = \left(\frac{k_3}{k_1}\right)\frac{(Z_{1_{out}} - Z_{3_{out}})(y_{B_0} - Z_{1_{out}} - Z_{3_{out}})}{(1 - Z_{1_{out}} - Z_{3_{out}})^2}\tau \tag{i}$$

Using the given data,

$$\left(\frac{k_2}{k_1 C_0}\right) = 0.2 \qquad \left(\frac{k_3}{k_1}\right) = 2$$

Substituting these values and noting that $y_{B_0} = 1 - y_{A_0}$, the design equations reduce to

$$Z_{1_{out}} - 2\left[\frac{(y_{A_0} - Z_{1_{out}})(1 - y_{A_0} - Z_{1_{out}} - Z_{3_{out}})}{(1 - Z_{1_{out}} - Z_{3_{out}})^2} - 0.2\frac{Z_{1_{out}} - Z_{3_{out}}}{1 - Z_{1_{out}} - Z_{3_{out}}}\right] = 0 \tag{j}$$

$$Z_{3_{out}} - 4\frac{(Z_{1_{out}} - Z_{3_{out}})(1 - y_{A_0} - Z_{1_{out}} - Z_{3_{out}})}{(1 - Z_{1_{out}} - Z_{3_{out}})^2} = 0 \tag{k}$$

Once we solve (j) and (k), we use Eq. 2.7.8 to obtain the species operating curve:

$$\frac{F_{A_{out}}}{(F_{tot})_0} = y_{A_0} - Z_{1_{out}} \tag{l}$$

$$\frac{F_{B_{out}}}{(F_{tot})_0} = y_{B_0} - Z_{1_{out}} - Z_{3_{out}} \tag{m}$$

$$\frac{F_{C_{out}}}{(F_{tot})_0} = Z_{1_{out}} - Z_{3_{out}} \tag{n}$$

$$\frac{F_{D_{out}}}{(F_{tot})_0} = Z_{3_{out}} \tag{o}$$

a. To determine the optimal feed composition, we solve (j) and (k) for different values of $y_{A_{in}}$ and calculate $F_{C_{out}}/(F_{tot})_0$ by (o). The following table provides the results of these calculations:

$y_{A_{in}}$	$Z_{1_{out}}$	$Z_{3_{out}}$	$F_{C_{out}}/(F_{tot})_0$
0.50	0.1956	0.1197	0.0759
0.55	0.1856	0.1039	0.0817
0.60	0.1726	0.0873	0.0853
0.65	0.1571	0.0708	0.0864
0.70	0.1398	0.0550	0.0847
0.75	0.1207	0.0405	0.0802
0.80	0.1000	0.0276	0.0724

The Figure E8.9.1 shows the plot of $F_{C_{out}}/(F_{tot})_0$ versus $y_{A_{in}}$. From the graph, maximum production rate of product C is achieved for $y_{A_{in}} = 0.65$.

b. The optimal feed rates of reactants A and B are

$$F_{A_{in}} = y_{A_{in}}(F_{tot})_0 = 20.8 \, mol/min$$

$$F_{B_{in}} = y_{B_{in}}(F_{tot})_0 = 11.2 \, mol/min$$

c. For the optimal reactant proportion, the solution of (j) and (k) is $Z_1 = 0.157$ and $Z_3 = 0.0708$. From (h) through (k), the flow rates of the individual species at the reactor outlet are $F_{A_{out}} = 15.8 \, mol/min$, $F_{B_{out}} = 3.91 \, mol/min$, $F_{C_{out}} = 2.76 \, mol/min$, and $F_{D_{out}} = 2.27 \, mol/min$.

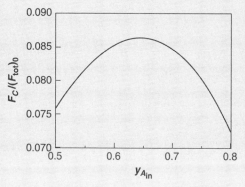

Figure E8.9.1 Optimal reactant proportion.

d. The conversions of reactants A and B are

$$f_{A_{out}} = \frac{F_{A_{in}} - F_{A_{out}}}{F_{A_{in}}} = 0.24$$

$$f_{B_{out}} = \frac{F_{B_{in}} - F_{B_{out}}}{F_{B_{in}}} = 0.651$$

Using Eq. 2.6.14, the yield of product C is

$$\eta_{B_{out}} = -\left(\frac{-1}{1}\right)\frac{F_{C_{out}} - F_{C_{in}}}{F_{A_{in}}} = 0.132$$

Example 8.10   The following reversible reactions (mechanism) were proposed for gas-phase cracking of hydrocarbons:

$$\begin{array}{ll} \text{Reactions 1 \& 2:} & \text{A} \rightleftharpoons \text{2B} \\ \text{Reactions 3 \& 4:} & \text{A} + \text{B} \rightleftharpoons \text{C} \\ \text{Reactions 5 \& 6:} & \text{A} + \text{C} \rightleftharpoons \text{D} \end{array}$$

The desired product is B. We want to design an isothermal CSTR to be operated at 489°C and 5 atm. A stream of reactant A at a rate of 1 mol/s is available in the plant.

a. Derive the design equations and plot the reaction and species operating curves.

b. Determine the volume of the reactor that provides the highest production rate of product B.

c. Determine the equilibrium composition.

d. Repeat parts (a) for a feed stream that consists of species D.

The rate expressions of the chemical reactions are:

$$r_1 = k_1 C_A \qquad r_2 = k_2 C_B^2 \qquad r_3 = k_3 C_A C_B \qquad r_4 = k_4 C_C$$

$$r_5 = k_5 C_A C_C \qquad r_6 = k_6 C_D$$

Data: At 489°C,

$$\begin{array}{lll} k_1 = 2\,\text{min}^{-1} & k_2 = 20\,\text{L/mol min}^{-1} & k_3 = 50\,\text{L/mol min}^{-1} \\ k_4 = 0.8\,\text{min}^{-1} & k_5 = 125\,\text{L/mol min}^{-1} & k_6 = 2\,\text{min}^{-1} \end{array}$$

Solution   We have considered this reaction scheme in Examples 4.1 and 7.8 but the values of the rate constants are different here. We select the three forward reactions as the set of independent reactions. Hence, the indices of the independent reactions are $m = 1, 3, 5$, and those of the dependent reactions are $k = 2, 4, 6$.

The factors α_{km}'s of the dependent reactions are $\alpha_{21} = -1$, $\alpha_{43} = -1$, and $\alpha_{65} = -1$, and all the others are zero. The stoichiometric coefficients of the independent reactions are

$$
\begin{array}{lllll}
s_{A_1} = -1 & s_{B_1} = 2 & s_{C_1} = 0 & s_{D_1} = 0 & \Delta_1 = 1 \\
s_{A_3} = -1 & s_{B_3} = -1 & s_{C_3} = 1 & s_{D_3} = 0 & \Delta_3 = -1 \\
s_{A_5} = -1 & s_{B_5} = 0 & s_{C_5} = -1 & s_{D_5} = 1 & \Delta_5 = -1
\end{array}
$$

We write Eq. 8.1.1 for each independent reaction:

$$
Z_{1_{out}} - Z_{1_{in}} = (r_{1_{out}} - r_{2_{out}})\tau\left(\frac{t_{cr}}{C_0}\right) \tag{a}
$$

$$
Z_{3_{out}} - Z_{3_{in}} = (r_{3_{out}} - r_{4_{out}})\tau\left(\frac{t_{cr}}{C_0}\right) \tag{b}
$$

$$
Z_{5_{out}} - Z_{5_{in}} = (r_{5_{out}} - r_{6_{out}})\tau\left(\frac{t_{cr}}{C_0}\right) \tag{c}
$$

We select the inlet stream as the reference stream; hence, $T_0 = 762$ K, $y_{A_0} = 1$, $y_{B_0} = y_{C_0} = y_{D_0} = 0$, and $Z_{1_{in}} = Z_{3_{in}} = Z_{5_{in}} = 0$. The reference concentration and volumetric flow rate are

$$
C_0 = \frac{P_0}{RT_0} = 0.08\,\text{mol/L}
$$

$$
v_0 = \frac{(F_{tot})_0}{C_0} = 750\,\text{L/min}
$$

We use Eq. 8.1.13 to express the species concentration, and the rates of the chemical reactions are

$$
r_1 = k_1 C_0 \frac{y_{A_0} - Z_{1_{out}} - Z_{3_{out}} - Z_{5_{out}}}{1 + Z_{1_{out}} - Z_{3_{out}} - Z_{5_{out}}} \tag{d}
$$

$$
r_2 = k_2 C_0^2 \left(\frac{y_{B_0} + 2Z_{1_{out}} - Z_{3_{out}}}{1 + Z_{1_{out}} - Z_{3_{out}} - Z_{5_{out}}}\right)^2 \tag{e}
$$

$$
r_3 = k_3 C_0^2 \frac{(y_{A_0} - Z_{1_{out}} - Z_{3_{out}} - Z_{5_{out}})(y_{B_0} + 2Z_{1_{out}} - Z_{3_{out}})}{(1 + Z_{1_{out}} - Z_{3_{out}} - Z_{5_{out}})^2} \tag{f}
$$

$$
r_4 = k_4 C_0 \frac{y_{C_0} + Z_{3_{out}} - Z_{5_{out}}}{1 + Z_{1_{out}} - Z_{3_{out}} - Z_{5_{out}}} \tag{g}
$$

$$
r_5 = k_5 C_0^2 \frac{(y_{A_0} - Z_{1_{out}} - Z_{3_{out}} - Z_{5_{out}})(y_{C_0} + Z_{3_{out}} - Z_{5_{out}})}{(1 + Z_{1_{out}} - Z_{3_{out}} - Z_{5_{out}})^2} \tag{h}
$$

$$
r_6 = k_6 C_0 \frac{y_{D_0} + Z_{5_{out}}}{1 + Z_{1_{out}} - Z_{3_{out}} - Z_{5_{out}}} \tag{i}
$$

We select Reaction 1 as the leading reaction; using Eq. 3.5.4 the characteristic reaction time is

$$t_{cr} = \frac{1}{k_1} = 0.5 \text{ min} \tag{j}$$

Substituting (d) through (i) and into (a), (b), and (c), and using (i), the design equations become

$$Z_{1_{out}} - \tau \left(\frac{y_{A_0} - Z_{1_{out}} - Z_{3_{out}} - Z_{5_{out}}}{1 + Z_{1_{out}} - Z_{3_{out}} - Z_{5_{out}}} \right) + \tau \left(\frac{k_2 C_0}{k_1} \right) \left(\frac{y_{B_0} + 2Z_{1_{out}} - Z_{3_{out}}}{1 + Z_{1_{out}} - Z_{3_{out}} - Z_{5_{out}}} \right)^2 = 0 \tag{k}$$

$$Z_{3_{out}} - \tau \left(\frac{k_3 C_0}{k_1} \right) \frac{(y_{A_0} - Z_{1_{out}} - Z_{3_{out}} - Z_{5_{out}})(y_{B_0} + 2Z_{1_{out}} - Z_{3_{out}})}{(1 + Z_{1_{out}} - Z_{3_{out}} - Z_{5_{out}})^2} +$$
$$\tau \left(\frac{k_4}{k_1} \right) \left(\frac{y_{C_0} + Z_{3_{out}} - Z_{5_{out}}}{1 + Z_{1_{out}} - Z_{3_{out}} - Z_{5_{out}}} \right) = 0 \tag{l}$$

$$Z_{5_{out}} - \tau \left(\frac{k_5 C_0}{k_1} \right) \frac{(y_{A_0} - Z_{1_{out}} - Z_{3_{out}} - Z_{5_{out}})(y_{C_0} + Z_{3_{out}} - Z_{5_{out}})}{(1 + Z_{1_{out}} - Z_{3_{out}} - Z_{5_{out}})^2} +$$
$$\tau \left(\frac{k_6}{k_1} \right) \left(\frac{y_{D_0} + Z_{5_{out}}}{1 + Z_{1_{out}} - Z_{3_{out}} - Z_{5_{out}}} \right) = 0 \tag{m}$$

The parameters are

$$\frac{k_2 C_0}{k_1} = 0.8 \quad \frac{k_3 C_0}{k_1} = 0.2 \quad \frac{k_4}{k_1} = 0.4 \quad \frac{k_5 C_0}{k_1} = 5 \quad \frac{k_6}{k_1} = 1$$

Once we solve the design equations, we use Eq. 2.7.8 to determine the species operating curves:

$$\frac{F_{A_{out}}}{(F_{tot})_0} = y_{A_0} - Z_{1_{out}} - Z_{3_{out}} - Z_{5_{out}} \tag{n}$$

$$\frac{F_{B_{out}}}{(F_{tot})_0} = y_{B_0} + 2Z_{1_{out}} - Z_{3_{out}} \tag{o}$$

$$\frac{F_{C_{out}}}{(F_{tot})_0} = y_{C_0} + Z_{3_{out}} - Z_{5_{out}} \tag{p}$$

$$\frac{F_{D_{out}}}{(F_{tot})_0} = y_{D_0} + Z_{5_{out}} \tag{q}$$

a. For the given feed, $y_{A_0} = 1$ and $y_{B_0} = y_{C_0} = y_{D_0} = 0$, We substitute these values into (k), (l), and (m) and solve them numerically for different values of τ. Figure E8.10.1 shows the reaction operating curves. We then use (n) through (q) to obtain the species curves, shown in Figure E8.10.2.

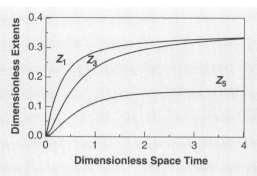

Figure E8.10.1 Reaction operating curves—feed of species A.

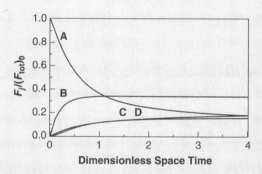

Figure E8.10.2 Species operating curves—feed of species A.

b. From the curve of product B, the highest $F_{B_{out}}$ is reached at $\tau = 1.48$, where $F_B/F_{(tot)0} = 0.3407$. At that space time, $Z_1 = 0.3061$, $Z_3 = 0.2704$, and $Z_5 = 0.1351$. The production rate of product B is $0.34\,\text{mol/s}$. Using Eq. 8.1.3 and (j), the reactor volume is

$$V_R = \tau \cdot v_0 \cdot t_{cr} = 8555\,\text{L}$$

c. To estimate the equilibrium composition, we solve the design equations for $\tau \to \infty$. For $\tau = 100$, the mole fractions of the various species are: $y_{Aeq} = 0.1371$; $y_{Beq} = 0.4086$; $y_{Ceq} = 0.2705$; $y_{Deq} = 0.1838$.

d. When the feed stream consists of species D, the reactor design proceeds in the same way as in parts (a). The only difference is that, now, $y_{A_0} = y_{B_0} = y_{C_0} = 0$ and $y_{D_0} = 1$. Substituting these values into (k), (l), and (m), we solve them numerically. Figure E8.10.3 shows the reaction operating curves. We then use (n) through (r) to obtain the species curves, shown in Figure E8.10.4. Note that in this case, the extents of Reactions 3 and 5 are negative, since they proceed reversely. To estimate the equilibrium composition, we solve the design equations for $\tau = 100$. The mole fractions of the various species are: $y_{Aeq} = 0.1344$; $y_{Beq} = 0.4039$; $y_{Ceq} = 0.2725$; $y_{Deq} = 0.1891$. The compositions are very close to those estimated in part (c).

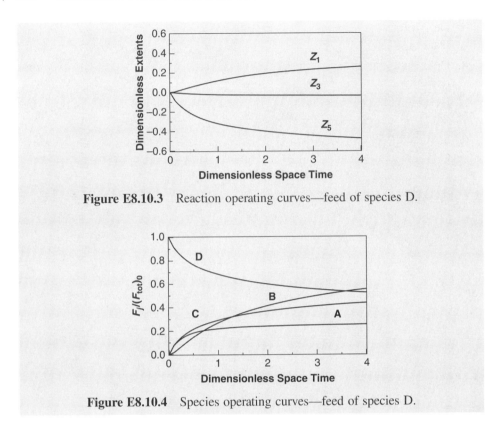

Figure E8.10.3   Reaction operating curves—feed of species D.

Figure E8.10.4   Species operating curves—feed of species D.

8.4   NONISOTHERMAL OPERATIONS

The design formulation of nonisothermal CSTRs with multiple reactions follows the same procedure outlined in the previous section—we write the design equation, Eq. 8.1.1, for each independent reaction. However, since the reactor temperature, T_{out}, is not known, we should solve the design equations simultaneously with the energy balance equation (Eq. 8.1.14).

Note that by definition, the temperature of a CSTR is the same everywhere; therefore, all CSTRs are, in principle, "isothermal." Usually, the inlet temperature is specified, and, as a part of the design, we have to determine the reactor temperature. When the reactor temperature is the same as the inlet temperature, $T_{out} = T_{in}$, we refer to a CSTR as an *isothermal reactor*. But when $T_{out} \neq T_{in}$, we have to determine the reactor temperature. Also, note that the energy balance equation contains another variable—the temperature of the heating (or cooling) fluid, θ_F. Because of the complex geometry of the heat-transfer surface in a CSTR, an average of the heating/cooling fluid's inlet and outlet temperatures is usually used.

The procedure for setting up the energy balance equation goes as follows:

1. Select and define the temperature of the reference stream, T_0.
2. Determine the specific molar heat capacity of the reference stream, $\hat{c}_{p_0}$.

3. Determine the dimensionless activation energies, γ_i's, of *all* chemical reactions.
4. Determine the dimensionless heat of reactions, DHR_m's, of the *independent* reactions.
5. Determine the correction factor of the heat capacity of the reacting fluid, CF (Z_m, θ).
6. Specify the dimensionless heat-transfer number, HTN (using Eq. 8.1.22).
7. Determine (or specify) the inlet temperature, θ_{in}.
8. Determine (or specify) the temperature of the heating/cooling fluid, θ_F.
9. Solve the energy balance equation simultaneously with the design equations to obtain Z_m's and θ_{out} as functions of the dimensionless space time, τ.

The design formulation of nonisothermal CSTRs consists of $(n_I + 1)$ simultaneous, nonlinear algebraic equations. We have to solve them for different values of dimensionless space time, τ. Below, we illustrate how to design nonisothermal CSTRs.

Example 8.11 The first-order chemical reaction

$$A \longrightarrow 2B$$

takes place in an aqueous solution. A solution of reactant A ($C_A = 0.8$ mol/L) is fed at a rate of 200 L/min into a cascade of two equal-size 100-L CSTRs connected in series. The feed temperature is 47°C. Based on the data below, for the indicated operations, determine the conversion of reactant A and the outlet temperature of each reactor:

a. Derive the reaction and species operating curves of each reactor for isothermal operation. Determine the feed flow rate if we would like to achieve 80% conversion.
b. Determine the heating load of each reactor in (a).
c. Determine the isothermal HTN of each reactor.
d. Derive the reaction and species operating curves of each reactor for adiabatic operation. Determine the feed flow rate if we would like to achieve 80% conversion.

Data: At 47°C, $k_1 = 0.4$ min^{-1}, $\Delta H_R(T_0) = -20$ kcal/mol B
$$E_a = 9000 \text{ cal/mol} \quad \rho = 1.0 \text{ kg/L} \quad \bar{c}_p = 1.0 \text{ kcal/kg K}$$

The temperature of the cooling fluid is 27°C.

Solution The stoichiometric coefficients of the reaction are

$$s_A = -1 \quad s_B = 2 \quad \Delta = 1$$

For the chemical formula used, the heat of reaction is

$$\Delta H_R(320\,\text{K}) = \left(\frac{-20\,\text{kcal}}{\text{mol B}}\right)\left(\frac{2\,\text{mol B}}{\text{mol extent}}\right) = -40\,\frac{\text{kcal}}{\text{mol extent}}$$

We select the inlet stream to the first reactor as the reference stream; hence, $y_{A_0} = 1$, $y_{B_0} = 0$, $T_0 = 320°\text{K}$, $y_{A_0} = 1$, $y_{B_0} = 0$, $T_0 = 320°\text{K}$, and $Z_{1_{in}} = 0$. The molar flow rate of the reference stream is

$$(F_{tot})_0 = v_0 C_0 = 160\,\text{mol/min} \tag{a}$$

Using Eq. 5.2.60, the specific molar heat capacity of the reference stream is

$$\hat{c}_{p0} = \frac{\rho}{C_0}\bar{c}_p = 1250\,\frac{\text{cal}}{\text{mol K}} \tag{b}$$

Using Eq. 5.2.62, for liquid-phase reactions and assuming constant heat capacity, $CF(Z_m, \theta) = 1.0$. The dimensionless heat of reaction is

$$\text{DHR} = \frac{\Delta H_R(T_0)}{T_0 \hat{c}_{p0}} = -0.10 \tag{c}$$

The dimensionless activation energy is

$$\gamma = \frac{E_a}{RT_0} = 14.15 \tag{d}$$

Using Eq. 8.1.1, the design equation for each reactor is

$$Z_{out} - Z_{in} = r_{out}\tau\left(\frac{t_{cr}}{C_0}\right) \tag{e}$$

Using Eq. 8.1.9 to express the species concentrations, the reaction rate is

$$r_{out} = k(T_0)C_0(1 - Z_{out})e^{\gamma(\theta-1)/\theta} \tag{f}$$

We define the characteristic reaction time as

$$t_{cr} = \frac{1}{k(T_0)} = 2.5\,\text{min} \tag{g}$$

Substituting (f) and (g) into (e), the design equation for the first reactor is

$$Z_{1_{out}} - (1 - Z_{1_{out}})\tau_1 e^{\gamma(\theta_1-1)/\theta_1} = 0 \tag{h}$$

and the design equation for the second reactor is

$$Z_{2_{out}} - Z_{1_{out}} - (1 - Z_{2_{out}})\tau_2 e^{\gamma(\theta_2-1)/\theta_2} = 0 \tag{i}$$

Using Eq. 8.1.14, the energy balance equation for the first reactor is

$$\text{HTN}_1\tau_1(\theta_F - \theta_1) = \text{DHR}(Z_{1_{out}} - 0) + CF(Z, \theta)_1(\theta_1 - 1) \tag{j}$$

The energy balance equation for the second reactor is

$$HTN_2\tau_2(\theta_F - \theta_2) = DHR(Z_{2_{out}} - Z_{1_{out}}) + CF(Z, \theta)_2(\theta_2 - \theta_1) \qquad (k)$$

Once we solve the design equations, we can determine the species curves, using Eq. 2.7.8:

$$\frac{F_{A_{out}}}{(F_{tot})_0} = 1 - Z_{out} \qquad (l)$$

$$\frac{F_{B_{out}}}{(F_{tot})_0} = 2Z_{out} \qquad (m)$$

a. For isothermal operation, $\theta_1 = \theta_2 = 1$, we solve (h) and (i) and obtain the reaction operating curve of the cascade, shown in Figure E8.11.1. We then use (l) and (m) to obtain the species curves, shown in Figure E8.11.2. From the curve of reactant A, 80% conversion $[F_A/(F_{tot})_0 = 0.2]$ is reached at $\tau = 2.48$. At that cascade space time, $Z_1 = 0.554$, and $Z_2 = 0.80$. Using Eq. 8.1.3 and (g), the flow rate of the reference stream is

$$v_0 = \frac{V_R}{\tau t_{cr}} = 32.4 \text{ L/min}$$

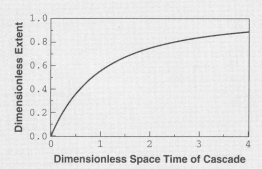

Figure E8.11.1 Reaction curve for the cascade—isothermal operation.

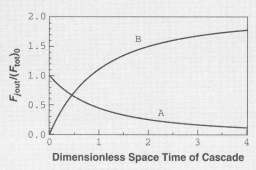

Figure E8.11.2 Species curves for the cascade—isothermal operation.

b. Combining Eqs. 8.1.14 and 8.1.16, the heating load of the first reactor is

$$\frac{\dot{Q}_1}{(F_{\text{tot}})_0 T_0 \hat{c}_{p_0}} = \text{DHR}(Z_{1_{\text{out}}} - 0) \tag{l}$$

and, the heating load of the second reactor is

$$\frac{\dot{Q}_2}{(F_{\text{tot}})_0 T_0 \hat{c}_{p_0}} = \text{DHR}(Z_{2_{\text{out}}} - Z_{1_{\text{out}}}) \tag{m}$$

Using (c), the heating loads of the reactors are

$$\dot{Q}_1 = (F_{\text{tot}})_0 \, \Delta H_R(T_0) Z_{\text{out}_1} = -2871 \text{ kcal/min}$$

$$\dot{Q}_2 = (F_{\text{tot}})_0 \, \Delta H_R(T_0)(Z_{\text{out}_2} - Z_{\text{out}_1}) = -1274 \text{ kcal/min}$$

The negative sign indicates that heat is being removed from the reactor.

c. To determine the isothermal HTN of each reactor, we use Eq. 8.1.20. For the first reactor,

$$\text{HTN}_{\text{iso}} = \frac{1}{(\theta_{F_1} - 1)\tau_1} \text{DHR}(Z_{1_{\text{out}}} - 0) \tag{n}$$

and for the second reactor

$$\text{HTN}_{\text{iso}} = \frac{1}{(\theta_{F_2} - 1)\tau_2} \text{DHR}(Z_{2_{\text{out}}} - Z_{1_{\text{out}}}) \tag{o}$$

For $\tau_1 = \tau_2 = 1.235$, we solve (n) and (o) and obtain $\text{HTN}_1 = 0.714$, and $\text{HTN}_2 = 0.319$.

d. For liquid-phase reaction, assuming constant heat capacity, CF $(z, \theta) = 1$. For adiabatic operations, (HTN = 0), the energy balance for the first reactor, (j), becomes

$$\text{DHR}(Z_{1_{\text{out}}} - 0) + (1)(\theta_1 - 1) = 0 \tag{p}$$

The energy balance equation of the second reactor, (k), becomes

$$\text{DHR}(Z_{2_{\text{out}}} - Z_{1_{\text{out}}}) + (1)(\theta_2 - \theta_1) = 0 \tag{q}$$

We solve (h) and (i) simultaneously with (p) and (q) for different values of τ. Figure E8.11.3 shows the reaction curve for the cascade, and Figure E8.11.4 shows the temperature curves. We then use (l) and (m) to obtain the species curves for the cascade, shown in Figure E8.11.5. From the curve of reactant A at the exit of the second reactor, 80% conversion $[F_A/(F_{\text{tot}})_0 = 0.2]$ is reached at $\tau = 1.04$. At that cascade space time, $Z_1 = 0.507$, $\theta_1 = 1.0507$,

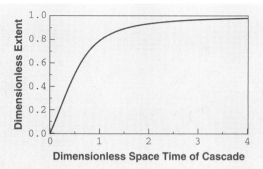

Figure E8.11.3   Reaction curve for the cascade—adiabatic operation.

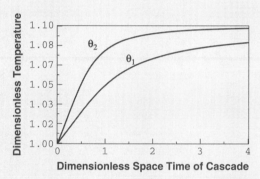

Figure E8.11.4   The temperature curves—adiabatic operation.

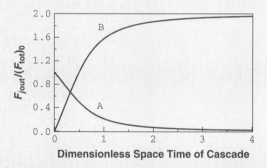

Figure E8.11.5   Species curves for the cascade—adiabatic operation.

and $Z_2 = 0.80$, $\theta_2 = 1.0802$. Using Eq. 8.1.3 and (g), the flow rate of the reference stream is

$$v_0 = \frac{V_R}{\tau t_{cr}} = 76.9 \text{ L/min}$$

The temperature of the first reactor is 336.2 K, and of the second reactor is 345.7 K.

Example 8.12 The gas-phase chemical reactions

$$\text{Reaction 1:} \quad A + B \longrightarrow C$$
$$\text{Reaction 2:} \quad C + B \longrightarrow D$$

are carried out in a CSTR. A gas stream (at 150°C and 2 atm) consisting of 40% reactant A, 40% reactant B, and 20% I (inert) is fed into a reactor at a rate of $0.4 \, \text{m}^3/\text{min}$. The reactor is cooled by condensing steam in the shell side at $T_F = 130°C$.

a. Derive the design equations and plot the reaction and species operating curves for isothermal operation.

b. Determine the volume of isothermal reactor that provides the highest production rate of product C.

c. Determine the heating rate in (b).

d. Determine the HTN for isothermal operation.

e. Derive the design equations and plot the reaction and species operating curves for adiabatic operation.

f. Determine the volume of adiabatic reactor that provides the highest production rate of product C. What is the operating temperature?

g. Determine the volume of the reactor that provides the highest production rate of product C if the HTN is half of the isothermal HTN. What is the operating temperature?

The rate expressions are: $r_1 = k_1 C_A C_B$, $r_2 = k_2 C_B C_C$

Data: At 150°C $k_1 = 0.2 \, \text{L/mol s}^{-1}$, $k_2 = 0.4 \, \text{L/mol s}^{-1}$

$$\Delta H_{R_1} = -6000 \, \text{cal/mol extent} \quad \Delta H_{R_2} = -4000 \, \text{cal/mol extent}$$

$$E_{a_1} = 4900 \, \text{cal/mol} \quad E_{a_2} = 7000 \, \text{cal/mol}$$

$$\hat{c}_{p_A} = 16 \, \text{cal/mol K} \quad \hat{c}_{p_B} = 8 \, \text{cal/mol K} \quad \hat{c}_{p_C} = 20 \, \text{cal/mol K}$$

$$\hat{c}_{p_D} = 26 \, \text{cal/mol K} \quad \hat{c}_{p_I} = 10 \, \text{cal/mol K}$$

Solution Since each reaction has a species that does not appear in the other reaction, the two reactions are independent, and there is no dependent reaction. The stoichiometric coefficients are

$$
\begin{array}{lllll}
s_{A_1} = -1 & s_{B_1} = -1 & s_{C_1} = 1 & s_{D_1} = 0 & \Delta_1 = -1 \\
s_{A_2} = 0 & s_{B_2} = -1 & s_{C_2} = -1 & s_{D_2} = 1 & \Delta_2 = -1
\end{array}
$$

We select the inlet stream as the reference stream; hence, $Z_{1_{in}} = Z_{2_{in}} = 0$, and $y_{A_0} = 0.40$, $y_{B_0} = 0.40$, $y_{I_0} = 0.20$, and $y_{C_0} = y_{D_0} = 0$. The reference concentration and flow rate are

$$C_0 = \frac{P}{RT_0} = 5.76 \times 10^{-2} \text{ mol/L}$$

$$(F_{\text{tot}})_0 = v_0 C_0 = 23.05 \text{ mol/min}$$

Using Eq. 5.2.58, the specific heat capacity of the reference stream is

$$\hat{c}_{P0} = \sum_j^J y_{j0}\hat{c}_{Pj}(1) = y_{A_0}\hat{c}_{PA}(T_0) + y_{B_0}\hat{c}_{PB}(T_0) + y_{I_0}\hat{c}_{PI}(T_0) = 11.6 \text{ cal/mol K}$$

The dimensionless activation energies of the two reactions are

$$\gamma_1 = \frac{E_{a_1}}{RT_0} = 5.83 \qquad \gamma_2 = \frac{E_{a_2}}{RT_0} = 8.33$$

The dimensionless heat of reactions are

$$\text{DHR}_1 = \frac{\Delta H_{R_1}(T_0)}{T_0\hat{c}_{P0}} = -1.222 \qquad \text{DHR}_2 = \frac{\Delta H_{R_2}(T_0)}{T_0\hat{c}_{P0}} = -0.815$$

Using Eq. 5.2.61, the correction factor of the heat capacity of the reacting fluid is

$$\text{CF}(Z_m, \theta) = 1 + \frac{1}{\hat{c}_{P0}} \sum_j^J \hat{c}_{Pj}(\theta) \sum_m^{n_I} (s_j)_m Z_m$$

$$= 1 + \frac{1}{\hat{c}_{P0}} [\hat{c}_{PA}(-Z_1) + \hat{c}_{PB}(-Z_1 - Z_2) + \hat{c}_{PC}(Z_1 - Z_2) + \hat{c}_{PD}Z_2]$$

$$= \frac{11.6 - 4Z_1 - 2Z_2}{11.6}$$

We write Eq. 8.1.1 for each independent reaction:

$$Z_{1_{\text{out}}} = r_{1_{\text{out}}}\tau\left(\frac{t_{\text{cr}}}{C_0}\right) \tag{a}$$

$$Z_{2_{\text{out}}} = r_{2_{\text{out}}}\tau\left(\frac{t_{\text{cr}}}{C_0}\right) \tag{b}$$

Using Eq. 8.1.13 to express the series concentrations, the rates of the two reactions are

$$r_{1_{\text{out}}} = k_1(T_0)C_0^2 \frac{(y_{A_0} - Z_{1_{\text{out}}})(y_{B_0} - Z_{1_{\text{out}}} - Z_{2_{\text{out}}})}{[(1 - Z_{1_{\text{out}}} - Z_{2_{\text{out}}})\theta]^2} e^{\gamma_1(\theta-1)/\theta} \tag{c}$$

$$r_{2_{\text{out}}} = k_2(T_0)C_0^2 \frac{(Z_{1_{\text{out}}} - Z_{2_{\text{out}}})(y_{B_0} - Z_{1_{\text{out}}} - Z_{2_{\text{out}}})}{[(1 - Z_{1_{\text{out}}} - Z_{2_{\text{out}}})\theta]^2} e^{\gamma_2(\theta-1)/\theta} \tag{d}$$

We select Reaction 1 as the leading reaction, and the characteristic reaction time is

$$t_{cr} = \frac{1}{k_1(T_0)C_0} = 86.85 = 1.45 \, \text{min} \tag{e}$$

Substituting (c), (d), and (e) into (a) and (b), the two design equations become

$$Z_{1_{out}} - \frac{(y_{A_0} - Z_{1_{out}})(y_{B_0} - Z_{1_{out}} - Z_{2_{out}})}{[(1 - Z_{1_{out}} - Z_{2_{out}})\theta]^2} \tau e^{\gamma_1(\theta-1)/\theta} = 0 \tag{f}$$

$$Z_{2_{out}} - \left[\frac{k_2(T_0)}{k_1(T_0)}\right] \frac{(Z_{1_{out}} - Z_{2_{out}})(y_{B_0} - Z_{1_{out}} - Z_{2_{out}})}{[(1 - Z_{1_{out}} - Z_{2_{out}})\theta]^2} \tau e^{\gamma_2(\theta-1)/\theta} = 0 \tag{g}$$

Substituting $CF(Z_m, \theta)$ into Eq. 8.1.14, the energy balance equation is

$$HTN\tau(\theta_F - \theta) - DHR_1 Z_{1_{out}} - DHR_2 Z_{2_{out}} -$$
$$\frac{11.6 - 4Z_{1_{out}} - 2Z_{2_{out}}}{11.6}(\theta - 1) = 0 \tag{h}$$

We have to solve (f), (g), and (h) simultaneously for different values of τ. Once we have the solution, we use Eq. 2.7.8 to obtain the species curves:

$$\frac{F_{A_{out}}}{(F_{tot})_0} = y_{A_0} - Z_{1_{out}} \tag{i}$$

$$\frac{F_{B_{out}}}{(F_{tot})_0} = y_{B_0} - Z_{1_{out}} - Z_{2_{out}} \tag{j}$$

$$\frac{F_{C_{out}}}{(F_{tot})_0} = y_{C_0} + Z_{1_{out}} - Z_{2_{out}} \tag{k}$$

$$\frac{F_{D_{out}}}{(F_{tot})_0} = y_{D_0} + Z_{2_{out}} \tag{l}$$

a. For isothermal operation ($\theta = 1$), the two design equations become

$$Z_{1_{out}} - \frac{(0.4 - Z_{1_{out}})(0.4 - Z_{1_{out}} - Z_{2_{out}})}{(1 - Z_{1_{out}} - Z_{2_{out}})^2} \tau = 0 \tag{m}$$

$$Z_{2_{out}} - (0.5)\frac{(Z_{1_{out}} - Z_{2_{out}})(0.4 - Z_{1_{out}} - Z_{2_{out}})}{(1 - Z_{1_{out}} - Z_{2_{out}})^2} \tau = 0 \tag{n}$$

We solve (m) and (n) numerically for different values of τ. Figure E8.12.1 shows the reaction operating curves. Using (i) through (l), we obtain the species operating curves, shown in Figure E8.12.2. From the curve of product C, the highest $F_C/(F_{tot})_0 = 0.0686$ and it is reached at $\tau = 2.84$. For that pace time, $Z_1 = 0.1664$ and $Z_2 = 0.0978$. Using (e) and Eq. 8.1.3, the reactor volume is

$$V_R = v_0 t_{cr}\tau = 1647 \, \text{L}$$

The production rate of product C is 1.58 mol/min.

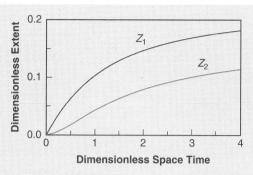

Figure E8.12.1 Reaction operating curves—isothermal operation.

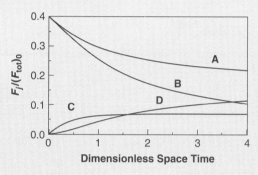

Figure E8.12.2 Species curves—isothermal operation.

b. For isothermal operation $\theta = 1$, and combining Eq. 8.1.14 and Eq. 8.1.17,

$$\frac{\dot{Q}}{T_0 (F_{\text{tot}})_0 \hat{c}_{p_0}} = \text{DHR}_1 Z_{1_{\text{out}}} + \text{DHR}_2 Z_{2_{\text{out}}} \tag{o}$$

which reduces to

$$\dot{Q} = (F_{\text{tot}})_0 \lfloor \Delta H_{R_1}(T_0) Z_{1_{\text{out}}} + \Delta H_{R_2}(T_0) Z_{2_{\text{out}}} \rfloor = -32.02 \text{ kcal/min}$$

The negative sign indicates that heat is removed from the reactor.

c. Using Eq. 8.1.20, the isothermal HTN is

$$\text{HTN}_{\text{iso}}(\tau) = \frac{1}{(\theta_F - 1)\tau} (\text{DHR}_1 Z_{1_{\text{out}}} + \text{DHR}_2 Z_{2_{\text{out}}}) \tag{p}$$

Figure E8.12.3 shows the HTN curve as a function of the reactor volume (space time). For the given reactor, $\tau = 2.84$, HTN $= 2.09$.

d. For adiabatic operation, HTN $= 0$, and (h) reduces to

$$\text{DHR}_1 Z_{1_{\text{out}}} + \text{DHR}_2 Z_{2_{\text{out}}} + \frac{11.6 - 4 Z_{1_{\text{out}}} - 2 Z_{2_{\text{out}}}}{11.6} (\theta - 1) = 0 \tag{q}$$

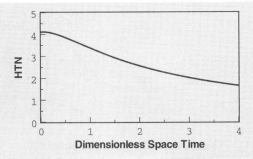

Figure E8.12.3 Isothermal HTN curve.

We solve (f), (g), and (q) numerically for different values of τ. Figure E8.12.4 shows the reaction curves, and Figure E8.12.5 shows the temperature curve. Using (i) through (l), we obtain the species operating curves, shown in Figure E8.12.6. From the curve of product C, the highest $F_C/(F_{tot})_0$ is 0.0516, and it is reached at $\tau = 0.66$. At that space time, $Z_1 = 0.1142$, $Z_2 = 0.0626$, and $\theta = 1.20$. Using (e) and Eq. 8.1.3, the reactor volume is

$$V_R = v_0 t_{cr} \tau = 382.8 \text{ L}$$

and the reactor temperature is $T = (1.2)(423) = 507.8$ K. The production rate of product C is 1.19 mol/min.

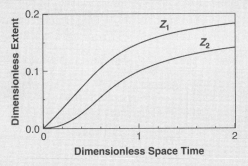

Figure E8.12.4 Reaction operating curves—adiabatic operation.

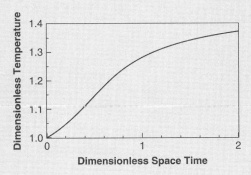

Figure E8.12.5 Temperature curve—adiabatic operation.

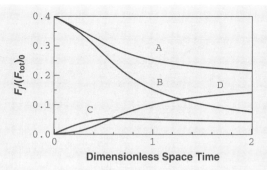

Figure E8.12.6   Species operating curves—adiabatic operation.

e. For nonisothermal operation with HTN $= 1.09$, we solve (f), (g), and (h) numerically for different values of τ. Figure E8.12.7 shows the reaction curves and Figure E8.12.8 shows the temperature curve. Using (i) through (l), we obtain the species operating curves, shown in Figure E8.12.9. The F_C curve does not have a maximum. If we use the same feed rate as in isothermal operation ($\tau = 2.84$), $Z_1 = 0.1722$, $Z_2 = 0.1073$, $\theta = 1.0375$, and $F_C/(F_{tot})_0 = 0.0649$. The reactor temperature is 439 K and the production rate of product C is 1.495 mol/min.

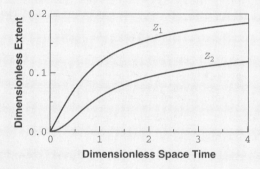

Figure E8.12.7   Reaction curves—nonisothermal operation.

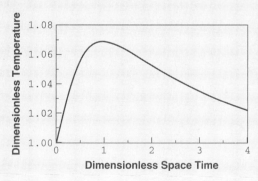

Figure E8.12.8   Temperature curve—nonisothermal operation.

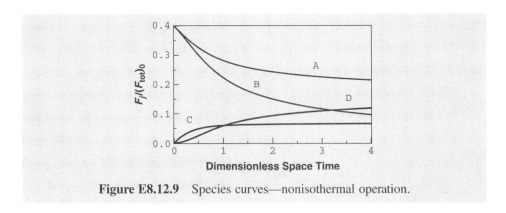

Figure E8.12.9 Species curves—nonisothermal operation.

8.5 SUMMARY

In this chapter we discussed the operation of continuous stirred-tank reactors. We covered the following topics:

1. The underlying assumptions of the CSTR model and when they are satisfied in practice.
2. The design equations, the energy balance equation, and the auxiliary relations for species concentrations.
3. The reaction operating curves and species operating curves.
4. Design and operation of isothermal CSTRs with single reactions.
5. Determination of the reaction rate expression and its parameters.
6. Operation and optimization of a cascade of CSTRs.
7. Design and operation of isothermal CSTRs with multiple reactions.
8. A procedure to estimate the range of the HTN.
9. Design and operation of nonisothermal CSTRs with multiple reactions.

PROBLEMS*

8.1₁ The gas-phase chemical reaction

$$2A \longrightarrow R$$

takes place in a CSTR. The rate expression is $r = 0.05C_A^2 \, \mathrm{mol/L\,s^{-1}}$. A stream consisting of 50% A–50% inert (by mole) is fed at a rate of

*Subscript 1 indicates simple problems that require application of equations provided in the text. Subscript 2 indicates problems whose solutions require some more in-depth analysis and modifications of given equations. Subscript 3 indicates problems whose solutions require more comprehensive analysis and involve application of several concepts. Subscript 4 indicates problems that require the use of a mathematical software or the writing of a computer code to obtain numerical solutions.

180 L/min into a 1-m^3 CSTR. Derive and plot the reaction and species operating curves. If $C_{A_{in}} = 0.3$ mol/L, what is the conversion of A? What is the production rate of product R?

8.2$_2$ The gas-phase chemical reaction

$$A + 2B \longrightarrow 3C$$

takes place in a CSTR. A gas stream at 2 atm and 677°K contains reactant A, and an inert at a proportion of 1/6 A and 5/6 inert is fed at a rate of 2 L/min. A second gas stream of reactant B at 1.95 atm and 330°K is fed at a rate of 0.5 L/min into the reactor. The volume of the reactor is 0.75 L, and it is kept at 440°K and 1.3 atm. The rate expression is $r = kC_A C_B$. The partial pressure of reactant A in the reactor and in the exit stream is 0.029 atm.

a. Derive and plot the reaction and species operating curves.

b. Determine the conversion of reactants A and B.

c. Determine the reaction rate constant of the reaction.

8.3$_1$ The liquid-phase reaction

$$A + B \longrightarrow R$$

takes place in a CSTR. An aqueous solution containing reactants A and B ($C_{A_{in}} = 100$ mmol/L and $C_{B_{in}} = 200$ mmol/L) is fed at a rate of 400 L/min to the reactor. The rate expression of the reaction is $r = 200 C_A C_B$ mol/L min^{-1}.

a. Derive and plot the reaction and species operating curves.

b. Determine the reactor volume needed for 90% conversion of A.

8.4$_2$ In the presence of an enzyme of fixed concentration, reactant A in aqueous solution decomposes according to the reaction

$$A \longrightarrow R + P$$

The following data were collected on an isothermal CSTR:

C_A (mol/L)	1	2	3	4	5	6	8	10
r (mol/L min)	1	2	3	4	4.7	4.9	5	5

We plan to carry out this reaction in a large-scale CSTR at the same fixed enzyme concentration. Use a graphical method (not requiring the derivation of the rate expression) to derive and plot the reaction and species operating curves for $C_{A_{in}} = 10$ mol/L.

a. Determine and plot the reaction and species operating curves.

b. Find the volumetric flow rate of the stream that we can feed to 250 L CSTR if we want to obtain 80% conversion.

8.5₂ The liquid-phase chemical reaction

$$A + B \longrightarrow R + S$$

is investigated on a CSTR. A feed stream containing reactants A and B is fed to a 1-L CSTR, and the following data are obtained:

Feed Composition (mol/L)	Flow Rate (L/min)	Output (mmol/L)
$C_{A_{in}} = C_{B_{in}} = 100$	$v = 1$	$C_{A_{out}} = 50$
$C_{A_{in}} = C_{B_{in}} = 200$	$v = 9$	$C_{A_{out}} = 150$
$C_{A_{in}} = 200, C_{B_{in}} = 100$	$v = 3$	$C_{A_{out}} = 150$

Determine the rate expression and the reaction rate constant.

8.6₂ The gas-phase decomposition reaction

$$A \longrightarrow B + 2C$$

is being studied in a CSTR. Reactant A at about 3 atm and 30°C ($C_{A_{in}} = 0.120$ mol/L) is fed into a 1-L CSTR at various flow rates, and its exit concentration is measured for each flow rate.

a. Derive a rate expression of the reaction.
b. Derive and plot the reaction and species operating curves for a CSTR.
c. A stream of reactant A ($C_{A_{in}} = 0.320$ mol/L) is fed into a 560-L CSTR. Determine volumetric feed flow rate if we want to achieve 50% conversion.

Data:

v_{in} (L/min)	0.06	0.48	1.5	8.1
$C_{A_{out}}$ (mmol/L)	30	60	80	105

8.7₂ The gas-phase reaction

$$A \longrightarrow 2B + 2C + D$$

is being investigated in a CSTR. A stream of a high-molecular-weight reactant A is fed continuously to the reactor. By changing the feed flow rate, different extents of cracking are obtained as follows:

$F_{A_{in}}$ (mmol/h)	300	1000	3000	5000
$C_{A_{out}}$ (mmol/L)	16	30	50	60

The volume of the reactor is 0.1 L, and, at the temperature of the reactor, the feed concentration is $C_{A_{in}} = 0.1$ mol/L. Determine the rate expression that fits the experimental data.

8.8₂ The first-order, liquid-phase reaction

$$A \longrightarrow R + P$$

is being studied in a CSTR. A liquid stream is fed into a CSTR, where reactant A decomposes according to the first-order reaction. The reactor is operated at different temperatures, and the feed flow rate is adjusted to keep the composition of the exit stream constant. Assuming constant feed concentration and constant density, determine the activation energy of the reaction from the following experimental results:

T (°C)	19	27	31	37
v_{in} (arbitrary units)	1	2	3	5

8.9$_2$ The second-order, gas-phase reaction

$$A \longrightarrow B + C$$

is carried out in a cascade of two isothermal CSTRs. Reactant A is fed at a rate of 100 mol/h into the first reactor, whose volume is 10 L. The molar flow rate of reactant A at the exit of the first reactor is 60 mol/h, and at the exit of the second reactor is 20 mol/h. The temperature in both reactors is 150°C, and the pressure is 2 atm. Determine the volume of Reactor 2 by:

a. Taking the inlet stream into the cascade as the reference stream.

b. Taking the inlet stream into Reactor 2 as the reference stream.

8.10$_2$ A biological reagent A decomposes by the liquid-phase reaction

$$A \longrightarrow R + P$$

The rate expression is

$$r = \frac{C_A}{0.2 + C_A} \text{ mol/L min}$$

We wish to treat 10 L/min of a waste liquid stream containing A ($C_{A_{in}} = 1$ mol/L), and we want to achieve 99% conversion of A. Two equal-size tanks (CSTR) are available. What is the best arrangement for the two tanks. Determine the size of the two units needed.

8.11$_4$ The first-order gas-phase reaction

$$A \longrightarrow B + C$$

takes place in a CSTR. A stream consisting of 90% of reactant A and 10% I (% mole) is fed into a 200-L reactor at a rate of 20 L/s. The feed is at 731 K and 3 atm.

a. Derive and plot the reaction and species curves for isothermal operation.

b. Determine the conversion of reactant A when the reactor is operated isothermally.

c. Determine the heating rate in part (b).

d. Determine the isothermal HTN.

e. Derive and plot the reaction, temperature, and species curves for adiabatic operation.

f. Determine the conversion of reactant A when the reactor is operated adiabatically.

g. Determine the reactor temperature in part (f).

h. Repeat parts (f) and (g) for nonisothermal operation with HTN half of the isothermal HTN.

Data: At 731 K, $k = 0.2 \, \text{s}^{-1}$, $E_a = 12,000 \, \text{cal/mol}$
$$\Delta H_R = -10,000 \, \text{cal/ mol extent}$$

$\hat{c}_{p_A} = 25 \, \text{cal/mol K}$ $\hat{c}_{p_B} = 15 \, \text{cal/mol K}$ $\hat{c}_{p_C} = 18 \, \text{cal/mol K}$

$\hat{c}_{p_I} = 9 \, \text{cal/mol K}$

8.12$_4$ The elementary liquid-phase reactions

$$A + B \longrightarrow C$$

$$C + B \longrightarrow D$$

take place in a cascade of two 100-L CSTRs. A solution $(C_{A_{in}} = 2 \, \text{mol/L}$ and $C_{B_{in}} = 2 \, \text{mol/L})$ at 80°C is fed into the first reactor at a rate of 200 L/min.

a. Derive and plot the reaction and species curves for isothermal operation.

b. Determine the conversion of reactant A and the production rate of product C at the exit of the second reactor when both reactors are operated isothermally.

c. Determine the heating rate of each reactor in part (b).

d. Determine the isothermal HTN of each reactor.

e. Derive and plot the reaction and species curves for adiabatic operation.

f. Determine the conversion of reactant A and the production rate of product C at the exit of the second reactor when both reactors are operated adiabatically.

g. Determine the temperature of each reactor in part (f).

h. Repeat parts (f) and (g) for nonisothermal operation with HTN half of the isothermal HTN.

Data: At 80°C, $k_1 = 0.1 \, \text{L/mol min}^{-1}$, $k_2 = 0.2 \, \text{L/mol min}^{-1}$

$$E_{a_1} = 12,000 \, \text{cal/mol} \quad E_{a_2} = 16,000 \, \text{cal/mol}$$

$$\Delta H_{R_1} = -15,000 \, \text{cal/mol extent} \quad \Delta H_{R_2} = -10,000 \, \text{cal/mol extent}$$

Density of the solution = 900 g/L. Heat capacity = 0.8 cal/g°C.

8.13$_4$ Methane is chlorinated in a gas phase at 400°C to produce mono-, di-, tri-, and tetrachloromethane. The desired products are CH_2Cl_2 and CCl_4. The following reactions take place:

$$CH_4 + Cl_2 \longrightarrow CH_3Cl + HCl$$

$$CH_3Cl + Cl_2 \longrightarrow CH_2Cl_2 + HCl$$

$$CH_2Cl_2 + Cl_2 \longrightarrow CHCl_3 + HCl$$

$$CHCl_3 + Cl_2 \longrightarrow CCl_4 + HCl$$

The reactions are elementary and the rate constants at $400°C$ are

$$k_1 = 30\,L/mol\ min \qquad k_2/k_1 = 3 \qquad k_3/k_1 = 1.5 \qquad k_4/k_1 = 0.375$$

A gas stream containing CH_4 and Cl_2 in the proportion of $1 : 1.2$ is fed at 1.2 atm into a CSTR operated at $400°C$. The feed rate is $1\ mol/min$.

a. Derive and plot the reaction and species operating curves.

b. What is the reactor volume needed to maximize the production of CH_2Cl_2?

c. What are the conversions of CH_4 and Cl_2?

d. What are the production rates of CH_3Cl, CH_2Cl_2, $CHCl_3$, and CCl_4 in (b)?

8.14₄ Methanol is produced by the gas-phase reaction, where the following reactions take place:

$$CO + 2H_2 \longrightarrow CH_3OH$$

$$CO + 3H_2 \longrightarrow CH_4 + H_2O$$

Both reactions are second order (each is first order with respect to each reactant) and $k_2/k_1 = 1.2$. A synthesis gas stream is fed into a CSTR operated at $450°C$ and 5 atm. Derive and plot the reaction and species operating curves for:

a. A feed consisting of $1/3$ CO and $2/3$ H_2 (mole basis).

b. A feed consisting of 50% CO and 50% H_2 (mole basis).

8.15₄ Below is a simplified kinetic model of the cracking of propane to produce ethylene.

$$C_3H_8 \longrightarrow C_2H_4 + CH_4 \qquad r_1 = k_1 C_{C_3H_8}$$

$$C_3H_8 \rightleftharpoons C_3H_6 + H_2 \qquad r_2 = k_2 C_{C_3H_8};\ \ r_{-2} = (k_2/K_2)C_{C_3H_6}C_{H_2}$$

$$C_3H_8 + C_2H_4 \longrightarrow C_2H_6 + C_3H_6 \qquad r_3 = k_3 C_{C_3H_8}C_{C_2H_4}$$

$$2C_3H_6 \longrightarrow 3C_2H_4 \qquad r_4 = k_4 C_{C_3H_6}$$

$$C_3H_6 \rightleftarrows C_2H_2 + CH_4 \quad r_5 = k_5 C_{C_3H_8}; \quad r_{-5}(k_5/K_5)C_{C_2H_2}C_{CH_4}$$

$$C_2H_4 + C_2H_2 \longrightarrow C_4H_6 \quad r_6 = k_6 C_{C_2H_4}C_{C_2H_2}$$

At 800°C, the values of the rate constants are $k_1 = 2.341$ s^{-1}, $k_2 = 2.12$ s^{-1}, $K_2 = 1000$, $k_3 = 23.63$ m^3/kmol s, $k_4 = 0.816$ s^{-1}, $k_5 = 0.305$ s^{-1}, $K_5 = 2000$, and $k_6 = 4.06 \times 10^3$ m^3/kmol s. You are asked to design a CSTR for the cracking of propane to be operated at 2 atm. Plot the reaction and species operating curves at 800°C.

9

OTHER REACTOR CONFIGURATIONS

In this chapter, the analysis of chemical reactors is expanded to additional reactor configurations that are commonly used to improve the yield and selectivity of the desirable products. In Section 9.1, we analyze semibatch reactors. Section 9.2 covers the operation of plug-flow reactors with continuous injection along their length. In Section 9.3, we examine the operation of one-stage distillation reactors, and Section 9.4 covers the operation of recycle reactors. In each section, we first derive the design equations, convert them to dimensionless forms, and then derive the auxiliary relations to express the species concentrations and the energy balance equation.

9.1 SEMIBATCH REACTORS

A semibatch reactor is a batch reactor into which one or more reactants are added continuously during the operation and no material is withdrawn. Semibatch reactors are used when it is desirable to maintain a low concentration of one reactant (the injected reactant). By adding this reactant slowly, as it is being consumed, its concentration is low throughout the operation. In some cases an inert species is injected to the reactor in order to supply (or remove) heat. For convenience, we divide semibatch reactors into two categories: reactors with liquid-phase reactions where the volume of the reacting fluid changes, shown schematically in the Figure 9.1a, and reactors with gas-phase reactions where the volume does not change, shown schematically in the Figure 9.1b.

Principles of Chemical Reactor Analysis and Design, Second Edition. By Uzi Mann
Copyright © 2009 John Wiley & Sons, Inc.

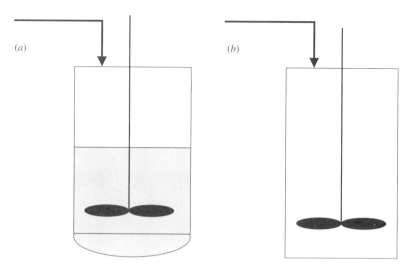

Figure 9.1   Semibatch reactors: (*a*) liquid phase (variable volume) and (*b*) gas phase (constant volume).

To derive the design equation of a semibatch reactor, we write a species balance for any species, say species *j*, that is *not* fed continuously into the reactor. Its molar balance equation is

$$\frac{dN_j}{dt} = (r_j)V_R(t) \tag{9.1.1}$$

where $V_R(t)$ is the volume of the reactor at time *t*. We follow the procedure used in Chapter 4 to derive the design equation of an ideal batch reactor for the *m*th-independent reaction:

$$\frac{dX_m}{dt} = \left(r_m + \sum_{k=1}^{n_D} a_{km}r_k \right)V_R(t) \tag{9.1.2}$$

where $X_m(t)$ is the extent of the *m*th-independent reaction in operating time *t*. Note that Eq. 9.1.2 is identical to the design equation of an ideal batch reactor (Eq. 4.3.8). The only differences between batch and semibatch operations are the way the reactor volume and the species concentrations vary during the operation.

Consider first a semibatch reactor with *liquid-phase* reactions. To derive a relation for the change in the volume of the reactor, we write an *overall* mass balance over the reactor:

$$\frac{d(\rho V_R)}{dt} = \rho_{\text{inj}}(t)v_{\text{inj}}(t) \tag{9.1.3}$$

where $v_{\text{inj}}(t)$ is the volumetric flow rate of the injected stream, and $\rho_{\text{inj}}(t)$ and ρ are the densities of the injected stream and the reacting liquid, respectively. For most liquid-phase reactions, the density of the feed is assumed to be the same as that of reacting liquid; hence, $\rho(t) = \rho_{\text{inj}}(t) = \rho$, and Eq. 9.1.3 reduces to

$$\frac{dV_R}{dt} = v_{\text{inj}}(t) \tag{9.1.4}$$

Since the volumetric injection flow rate $v_{\text{inj}}(t)$ may vary with time, we have to solve Eq. 9.1.4 simultaneously with Eq. 9.1.2.

The reactor volume at time t is

$$V_R(t) = V_R(0) + \int_0^t v_{\text{inj}}(x)\, dx \tag{9.1.5}$$

Substituting Eq. 9.1.5 into Eq. 9.1.2, the design equation of a semibatch reactor is

$$\frac{dX_m}{dt} = \left(r_m + \sum_{k=1}^{n_D} \alpha_{km} r_k \right) \left[V_R(0) + \int_0^t v_{\text{inj}}(x)\, dx \right] \tag{9.1.6}$$

To reduce the design equation to dimensionless form, we have to select a reference state and define dimensionless extents and dimensionless time. The reference state should apply to all operations, including those with an initially empty reactor, and should enable us to compare the operation of a semibatch reactor to that of a batch reactor. Therefore, we select the molar content of the reference state, $(N_{\text{tot}})_0$, as the *total* moles of species added to the reactor. The dimensionless extent is defined by

$$Z_m(t) \equiv \frac{\text{Extent of the } m\text{th-independent reaction at time } t}{\text{Total number of moles added to the reactor}} = \frac{X_m(t)}{(N_{\text{tot}})_0} \tag{9.1.7}$$

The total number of moles added to a semibatch reactor during the *entire* operation is

$$(N_{\text{tot}})_0 \equiv N_{\text{tot}}(0) + (N_{\text{tot}})_{\text{inj}} = N_{\text{tot}}(0) + \int_0^{t_{\text{op}}} v_{\text{inj}}(x)(C_0)_{\text{inj}}\, dx \tag{9.1.8}$$

where $N_{\text{tot}}(0)$ is the total number of moles initially charged to the reactor, $(N_{\text{tot}})_{\text{inj}}$ is the total moles injected, $(C_0)_{\text{inj}}$ is the *total* concentration of the injected stream, and t_{op} is the total operating time.

The reference volume, V_{R_0}, is defined as the total volume of reacting liquid added to the reactor during the entire operation:

$$V_{R_0} \equiv V_R(0) + V_{\text{inj}} = V_R(0) + \int_0^{t_{\text{op}}} v_{\text{inj}}(x)\, dx \tag{9.1.9}$$

where V_{inj} is the total volume injected during the operation. The concentration of the reference state, C_0, is defined by

$$C_0 \equiv \frac{(N_{tot})_0}{V_{R_0}} \qquad (9.1.10)$$

The dimensionless time is defined by

$$\tau \equiv \frac{t}{t_{cr}} \qquad (9.1.11)$$

where t_{cr} is the characteristic reaction time, defined in Section 3.5 (Eq. 3.5.1). Differentiating Eq. 9.1.7 and Eq. 9.1.11,

$$dX_m = (N_{tot})_0 \, dZ_m$$

$$dt = t_{cr} \, d\tau$$

and substituting these into Eq. 9.1.2, the dimensionless design equation for semi-batch reactor is

$$\frac{dZ_m}{d\tau} = \left(r_m + \sum_{k=1}^{n_D} \alpha_{km} r_k \right) \left[\frac{V_R(\tau)}{V_{R_0}} \right] \left(\frac{t_{cr}}{C_0} \right) \qquad 0 \leq \tau \leq \tau_{op} \qquad (9.1.12)$$

where $V_R(\tau)$ is given by Eq. 9.1.5 with $t = t_{cr}\tau$. Equation 9.1.12 is the general design equation of liquid-phase semibatch reactors with any given injection rate, written for the mth-independent reaction. To describe the reactor operation, we have to write the design equation for each *independent* chemical reaction. Note that Eq. 9.1.12 is valid only for $0 \leq \tau \leq \tau_{op}$, where τ_{op} is the dimensionless operating time defined by

$$\tau_{op} \equiv \frac{t_{op}}{t_{cr}} \qquad (9.1.13)$$

To solve the design equations, we should express the rates of all the chemical reactions (r_m's and r_k's) in terms of the extents of the independent reactions. The concentration of species j at time t is

$$C_j(t) \equiv \frac{N_j(t)}{V_R(t)} \qquad (9.1.14)$$

Using Eq. 2.3.3, the molar content of species j at time t is

$$N_j(t) = N_j(0) + \int_0^t v_{inj}(x)(C_j)_{inj} \, dx + \sum_m^{n_I} (s_j)_m X_m(t) \qquad (9.1.15)$$

where the first term on the right, $N_j(0)$, indicates the number of moles of species j charged to the reactor initially, the second term indicates the moles of species j

injected during time t, and the third term indicates the mole of species j formed by chemical reactions. Substituting Eq. 9.1.15 into Eq. 9.1.14,

$$C_j(t) = \frac{N_j(0) + \int_0^t v_{inj}(x)(C_j)_{inj}\, dx + \sum_m^{n_I} (s_j)_m X_m(t)}{V_R(0) + \int_0^t v_{inj}(x)\, dx} \tag{9.1.16}$$

Using Eq. 9.1.7 and Eq. 9.1.11, the concentration of species j at dimensionless time τ is

$$C_j(\tau) = C_0 \frac{[1/(N_{tot})_0]\left[N_j(0) + \int_0^{t_{cr}\tau} v_{inj}(x)(C_j)_{inj}\, dx\right] + \sum_m^{n_I} (s_j)_m Z_m(\tau)}{[1/V_{R_0}]\left[V_R(0) + \int_0^{t_{cr}\tau} v_{inj}(x)\, dx\right]} \tag{9.1.17}$$

For a species that is not charged initially into the reactor, $N_j(0) = 0$, and for a species that is not injected continuously, $(C_j)_{inj} = 0$.

When the injection rate is *uniform*, $v_{inj}(t) = (v_0)_{inj}$, Eq. 9.1.5 becomes

$$V_R(t) = V_R(0) + (V_0)_{inj}t \tag{9.1.18}$$

and using Eq. 9.1.9, the injection rate is

$$(v_0)_{inj} = \frac{V_{inj}}{t_{op}} = \frac{V_{R_0} - V_R(0)}{t_{op}} \tag{9.1.19}$$

where V_{inj} is the total injected volume. Substituting Eq. 9.1.18 and Eq. 9.1.19 into Eq. 9.1.12, the design equation of semibatch reactors with a uniform injection rate is

$$\frac{dZ_m}{d\tau} = \left(r_m + \sum_k^{n_D} \alpha_{km} r_k\right)\left[\frac{V_R(0)}{V_{R_0}} + \left(\frac{V_{inj}}{V_{R_0}}\right)\frac{\tau}{\tau_{op}}\right]\left(\frac{t_{cr}}{C_0}\right) \qquad 0 \le \tau \le \tau_{op} \tag{9.1.20}$$

and Eq. 9.1.17 reduces to

$$C_j(\tau) = C_0 \frac{\dfrac{N_j(0)}{(N_{tot})_0} + \left[\dfrac{(N_j)_{inj}}{(N_{tot})_0}\right]\left(\dfrac{\tau}{\tau_{op}}\right) + \sum_m^{n_I} (s_j)_m Z_m(\tau)}{\dfrac{V_R(0)}{V_{R_0}} + \left(\dfrac{V_{inj}}{V_{R_0}}\right)\dfrac{\tau}{\tau_{op}}} \tag{9.1.21}$$

For constant volume, *gas-phase* semibatch reactors, $V_R(t) = V_R(0) = V_{R_0}$, and the design equation (Eq. 9.1.2) reduces to

$$\frac{dX_m}{dt} = \left(r_m + \sum_{k=1}^{n_D} \alpha_{km} r_k \right) V_{R_0} \tag{9.1.22}$$

Substituting Eq. 9.1.7 and Eq. 9.1.11 into Eq. 9.1.22, the dimensionless design equation is

$$\frac{dZ_m}{d\tau} = \left(r_m + \sum_{k=1}^{n_D} \alpha_{km} r_k \right) \left(\frac{t_{cr}}{C_0} \right) \tag{9.1.23}$$

Using Eq. 9.1.17, the concentration of species j at time τ is

$$C_j(\tau) = C_0 \left[\frac{N_j(0)}{(N_{tot})_0} + \frac{1}{(N_{tot})_0} \int_0^{t_{cr}\tau} v_{inj}(x)(C_j)_{inj}\, dx + \sum_m^{n_I} (s_j)_m Z_m(\tau) \right] \tag{9.1.24}$$

For uniform injection rate, $v_{inj}(t) = (v_0)_{inj}$, and using Eq. 9.1.19, the concentration of species j is

$$C_j(\tau) = C_0 \left[\frac{N_j(0)}{(N_{tot})_0} + \left(\frac{(N_j)_{inj}}{(N_{tot})_0} \right) \left(\frac{\tau}{\tau_{op}} \right) + \sum_m^{n_I} (s_j)_m Z_m(\tau) \right] \tag{9.1.25}$$

To express the temperature changes, we write the energy balance equation. For semibatch reactors, the expansion work is usually negligible, and assuming isobaric operation, the energy balance equation is

$$\Delta H(t) = H(t) - H(0) = Q(t) + \int_0^t (\dot{m}_{inj} h_{inj})\, dt - W_{sh}(t) \tag{9.1.26}$$

where $(\dot{m}_{inj} h_{inj})$ is the rate enthalpy added to the reactor by the injection stream. Assuming no phase change, the change in the enthalpy of the reacting fluid is

$$H(t) - H(0) = M(t)\bar{c}_p[T(t) - T_0] - M(0)\bar{c}_p[T(0) - T_0]$$
$$+ \sum_m^{n_I} \Delta H_{R_m}(T_0) X_m(t) \tag{9.1.27}$$

Differentiating Eq. 9.1.26 and Eq. 9.1.27 with respect to time and noting that $dM/dt = \dot{m}_{inj}$,

$$\frac{dH}{dt} = \dot{Q} + \dot{m}_{inj}h_{inj} - \dot{W}_{sh}$$

$$\frac{dH}{dt} = [M(t)\bar{c}_p]\frac{dT}{dt} + \dot{m}_{inj}\bar{c}_p[T(t) - T_0] + \sum_m^{n_I} \Delta H_{R_m}(T_0)\frac{dX_m}{dt} \qquad (9.1.28)$$

where $[M(t)\bar{c}_p]$ is the heat capacity of the reacting fluid at time t. The rate heat is transferred to the reactor, $\dot{Q}$, and is expressed by

$$\dot{Q} = U\left(\frac{S}{V}\right)V_R(t)(T_F - T) \qquad (9.1.29)$$

Combining Eqs. 9.1.27, 9.1.28, and 9.1.29, and noting that from overall material balance

$$\frac{dM}{dt} = \dot{m}_{inj}$$

for reactors with negligible shaft work,

$$\frac{dT}{dt} =$$

$$\frac{1}{M(t)\bar{c}_p}\left[U\left(\frac{S}{V}\right)V_R(t)(T_F - T) + \dot{m}_{inj}\bar{c}_p[T - T_0] - \dot{m}_{inj}h_{inj} - \sum_m^{n_I} \Delta H_{R_m}(T_0)\frac{dX_m}{dt}\right]$$

$$(9.1.30)$$

Equation 9.1.30 describes the change in the reactor temperature.

To reduce Eq. 9.1.30 to dimensionless form, we relate the heat capacity of the reacting fluid to that of the reference state by defining a correction factor

$$M(t)\bar{c}_p \equiv CF(N_{tot})_0\hat{c}_{p0} \qquad (9.1.31)$$

where CF is the correction factor of the heat capacity at time t, and define the dimensionless temperature by

$$\theta \equiv \frac{T}{T_0} \qquad (9.1.32)$$

where T_0 is the temperature of the reference state.

For *liquid-phase* semibatch reactors, the enthalpy of the injected stream is

$$\dot{m}_{inj}h_{inj} = \dot{m}_{inj}\bar{c}_{p_{inj}}(T_{inj} - T_0) \qquad (9.1.33)$$

where $\bar{c}_{p_{inj}}$ is the specific mass-based heat capacity of the injection stream. The total mass of reacting fluid added to the reactor during the entire operation, M_{tot}, is

$$M_{tot} \equiv V_R(0)\rho + \int_0^{t_{op}} v_{inj}(x)\rho_{inj}\, dx \tag{9.1.34}$$

The specific molar-based heat capacity, $\hat{c}_{p_0}$, of the reference state is defined by

$$\hat{c}_{p_0} \equiv \frac{M_{tot}\bar{c}_p}{(N_{tot})_0} \tag{9.1.35}$$

where $(N_{tot})_0$ is defined by Eq. 9.1.8. For liquid-phase reactions, assuming constant density and constant specific heat capacity, the correction factor of the heat capacity is

$$CF = \frac{M(\tau)\bar{c}_p}{M_{tot}\bar{c}_p} = \frac{V_R(\tau)}{V_{R_0}} \tag{9.1.36}$$

where $V_R(\tau)$ is given by Eq. 9.1.5 with $t = t_{cr}\tau$. When there is no change in phase,

$$\dot{m}_{inj}h_{inj} = \dot{m}_{inj}\bar{c}_p\lfloor T_{inj} - T_0 \rfloor$$

Using Eqs. 9.1.7, 9.1.10, 9.1.11, 9.1.32, and 9.1.36, Eq. 9.1.30 reduces to

$$\frac{d\theta}{d\tau} = \frac{1}{CF}\left[HTN\left(\frac{V_R}{V_{R_0}}\right)(\theta_F - \theta) + \frac{\dot{m}_{inj}t_{cr}\bar{c}_p}{(N_{tot})_0\hat{c}_{p_0}}(\theta_{inj} - \theta) - \sum_m^{n_I} DHR_m \frac{dZ_m}{d\tau}\right] \tag{9.1.37}$$

where HTN is the heat-transfer number, defined by Eq. 5.2.22:

$$HTN \equiv \frac{Ut_{cr}}{C_0\hat{c}_{p_0}}\left(\frac{S}{V}\right)$$

and DHR_m is the dimensionless heat of reaction of the mth-independent reaction, defined by Eq. 5.2.23,

$$DHR_m \equiv \frac{\Delta H_{R_m}(T_0)}{T_0\hat{c}_{p_0}}$$

When the injection rate is uniform,

$$\dot{m}_{inj} = \frac{M_{inj}}{t_{op}}$$

Eq. 9.1.37 reduces to

$$\frac{d\theta}{d\tau} = \frac{1}{CF}\left[HTN\left(\frac{V_R}{V_{R_0}}\right)(\theta_F - \theta) + \frac{1}{\tau_{op}}\left(\frac{V_{inj}}{V_{R_0}}\right)(\theta_{inj} - \theta) - \sum_m^{n_I} DHR_m \frac{dZ_m}{d\tau}\right]$$

(9.1.38)

Eq. 9.1.36 reduces to

$$CF = \frac{V_R(0)}{V_{R_0}} + \left(\frac{V_{inj}}{V_{R_0}}\right)\frac{\tau}{\tau_{op}}$$

(9.1.39)

For *gas-phase* reactions, the heat capacity of the reacting fluid may vary with both the composition and temperature, and specific molar heat capacities are used. The rate enthalpy added by the injection stream is

$$\dot{m}_{in}h_{in} = \left(\sum_j^J (F_j)_{inj}\hat{c}_{p_j}(T_{inj})\right)(T_{inj} - T_0)$$

(9.1.40)

where $(F_j)_{inj}$ is the injection molar flow rate of species j. The heat capacity of the reference state is

$$(N_{tot})_0\hat{c}_{p_0} \equiv \sum_J^J [N_j(0) + (N_j)_{inj}]\hat{c}_{p_j}(T_0)$$

(9.1.41)

and the specific molar-based heat capacity of the reference state is

$$\hat{c}_{p_0} \equiv \frac{1}{(N_{tot})_0}\sum_j^J [N_j(0) + (N_j)_{inj}]\hat{c}_{p_j}(T_0)$$

(9.1.42)

For gas-phase reactions, the heat capacity of the reacting fluid is written in terms of the species molar contents:

$$M\bar{c}_p \equiv \sum_j^J N_j\hat{c}_{p_j}(T)$$

(9.1.43)

Differentiating with respect to time

$$\frac{dM}{dt}\bar{c}_p = \sum_j^J \frac{dN_j}{dt}\hat{c}_{p_j}(T)$$

(9.1.44)

From material balance over species j,

$$\frac{dN_j}{dt} = (F_j)_{inj} + \sum_m^{n_I} (s_j)_m \frac{dX_m}{dt}$$

(9.1.45)

Substituting Eqs. 9.1.31, 9.1.40, 9.1.43, 9.1.44, and 9.1.45 into Eq. 9.1.30 and then reducing the energy balance equation to dimensionless form, we obtain

$$
\frac{d\theta}{d\tau} = \frac{1}{CF(Z_m, \theta)} \left[HTN(\theta_F - \theta) + \frac{t_{cr} \sum_{j}^{J} (F_j)_{inj} \hat{c}_{p_j}(\theta_{inj})}{(N_{tot})_0 \hat{c}_{p_0}} (\theta_{inj} - 1) \right]
$$

$$
- \frac{1}{CF(Z_m, \theta)} \left[\sum_{j}^{J} \left(\sum_{m}^{n_I} (s_j)_m \frac{dZ_m}{d\tau} \right) \hat{c}_{p_j}(\theta)[\theta - 1] + \sum_{m}^{n_I} DHR_m \frac{dZ_m}{d\tau} \right] \quad (9.1.46)
$$

Equation 9.1.46 is the dimensionless energy balance equation for gas-phase, *constant-volume* semibatch reactors. The correction factor of the heat capacity is

$$
CF(Z_m, \theta) = \sum_{j}^{J} \left[\frac{N_j(0)}{(N_{tot})_0} + \frac{1}{(N_{tot})_0} \int_0^{t_{cr}\tau} (F_j)_{inj} \, dx + \sum_{m}^{n_I} (s_j)_m Z_m(\tau) \right] \frac{\hat{c}_{p_j}(\theta)}{\hat{c}_{p_0}} \quad (9.1.47)
$$

For *liquid-phase* isothermal semi-batch operations with uniform injection rate, we can use the energy balance equation (Eq. 3.1.38) to determine the needed heating (or cooling) load and to estimate the isothermal *HTN*. Recall that the first term in Eq. 9.1.38 expresses the rate of heat transfer to the reactor

$$
\frac{d}{d\tau} \left(\frac{Q}{(N_{tot})_0 \cdot \hat{c}_{p_0} \cdot T_0} \right) = HTN \left(\frac{V_R(\tau)}{V_{R_0}} \right)(\theta_F - \theta) \quad (9.1.48)
$$

where

$$
\frac{Q}{(N_{tot})_0 \cdot \hat{c}_{p_0} \cdot T_0} \quad (9.1.49)
$$

is the dimensionless heat transferred to the reactor. Hence, for isothermal operation $(d\theta/d\tau = 0)$,

$$
\frac{d}{d\tau} \left(\frac{Q}{(N_{tot})_0 \cdot \hat{c}_{p_0} \cdot T_0} \right) = -\frac{1}{\tau_{op}} \left(\frac{V_{inj}}{V_{R_0}} \right)(\theta_{inj} - \theta) + \sum_{m}^{n_I} DHR_m \frac{dZ_m}{d\tau} \quad (9.1.50)
$$

We determine the *HTN* at any instance by combining Eqs. 9.1.48 and 9.1.50 (selecting the operating temperature as the reference temperature, $\theta = 1$).

$$
HTN_{iso}(\tau_{op}) = \left(\frac{V_{R_0}}{V_R(\tau)} \right)
$$

$$
\times \frac{\tau_{op}}{(\theta_F - 1)} \left[-\frac{1}{\tau_{op}} \cdot \left(\frac{V_{inj}}{V_{R_0}} \right) \cdot (\theta_{inj} - 1) + \sum_{m}^{n_I} DHR_m \cdot \frac{dZ_m}{d\tau} \right] \quad (9.1.51)
$$

We define an average *HTN* for isothermal operation by

$$HTN_{ave} \equiv \frac{1}{\tau_{op}} \int_0^{\tau_{op}} HTN(\tau)d\tau \tag{9.2.52}$$

Example 9.1 A valuable product V is produced in a reactor where the following simultaneous chemical reactions take place:

$$\text{Reaction 1:} \quad \text{A} \longrightarrow \text{V}$$
$$\text{Reaction 2:} \quad 2\text{A} \longrightarrow \text{W}$$

Reaction 1 is first order and Reaction 2 is second order. To improve the yield of the desirable product, it was suggested to carry out the operation in a semibatch reactor. A 200-L aqueous solution of A with a concentration of 4 mol/L at 60°C is to be processed.

a. Show qualitatively the advantage of semibatch operation.

b. Derive the design equations, and plot the reaction and species operating curves for isothermal semibatch operation.

c. Compare the semibatch operations in (b) to the corresponding batch operation. Determine the operating time needed to achieve 90% conversion of A in (b) and (c) and the amounts of products V and W generated in each operation.

d. Determine the heating load, and estimate the isothermal HTN (with $T_F = 40°C$).

e. Repeat (b), (c), and (d) for adiabatic operation and compare the production of products V and W for $\tau_{op} = 8$.

Data: At 60°C, $k_1 = 0.1 \text{ min}^{-1}$, $k_2 = 0.15 \text{ L/mol min}^{-1}$

$$\Delta H_{R_1} = -8000 \text{ cal/mol} \qquad \Delta H_{R_2} = -12,000 \text{ cal/mol}$$

$$E_{a_1} = 9000 \text{ cal/mol} \qquad E_{a_2} = 12,000 \text{ cal/mol}$$

The density is 1000 g/L, and heat capacity is $\bar{c}_p = 1$ cal/g K.

Solution

a. To identify the preferable reactor operation mode, we examine the ratio of the formation rates of the desired and undesired products:

$$\frac{r_V}{r_W} = \frac{k_1(T)C_A}{k_2(T)C_A^2} = \left(\frac{k_1}{k_2}\right)\left(\frac{1}{C_A}\right)\exp\left(-\frac{E_{a_1} - E_{a_2}}{RT}\right)$$

Hence, we would like to maintain low concentration of reactant A. This is achieved by starting with an empty reactor and adding reactant A slowly, so that A is diluted with the products. Also, since $E_{a_1} < E_{a_2}$, it is preferable to operate the reactor at a low temperature, while achieving a reasonable reaction rate.

b. The stoichiometric coefficients of the chemical reactions are

$$s_{A_1} = -1 \qquad s_{V_1} = 1 \qquad s_{W_1} = 0 \qquad \Delta_1 = 0$$
$$s_{A_2} = -2 \qquad s_{V_2} = 0 \qquad s_{W_2} = 1 \qquad \Delta_2 = -1$$

The two chemical reactions are independent, and there is no dependent reaction. We write Eq. 9.1.20 for each independent reaction:

$$\frac{dZ_1}{d\tau} = r_1 \left(\frac{\tau}{\tau_{op}}\right)\left(\frac{t_{cr}}{C_0}\right) \qquad 0 \le \tau \le \tau_{op} \tag{a}$$

$$\frac{dZ_2}{d\tau} = r_2 \left(\frac{\tau}{\tau_{op}}\right)\left(\frac{t_{cr}}{C_0}\right) \qquad 0 \le \tau \le \tau_{op} \tag{b}$$

We select the total amount introduced into the reactor during the operation as the reference state. Since the reactor is empty initially, $V_R(0) = 0$, $N_{tot}(0) = N_A(0) = N_V(0) = 0$, and $N_W(0) = 0$. For uniform injection rate, $V_{R_0} = (v_0)_{inj} t_{op} = 200\,L$, and, since only reactant A is injected, $(C_0)_{inj} = (C_A)_{inj} = 4$ mol/L and $(N_{tot})_0 = 800$ mol. Using Eq. 9.1.10, the reference concentration is

$$C_0 = \frac{(N_{tot})_0}{V_{R_0}} = 4\,mol/L$$

Using Eq. 9.1.21, the concentration of reactant A in the reactor is

$$C_A(\tau) = C_0 \left(\frac{\tau_{op}}{\tau}\right)\left[\frac{\tau}{\tau_{op}} + (-1)Z_1(\tau) + (-2)Z_2(\tau)\right] \tag{c}$$

The rates of the two chemical reactions are

$$r_1 = k_1(T_0)C_0 \left(\frac{\tau_{op}}{\tau}\right)\left(\frac{\tau}{\tau_{op}} - Z_1 - 2Z_2\right)e^{\gamma_1(\theta-1)/\theta} \tag{d}$$

$$r_2 = k_2(T_0)C_0^2 \left(\frac{\tau_{op}}{\tau}\right)^2\left(\frac{\tau}{\tau_{op}} - Z_1 - 2Z_2\right)^2 e^{\gamma_2(\theta-1)/\theta} \tag{e}$$

We select Reaction 1 as the leading reaction; hence, the characteristic reaction time is

$$t_{cr} = \frac{1}{k_1(T_0)} = 10\,min \tag{f}$$

Substituting (d), (e), and (f) into (a) and (b), the design equations reduce to

$$\frac{dZ_1}{d\tau} = \left(\frac{\tau}{\tau_{op}} - Z_1 - 2Z_2\right)e^{\gamma_1(\theta-1)/\theta} \qquad 0 \le \tau \le \tau_{op} \qquad (g)$$

$$\frac{dZ_2}{d\tau} = \left[\frac{k_2(T_0)C_0}{k_1(T_0)}\right]\left(\frac{\tau_{op}}{\tau}\right)\left(\frac{\tau}{\tau_{op}} - Z_1 - 2Z_2\right)^2 e^{\gamma_2(\theta-1)/\theta} \qquad 0 \le \tau \le \tau_{op} \quad (h)$$

For isothermal operation, $\theta = 1$, and the design equations reduce to

$$\frac{dZ_1}{d\tau} = \frac{\tau}{\tau_{op}} - Z_1 - 2Z_2 \qquad 0 \le \tau \le \tau_{op} \qquad (i)$$

$$\frac{dZ_2}{d\tau} = \left[\frac{k_2(T_0)C_0}{k_1(T_0)}\right]\left(\frac{\tau_{op}}{\tau}\right)\left(\frac{\tau}{\tau_{op}} - Z_1 - 2Z_2\right)^2 \qquad 0 \le \tau \le \tau_{op} \qquad (j)$$

We solve (i) and (j) for different values of operating times, τ_{op}, subject to the initial conditions that at $\tau = 0$, $Z_1 = Z_2 = 0$, and determine the two extents for each τ_{op}. Figure E9.1.1 shows the two reaction operating curves. Using Eq. 9.1.15, we determine the species operating curves, $N_j(\tau_{op})/(N_{tot})_0$, shown in the Figure E9.1.2.

c. We compare the species curves of the two products to those of an ideal batch reactor (design equations derived below). Figure E9.1.3 shows the curves of the desired product V, and Figure E9.1.4 shows the curves of the undesirable product W. An examination of the two figures indicates that additional amount of desirable product V can be obtained by semibatch operation over a longer operating times. From the species operating curves, $N_A(\tau_{op})/(N_{tot})_0 = 0.1$ (90% conversion of A) is reached at $\tau_{op} = 4.4$, which corresponds to 44 min. At this τ_{op}, $N_V(\tau_{op})/(N_{tot})_0 = 0.309$ and $N_W(\tau_{op})/(N_{tot})_0 = 0.296$ which correspond to 247.2 and 236.8 mol, respectively. Similarly, for isothermal batch operation, $N_A(\tau)/(N_{tot})_0 = 0.1$ is reached at $\tau_{op} = 0.56$, which corresponds to 5.6 min. At this operating

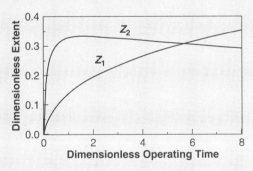

Figure E9.1.1 Reaction operating curves—semibatch.

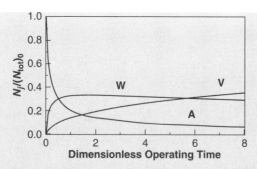

Figure E9.1.2 Species operating curves—semibatch; isothermal operation.

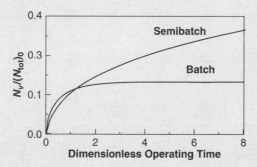

Figure E9.1.3 Comparison of V production—isothermal operation.

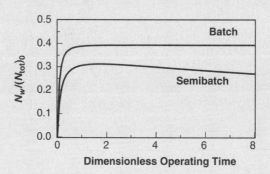

Figure E9.1.4 Comparison of W production—isothermal operation.

time, $N_V(\tau)/(N_{tot})_0 = 0.151$ and $N_W(\tau)/(N_{tot})_0 = 0.388$, which correspond to 120.8 and 302.4 mol, respectively. Hence, the operating time of semibatch is about 8 times longer than that of a batch operation, but more than twice the amount of product V, and a smaller amount of undesirable product W are generated.

d. The heating load curve is calculated by Eq. 9.1.50, and is shown in Figure E9.1.5. For 90% conversion, $\tau_{op} = 4.4$, and $Q/[(N_{tot})_0 \hat{c}_{p_0} T_0] =$

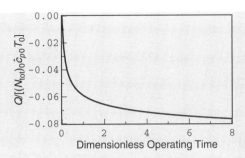

Figure E9.1.5 Heat-transfer curve—semibatch; isothermal operation.

-0.0723, which corresponds to -4817.3 kcal. The instantaneous isothermal HTN is calculated by Eq. 9.1.51.

e. For adiabatic semibatch operation, we have to solve design equations (g) and (h) simultaneously with the energy balance equation. Using Eq. 9.1.35, the specific molar heat capacity of the reference state is

$$\hat{c}_{p0} = \frac{(200\,\text{L})(1000\,\text{g/L})(1\,\text{cal/g K})}{800\,\text{mol}} = 250\,\text{cal/mol K}$$

The reference temperature is $T_0 = 333$ K, and the dimensionless heats of reactions are

$$\text{DHR}_1 = \frac{\Delta H_{R_1}(T_0)}{T_0 \hat{c}_{p0}} = -0.0961 \qquad \text{DHR}_2 = \frac{\Delta H_{R_2}(T_0)}{T_0 \hat{c}_{p0}} = -0.1441$$

The dimensionless activation energies of the two chemical reactions are

$$\gamma_1 = \frac{E_{a_1}}{RT_0} = 13.6 \qquad \gamma_2 = \frac{E_{a_2}}{RT_0} = 18.13$$

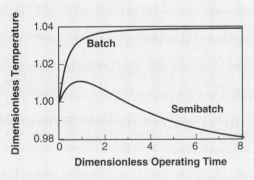

Figure E9.1.6 Temperature curve—adiabatic operation.

For adiabatic operation, $HTN = 0$. For uniform injection rate, using Eq. 9.1.39, the correction factor of the heat capacity is

$$CF(Z_m, \theta) = \frac{\tau}{\tau_{op}} \qquad (r)$$

Since $T_{inj} = T_0(\theta_{inj} = 1)$, the energy balance equation (Eq. 9.1.38) reduces to

$$\frac{d\theta}{d\tau} = \left(\frac{\tau_{op}}{\tau}\right)\left[\frac{1}{\tau_{op}}\left(\frac{V_{inj}}{V_{R_0}}\right)(1 - \theta) - DHR_1\frac{dZ_1}{d\tau} - DHR_2\frac{dZ_2}{d\tau}\right] \qquad (s)$$

We solve (s) simultaneously with (g) and (h) for different values of operating times, τ_{op}, subject to the initial conditions that at $\tau = 0$, $Z_1 = Z_2 = 0$, and $\theta = 1$, and we determine Z_1, Z_2, and θ for each τ_{op}. Once we obtain the reaction operating curves (Z_1 and Z_2 versus τ_{op}), we use Eq. 9.1.15 to obtain the species operating curves. Figure E9.1.7 shows the curve for product V and compares it to that of an adiabatic batch reactor. Figure E9.1.8 shows the curve for product W and compares it to that of an adiabatic batch reactor. Figure E9.1.6 shows the temperature curve and compares it to that of a

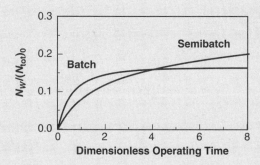

Figure E9.1.7   Comparison of V production—adiabatic operation.

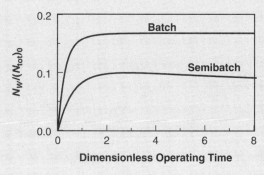

Figure E9.1.8   Comparison of W production—adiabatic operation.

batch reactor. Note that since the activation energy of the second reaction is higher than that of the first, and both reactions are exothermic, isothermal operation is preferable.

For a batch reactor (constant volume), the design equations (Eq. 7.1.1) are

$$\frac{dZ_1}{d\tau} = r_1 \left(\frac{t_{cr}}{C_0} \right) \tag{t}$$

$$\frac{dZ_2}{d\tau} = r_2 \left(\frac{t_{cr}}{C_0} \right) \tag{u}$$

Using Eq. 6.1.5 to express the species concentrations, the reaction rates are

$$r_1 = k_1(T_0)C_0(1 - Z_1 - 2Z_2)e^{\gamma_1(\theta-1)/\theta} \tag{v}$$

$$r_2 = k_2(T_0)C_0^2(1 - Z_1 - 2Z_2)^2 e^{\gamma_2(\theta-1)/\theta} \tag{w}$$

Substituting (v), (w), and (f) into (t) and (u), the design equations become

$$\frac{dZ_1}{d\tau} = (1 - Z_1 - 2Z_2)e^{\gamma_1(\theta-1)/\theta} \tag{x}$$

$$\frac{dZ_2}{d\tau} = \left[\frac{k_2(T_0)C_0}{k_1(T_0)} \right] (1 - Z_1 - 2Z_2)^2 e^{\gamma_2(\theta-1)/\theta} \tag{y}$$

For isothermal operation, $\theta = 1$, and (x) and (y) are solved simultaneously. For adiabatic operation, the energy balance equation is

$$\frac{d\theta}{d\tau} = -\text{DHR}_1 \frac{dZ_1}{d\tau} - \text{DHR}_2 \frac{dZ_2}{d\tau} \tag{z}$$

We solve (x), (y), and (z) for different values of operating times, τ_{op}, subject to the initial conditions that $\tau = 0$, $Z_1 = Z_2 = 0$, and $\theta = 1$. The species curves of the product V, and the undesirable product W, are compared in the figures above.

Example 9.2 Valuable product V is produced in a semibatch reactor where the following simultaneous, liquid-phase chemical reactions take place.

$$\text{Reaction 1:} \quad \text{A} + \text{B} \longrightarrow \text{V}$$
$$\text{Reaction 2:} \quad 2\text{A} \longrightarrow \text{W}$$

The rate expressions are $r_1 = k_1 C_A C_B$, and $r_2 = k_2 C_A^2$. The reactor is initially charged with a 200-L solution of reactant B with a concentration of $C_B(0) = 3$ mol/L, and 200-L solution of reactant A ($C_A = 3$ mol/L) is fed continuously into the reactor at a constant rate. Both the injected stream temperature and the initial reactor temperature are 60°C.

a. Show qualitatively the advantage of semi-batch operation over batch operation.

b. Derive the design equations, and plot the reaction and species curves for isothermal semi-batch operation. Compare the product curves to those of isothermal batch operation. What is the operating time for each mode of operation for 80% conversion of reactant A? What is the amount of product V and W generated in each mode?

c. Derive the design equations, and plot the reaction and species curves for adiabatic semi-batch operation. Compare the product curves to those of adiabatic batch operation with $T(0) = 60°C$, and $T_{inj} = 50°C$.

d. Derive the design and energy balance equations for batch operation.

Data: At 60°C, $k_1 = 0.02\,L/mol\ min^{-1}$, $k_2 = 0.04\,L/mol\ min^{-1}$

$$\Delta H_{R_1} = -9000\,cal/mol \qquad \Delta H_{R_2} = -13,000\,cal/mol$$

$$E_{a_1} = 12,000\,cal/mol \qquad E_{a_2} = 20,000\,cal/mol$$

The heat capacity of the solution is $\bar{c}_p = 0.9\,cal\ g^{-1}\ K^{-1}$, and its density is $0.85\,kg/L$.

Solution

a. To identify the preferable reactor operation mode, we write the ratio of the formation rates of the desired and undesired products:

$$\frac{r_V}{r_W} = \frac{k_1(T)C_A C_B}{k_2(T)C_A^2} = \left(\frac{k_1}{k_2}\right)\left(\frac{C_B}{C_A}\right)\exp\left(-\frac{E_{a_1} - E_{a_2}}{RT}\right)$$

Hence, we would like to maintain high concentration of reactant B and low concentration of reactant A. This is achieved by charging the reactor with reactant B and then injecting reactant A slowly. Also, since $E_{a_1} < E_{a_2}$, it is preferable to operate the reactor at a lower temperature.

b. The stoichiometric coefficients of the chemical reactions are

$$
\begin{array}{lllll}
s_{A_1} = -1 & s_{B_1} = -1 & s_{V_1} = 1 & s_{W_1} = 0 & \Delta_1 = -1 \\
s_{A_2} = -2 & s_{B_2} = 0 & s_{V_2} = 0 & s_{W_2} = 1 & \Delta_2 = -1
\end{array}
$$

Since each reaction has a species that does not appear in the other, the two reactions are independent, and there is no dependent reaction. We select the total amount introduced into the reactor during the operation as the reference state. Hence, $V_R(0) = 200\,\text{L}$, and, for constant injection rate, $(V_{\text{inj}})_0 = v_0 t_{\text{op}} = 200\,\text{L}$, and

$$V_{R_0} = V_R(0) + (V_{\text{inj}})_0 = 200 + 200 = 400\,\text{L}$$

Since only reactant B is charged initially, $N_{\text{tot}}(0) = N_B(0) = 600\,\text{mol}$, $N_A(0) = 0$, $N_V(0) = 0$, and $N_W(0) = 0$. Also, since only reactant A is injected, $(N_{\text{tot}})_{\text{inj}} = (C_A)_{\text{inj}}(V_{\text{inj}})_0 = 600\,\text{mol}$, and the total molar content of the reference state is

$$(N_{\text{tot}})_0 = N_{\text{tot}}(0) + (N_{\text{inj}})_0 = 1200\,\text{mol}$$

The reference concentration is

$$C_0 = \frac{(N_{\text{tot}})_0}{V_{R_0}} = \frac{1200\,\text{mol}}{400\,\text{L}} = 3\,\text{mol/L}$$

We write Eq. 9.1.20 for each independent reaction:

$$\frac{dZ_1}{d\tau} = r_1 \left[\frac{1}{2}\left(1 + \frac{\tau}{\tau_{\text{op}}}\right) \right] \left(\frac{t_{\text{cr}}}{C_0} \right) \qquad 0 \le \tau \le \tau_{\text{op}} \qquad (a)$$

$$\frac{dZ_2}{d\tau} = r_2 \left[\frac{1}{2}\left(1 + \frac{\tau}{\tau_{\text{op}}}\right) \right] \left(\frac{t_{\text{cr}}}{C_0} \right) \qquad 0 \le \tau \le \tau_{\text{op}} \qquad (b)$$

Using Eq. 9.1.21, the concentrations of the reactants in the reactor are

$$C_A(\tau) = C_0 \left(\frac{2\tau_{\text{op}}}{\tau_{\text{op}} + \tau} \right) \left[(0.5)\frac{\tau}{\tau_{\text{op}}} + (-1)Z_1(\tau) + (-2)Z_2(\tau) \right] \qquad (c)$$

$$C_B(\tau) = C_0 \left(\frac{2\tau_{\text{op}}}{\tau_{\text{op}} + \tau} \right) [(0.5) + (-1)Z_1(\tau)] \qquad (d)$$

The rates of the two reactions are

$$r_1 = k_1(T_0)C_0^{\,2} \left(\frac{2\tau_{\text{op}}}{\tau_{\text{op}} + \tau} \right)^2 \left((0.5)\frac{\tau}{\tau_{\text{op}}} - Z_1 - 2Z_2 \right)(0.5 - Z_1)e^{\gamma_1(\theta - 1)/\theta} \qquad (e)$$

$$r_2 = k_2(T_0)C_0^{\,2} \left(\frac{2\tau_{\text{op}}}{\tau_{\text{op}} + \tau} \right)^2 \left((0.5)\frac{\tau}{\tau_{\text{op}}} - Z_1 - 2Z_2 \right)^2 e^{\gamma_2(\theta - 1)/\theta} \qquad (f)$$

We select Reaction 1 as the leading reaction; hence, the characteristic reaction time is

$$t_{\text{cr}} = \frac{1}{k_1(T_0)C_0} = 16.67\,\text{min} \qquad (g)$$

Substituting (e), (f), and (g) into (a) and (b), the design equations reduce to

$$\frac{dZ_1}{d\tau} = \left(\frac{2\tau_{op}}{\tau_{op}+\tau}\right)\left[(0.5)\frac{\tau}{\tau_{op}}-Z_1-2Z_2\right](0.5-Z_1)e^{\gamma_1(\theta-1)/\theta} \quad 0\leq\tau\leq\tau_{op} \qquad (h)$$

$$\frac{dZ_2}{d\tau} = \left[\frac{k_2(T_0)}{k_1(T_0)}\right]\left(\frac{2\tau_{op}}{\tau_{op}+\tau}\right)\left[(0.5)\frac{\tau}{\tau_{op}}-Z_1-2Z_2\right]^2 e^{\gamma_2(\theta-1)/\theta} \quad 0\leq\tau\leq\tau_{op} \qquad (i)$$

For isothermal operation, $\theta = 1$, and the design equations reduce to

$$\frac{dZ_1}{d\tau} = \left(\frac{2\tau_{op}}{\tau_{op}+\tau}\right)\left[(0.5)\frac{\tau}{\tau_{op}}-Z_1-2Z_2\right](0.5-Z_1) \quad 0\leq\tau\leq\tau_{op} \qquad (j)$$

$$\frac{dZ_2}{d\tau} = \left[\frac{k_2(T_0)}{k_1(T_0)}\right]\left(\frac{2\tau_{op}}{\tau_{op}+\tau}\right)\left[(0.5)\frac{\tau}{\tau_{op}}-Z_1-2Z_2\right]^2 \quad 0\leq\tau\leq\tau_{op} \qquad (k)$$

We solve (j) and (k) for different values of operating times, τ_{op}, subject to the initial conditions that at $\tau = 0$, $Z_1 = Z_2 = 0$, and then determine the final extents for each τ_{op}. Figure E9.2.1 shows the reaction curves for isothermal semibatch operation. Using Eq. 9.1.15, we determine the species curves, shown in Figure E9.2.2. The curves of products V and W are compared to those of isothermal batch operation (derived below) in Figures E9.2.3 and E9.2.4, respectively.

For semibatch operation, 80% conversion of reactant A ($N_A/(N_{tot})_0 = 0.1$), is reached at $\tau_{op} = 6.95$, which corresponds to 116 min. At that operating time, $Z_1 = 0.180$, $Z_2 = 0.08$, $N_V/(N_{tot})_0 = 0.20$, and $N_W/(N_{tot})_0 = 0.10$. Hence, 240 mol of product V and 120 mol of product W were produced. For batch operation, 80% conversion is reached at $\tau_{op} = 1.26$, which corresponds to 21 min. At that operating time, $N_V/(N_{tot})_0 = 0.118$, and $N_W/(N_{tot})_0 = 0.141$. Hence, 141.6 mole of product V and 169.2 mol of product W were produced.

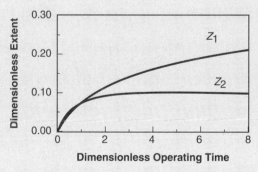

Figure E9.2.1 Reaction operating curves; isothermal semibatch operation.

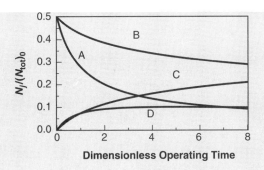

Figure E9.2.2   Species operating curves; isothermal semibatch operation.

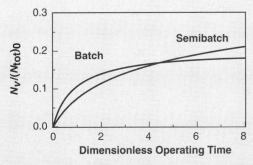

Figure E9.2.3   Comparison with batch reactor—product V (isothermal operation).

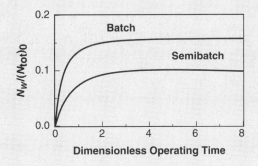

Figure E9.2.4   Comparison with batch reactor—product W (isothermal operation).

c. For nonisothermal semibatch operations, we have to solve design equations (h) and (i) simultaneously with the energy balance equation. Using Eq. 9.1.35,

$$\hat{c}_{p0} = \frac{(400 \, \text{L})(850 \, \text{g/L})(0.9 \, \text{cal/g K})}{(1200 \, \text{mol})} = 255 \, \text{cal/mol K}$$

The reference temperature is $T_0 = 333$ K, and the dimensionless heats of reactions are

$$\text{DHR}_1 = \frac{\Delta H_{R_1}(T_0)}{T_0 \hat{c}_{p_0}} = -0.106 \qquad \text{DHR}_2 = \frac{\Delta H_{R_2}(T_0)}{T_0 \hat{c}_{p_0}} = -0.153$$

The dimensionless activation energies of the two reactions are

$$\gamma_1 = \frac{E_{a_1}}{RT_0} = 18.14 \qquad \gamma_2 = \frac{E_{a_2}}{RT_0} = 30.21$$

For adiabatic operation, HTN $= 0$. Using Eq. 9.1.39,

$$\text{CF}(Z_m, \theta) = \frac{1}{2}\left(1 + \frac{\tau}{\tau_{\text{op}}}\right) \tag{l}$$

The energy balance equation (Eq. 9.1.37) reduces to

$$\frac{d\theta}{d\tau} = \left(\frac{2\tau_{\text{op}}}{\tau_{\text{op}} + \tau}\right)\left[\frac{(N_{\text{tot}})_{\text{inj}}}{(N_{\text{tot}})_0}\frac{1}{\tau_{\text{op}}}(\theta_{\text{inj}} - \theta) + \text{DHR}_1\frac{dZ_1}{d\tau} + \text{DHR}_2\frac{dZ_2}{d\tau}\right] \tag{m}$$

In this case, $\theta_{\text{inj}} = 0.97$. We solve (m) simultaneously with (h) and (i) for different values of τ_{op}, subject to the initial conditions that at $\tau = 0$, $Z_1 = Z_2 = 0$, $\theta = 1$, and then determine the extents and θ at each τ_{op}. Figure E9.2.5 shows the reaction operating curves and Figure E9.2.6 shows the temperature curve. Using Eq. 9.1.15 we determine the species curves. The curves of products V and W are shown in Figures E9.2.7 and E9.2.8, respectively, and are compared to those of adiabatic batch operation.

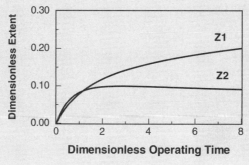

Figure E9.2.5 Reaction curves—semibatch; adiabatic operation.

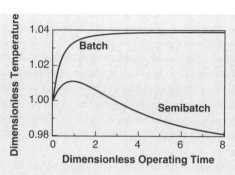

Figure E9.2.6   Temperature curve; adiabatic operation.

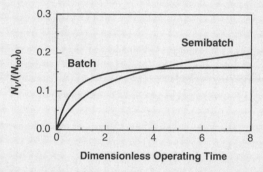

Figure E9.2.7   Comparison with batch reactor—V production; adiabatic operation.

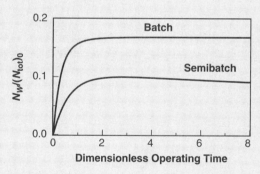

Figure E9.2.8   Comparison with batch reactor—W production; adiabatic operation.

d. For a constant-volume batch reactor, the design equations (Eq. 6.1.1) are

$$\frac{dZ_1}{d\tau} = r_1\left(\frac{t_{cr}}{C_0}\right) \tag{n}$$

$$\frac{dZ_2}{d\tau} = r_2\left(\frac{t_{cr}}{C_0}\right) \tag{o}$$

When both reactant solutions are charged initially into the reactor, we use the design equation of batch reactor with $C_0 = 3.0 \, \text{mol/L}$ and $y_A(0) = y_B(0) = 0.5$. Using Eq. 6.1.10 to express the species concentrations, the rates of the reactions are

$$r_1 = k_1(T_0)C_0^2(0.5 - Z_1 - 2Z_2)(0.5 - Z_1)e^{\gamma_1(\theta-1/\theta)} \tag{p}$$

$$r_2 = k_2(T_0)C_0^2(0.5 - Z_1 - 2Z_2)^2 e^{\gamma_2(\theta-1/\theta)} \tag{q}$$

The design equations are

$$\frac{dZ_1}{d\tau} = (0.5 - Z_1 - 2Z_2)(0.5 - Z_1)e^{\gamma_1(\theta-1/\theta)} \tag{r}$$

$$\frac{dZ_2}{d\tau} = \left[\frac{k_2(T_0)}{k_1(T_0)}\right](0.5 - Z_1 - 2Z_2)^2 e^{\gamma_2(\theta-1/\theta)} \tag{s}$$

For isothermal operation, $\theta = 1$, and we solve these equations and calculate the dimensionless species operating curves. To compare semibatch and batch operations, we plot the species operating curves for V and W, shown above. For adiabatic batch operation, the energy balance equation is

$$\frac{d\theta}{d\tau} = \text{DHR}_1 \frac{dZ_1}{d\tau} + \text{DHR}_2 \frac{dZ_2}{d\tau} \tag{t}$$

We solve (t) simultaneously with (r) and (s) subject to the initial conditions that $Z_1(0) = Z_2(0) = 0$ and $\theta(0) = 1$, and obtain the reaction and species operating curves for batch operation. An examination of the product operating curves indicates the advantage of selecting the semibatch mode to reduce the production of undesirable product W.

9.2 PLUG-FLOW REACTOR WITH DISTRIBUTED FEED

In this section, we discuss the operation of a plug-flow reactor with a distributed feed along the reactor length, shown schematically in Figure 9.2. This reactor configuration is the flow reactor analog to semibatch operation. While in semibatch reactors products are generated over time, here products are generated over space

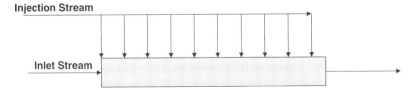

Figure 9.2 Plug-flow reactor with distributed feed.

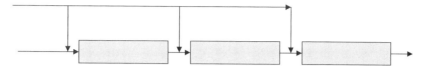

Figure 9.3   Actual implementation of distributed feed.

(volume). As for semibatch reactors, a plug-flow reactor with a distributed feed is used when it is desirable to maintain one reactant at a high concentration and another reactant at low concentration in order to improve the yield of a desirable product. In practice, it is not easy to control the injection rate along the reactor. Rather, the plug-flow reactor is divided into distinct sections, each with an injection stream as shown in Figure 9.3. Nevertheless, a plug-flow reactor with a distributed feed serves as the limiting model when the number of sections approaches infinity.

To derive the design equation of a plug-flow reactor with a distributed feed, we write a species balance equation for any species, say species j, that is *not* injected along the reactor over reactor element dV. Since the species is not fed or withdrawn, its molar balance equation is

$$dF_j = (r_j)\,dV \tag{9.2.1}$$

We follow the same procedure as the one used in Chapter 4 to derive the reaction-based design equation of a plug-flow reactor and obtain

$$\frac{d\dot{X}_m}{dV} = r_m + \sum_{k}^{n_D} \alpha_{km} r_k \tag{9.2.2}$$

where $\dot{X}_m$ is the extent per unit time of the mth-independent reaction. Equation 9.2.2 is the differential design equation for a plug-flow reactor with a distributed feed, written for the mth-independent chemical reaction. Note that it is identical to the design equation of a plug-flow reactor. The only difference between the two is the way the volumetric flow rate and the species concentrations vary along the reactor.

To reduce the design equation to a dimensionless form, we select a reference stream that combines the inlet steam and the injection stream. Hence, its total volumetric flow rate is

$$v_0 \equiv v_{\text{in}} + v_{\text{inj}} \tag{9.2.3}$$

Its molar flow rate is

$$(F_{\text{tot}})_0 \equiv (F_{\text{tot}})_{\text{in}} + (F_{\text{tot}})_{\text{inj}} \tag{9.2.4}$$

The reference concentration is defined by

$$C_0 \equiv \frac{(F_{\text{tot}})_0}{v_0} \qquad (9.2.5)$$

The dimensionless extent of the m-th independent reaction is defined by

$$Z_m \equiv \frac{\dot{X}_m}{(F_{\text{tot}})_0} \qquad (9.2.6)$$

Also, the dimensionless space time (or dimensionless volume) is defined by

$$\tau \equiv \frac{V}{v_0 t_{\text{cr}}} \qquad (9.2.7)$$

Differentiating Eqs. 9.2.6 and 9.2.7, $d\dot{X}_m = (F_{\text{tot}})_0 \, dZ_m$, and $dV = (v_0 t_{\text{cr}}) \, d\tau$, and substituting these into Eq. 9.2.2, the design equation reduces to

$$\frac{dZ_m}{d\tau} = \left(r_m + \sum_{k}^{n_D} \alpha_{km} r_k \right)\left(\frac{t_{\text{cr}}}{C_0} \right) \qquad (9.2.8)$$

Equation 9.2.8 is the dimensionless design equation, written for the mth-independent reaction. To describe the reactor operation, we have to write Eq. 9.2.8 for each independent reaction.

To solve the design equations, we have to express reaction rates in terms of the extents of the independent reactions. Using stoichiometric relation (Eq. 2.3.3), the local molar flow rate of species j at volume V from the inlet is

$$F_j = (F_j)_{\text{in}} + \int_0^V \lambda(x)(C_j)_{\text{inj}} \, dx + \sum_{m}^{n_I} (s_j)_m \dot{X}_m \qquad (9.2.9)$$

where $\lambda(x)$ is the local volumetric injection rate per unit reactor volume, $(C_j)_{\text{inj}}$ is the concentration of species j in the injection stream, and $(F_j)_{\text{in}}$ is the rate species j is fed at the reactor inlet. In general, the mode of the injection rate along the reactor, $\lambda(x)$, should be specified. The local concentration of species j is

$$C_j = \frac{F_j}{v} = \frac{1}{v}\left[(F_j)_{\text{in}} + \int_0^V \lambda(x)(C_j)_{\text{inj}} \, dx + \sum_{m}^{n_I} (s_j)_m \dot{X}_m \right] \qquad (9.2.10)$$

where v is the local volumetric flow rate. The first term in the bracket indicates the molar flow rate of species j fed at the reactor inlet, the second indicates the amount injected, and the third indicates the amount generated by the chemical reactions.

For *liquid-phase* reactions, the density of the reacting fluid does not vary much (and it is assumed equal to that of the inlet stream and the injected stream), the volumetric flow rate at any point in the reactor is

$$v = v_{\text{in}} + \int_0^V \lambda(x)\, dx \qquad (9.2.11)$$

When the injection flow rate is uniform,

$$\lambda(x) = \lambda_0 = \frac{v_{\text{inj}}}{V_{R_{\text{tot}}}} \qquad (9.2.12)$$

where $V_{R_{\text{tot}}}$ is the total volume of the reactor, and the local volumetric flow rate is

$$v = v_{\text{in}} + v_{\text{inj}} \left(\frac{V}{V_{R_{\text{tot}}}} \right) \qquad (9.2.13)$$

Substituting Eq. 9.2.13 into Eq. 9.2.10, the local species concentration is

$$C_j = C_0 \frac{(v_{\text{in}}/v_0)(C_j)_{\text{in}}/C_0 + (v_{\text{inj}}/v_0)(\tau/\tau_{\text{tot}})(C_j)_{\text{inj}}/C_0 + \sum_{m}^{n_I} (s_j)_m Z_m}{(v_{\text{in}}/v_0) + (v_{\text{inj}}/v_0)(\tau/\tau_{\text{tot}})} \qquad (9.2.14)$$

where τ_{tot} is the total dimensionless space time of the reactor, defined by

$$\tau_{\text{tot}} \equiv \frac{V_{R_{\text{tot}}}}{v_0 t_{\text{cr}}} \qquad (9.2.15)$$

For *gas-phase* reactions, assuming ideal-gas behavior, the local volumetric flow rate is

$$v = v_0 \left[\frac{F_{\text{tot}}}{(F_{\text{tot}})_0} \right] \left(\frac{T}{T_0} \right) \left(\frac{P_0}{P} \right) \qquad (9.2.16)$$

where F_{tot} is the local total molar flow rate given by

$$F_{\text{tot}} = (F_{\text{tot}})_{\text{in}} + \int_0^V \lambda(x)(C_{\text{tot}})_{\text{inj}}\, dx + \sum_{m}^{n_I} \Delta_m \dot{x}_m \qquad (9.2.17)$$

For uniform injection along the reactor, and assuming negligible pressure drop, the local volumetric flow rate is

$$v = v_0 \left[\frac{(F_{tot})_{in}}{(F_{tot})_0} + \frac{(F_{tot})_{inj}}{(F_{tot})_0} \left(\frac{\tau}{\tau_{tot}} \right) + \sum_m^{n_I} \Delta_m Z_m \right] \theta \qquad (9.2.18)$$

where $(F_{tot})_{in}$ and $(F_{tot})_{inj}$ are the molar flow rates of the inlet and the injection streams, respectively. Combining Eqs. 9.2.18 and 9.2.10, the local species concentration is

$$C_j = C_0 \frac{\dfrac{(F_j)_{in}}{(F_{tot})_0} + \dfrac{(F_j)_{inj}}{(F_{tot})_0} \left(\dfrac{\tau}{\tau_{tot}} \right) + \sum_m^{n_I} (s_j)_m Z_m}{\left[\dfrac{(F_{tot})_{in}}{(F_{tot})_0} + \dfrac{(F_{tot})_{inj}}{(F_{tot})_0} \left(\dfrac{\tau}{\tau_{tot}} \right) + \sum_m^{n_I} \Delta_m Z_m \right] \theta} \qquad (9.2.19)$$

For nonisothermal operations, we have to use the energy balance equation to express temperature variation along the reactor. Consider a section of volume V from the reactor inlet. Assuming negligible kinetic and potential energies, the energy balance equation is

$$\dot{H}(V) - \dot{H}_{in} = \dot{Q}(V) + \dot{m}_{inj}(V) h_{inj} \qquad (9.2.20)$$

where $\dot{H}(V) - \dot{H}_{in}$ is the rate enthalpy flows in and out of a reactor section of volume V, $\dot{Q}(V)$ is the rate heat is added to the section, and $\dot{m}_{inj}(V) h_{inj}$ is the rate enthalpy added to the section by the injection stream (h_{inj} is the specific mass-based enthalpy of the injection stream). For reacting fluids,

$$\dot{H}(V) - \dot{H}_{in} = \dot{m}(V) \bar{c}_p (T - T_0) - \dot{m}_{in} \bar{c}_p (T_{in} - T_0) + \sum_m^{n_I} \Delta H_{R_m} \dot{X}_m \qquad (9.2.21)$$

The first two terms on the right-hand side indicate the change in the sensible heat, and the third indicates the change in enthalpy due to the chemical reactions. Differentiating Eq. 9.2.20 with respect to the reactor volume,

$$\frac{d\dot{H}}{dV} = \frac{d\dot{Q}}{dV} + (\lambda \rho_{inj}) h_{inj} \qquad (9.2.22)$$

Differentiating Eq. 9.2.21 with respect to the reactor volume,

$$\frac{d\dot{H}}{dV} = \dot{m} \bar{c}_p \frac{dT}{dV} + \frac{d\dot{m}}{dV} \bar{c}_p (T - T_0) + \sum_m^{n_I} \Delta H_{R_m} \frac{d\dot{X}_m}{dV} \qquad (9.2.23)$$

From overall material balance,

$$\frac{d\dot{m}}{dV} = \lambda \rho_{inj}$$

and the rate heat transferred per unit volume is

$$\frac{d\dot{Q}}{dV} = U(T_F - T)\left(\frac{S}{V}\right)$$

Combining Eqs. 9.2.22 and 9.2.23, and using these relations, the energy balance equation becomes

$$\frac{dT}{dV} = \frac{1}{\dot{m}\bar{c}_p}\left[U\left(\frac{S}{V}\right)(T_F - T) + (\lambda\rho_{inj})h_{inj} - (\lambda\rho_{inj})\bar{c}_p(T - T_0) - \sum_{m}^{n_I}\Delta H_{R_m}\frac{d\dot{X}_m}{dV}\right]$$

$$(9.2.24)$$

Equation 9.2.24 describes the change of temperature along the reactor. To reduce it to dimensionless form, recall that

$$dV = v_0 t_{cr}\, d\tau \qquad dT = T_0\, d\theta \qquad d\dot{X}_m = (F_{tot})_0\, dZ$$

$$\text{HTN} = \frac{U t_{cr}}{C_0 \hat{c}_{p_0}}\left(\frac{S}{V}\right) \qquad \text{DHR}_m = \frac{\Delta H_{R_m}(T_0)}{T_0 \hat{c}_{p_0}}$$

and using a correction factor of the heat capacity, defined by

$$\text{CF} \equiv \frac{\dot{m}\bar{c}_p}{(F_{tot})_0 \hat{c}_{p_0}}$$

Equation 9.2.24 reduces to

$$\frac{d\theta}{d\tau} = \frac{1}{\text{CF}}\left[\text{HTN}(\theta_F - \theta) + \frac{(\lambda\rho_{inj})t_{cr}}{C_0\hat{c}_{p_0}T_0}h_{inj} - \frac{(\lambda\rho_{inj})\bar{c}_p t_{cr}}{C_0\hat{c}_{p_0}}(\theta - 1) - \sum_{m}^{n_I}\text{DHR}_m\frac{dZ_m}{d\tau}\right]$$

$$(9.2.25)$$

For uniform injection rate,

$$\lambda = \frac{v_{inj}}{V_{R_{tot}}} = \left(\frac{v_{inj}}{v_0}\right)\left(\frac{v_0}{V_{R_{tot}}}\right)$$

and the energy balance equation reduces to

$$\frac{d\theta}{d\tau} =$$

$$\frac{1}{\text{CF}}\left[\text{HTN}(\theta_F - \theta) + \frac{1}{\tau_{\text{tot}}}\left(\frac{v_{\text{inj}}}{v_0}\right)\frac{\rho_{\text{inj}}h_{\text{inj}}}{C_0\hat{c}_{p_0}T_0} - \frac{1}{\tau_{\text{tot}}}\left(\frac{v_{\text{inj}}}{v_0}\right)\frac{\rho_{\text{inj}}\bar{c}_p}{C_0\hat{c}_{p_0}}(\theta - 1) - \sum_m^{n_I}\text{DHR}_m\frac{dZ_m}{d\tau}\right]$$

$$(9.2.26)$$

where τ_{tot}, the total dimensionless space time, is defined in Eq. 9.2.15. When the injection stream does not involve phase change,

$$h_{\text{inj}} = \bar{c}_{p_{\text{inj}}}(T_{\text{inj}} - T_0) \tag{9.2.27}$$

For most *liquid-phase* reactions, the density and the mass-based specific heat capacity of the reacting fluid and the injection stream are assumed to be the same and constant. For uniform injection rate, Eq. 9.2.26 reduces to

$$\frac{d\theta}{d\tau} = \frac{1}{\text{CF}}\left[\text{HTN}(\theta_F - \theta) + \frac{1}{\tau_{\text{tot}}}\left(\frac{v_{\text{inj}}}{v_0}\right)\frac{\rho\bar{c}_p}{C_0\hat{c}_{p_0}}[\theta_{\text{inj}} - \theta] - \sum_m^{n_I}\text{DHR}_m\frac{dZ_m}{d\tau}\right]$$

$$(9.2.28)$$

where the specific molar heat capacity of the reference stream is

$$\hat{c}_{p_0} \equiv \frac{(\dot{m}_{\text{in}} + \dot{m}_{\text{inj}})\bar{c}_p}{(F_{\text{tot}})_0} \tag{9.2.29}$$

and the correction factor of the heat capacity is

$$\text{CF} = \frac{\bar{c}_p}{(F_{\text{tot}})_0\hat{c}_{p_0}}\left[\dot{m}_{\text{in}} + \dot{m}_{\text{inj}}\left(\frac{\tau}{\tau_{\text{tot}}}\right)\right] \tag{9.2.30}$$

Note that although the specific heat capacity of the reacting liquid is constant, the correction factor varies along the reactor because the mass flow rate changes.

For *gas-phase* reactions, the heat capacity of the reacting fluid depends on both composition and temperature. Also, the specific heat capacity of the injection stream is usually different from that of the inlet stream. To account for the effect of composition and temperature, molar-based specific heat capacities are used. The molar-based and the mass-based specific heat capacities are related by

$$\dot{m}\bar{c}_p \equiv \sum_j^J F_j\hat{c}_{p_j}(\theta) \qquad \dot{m}_{\text{inj}}\bar{c}_p \equiv \sum_j^J (F_j)_{\text{inj}}\hat{c}_{p_j}(\theta_{\text{inj}}) \tag{9.2.31}$$

Substituting Eq. 9.2.31 into Eq. 9.2.20,

$$\dot{H} - \dot{H}_{\text{in}} \equiv \dot{Q} + \left[\sum_j^J F_{j_{\text{inj}}} \hat{c}_{p_j}(T) \right] (T_{\text{inj}} - T_0) \qquad (9.2.32)$$

Substituting Eq. 9.2.31 into Eq. 9.2.21,

$$\dot{H} - \dot{H}_{\text{in}} \equiv \left[\sum_j^J F_j \hat{c}_{p_j}(T) \right] (T - T_0) - \left[\sum_j^J (F_j)_{\text{in}} \hat{c}_{p_j}(T_{\text{in}}) \right] (T_{\text{in}} - T_0) + \sum_m^{n_I} \Delta H_{Rm}(T_0) \dot{X}_m$$

$$(9.2.33)$$

Differentiating these relations with respect to the volume V,

$$\frac{d\dot{H}}{dV} = \frac{d\dot{Q}}{dV} + \left[\sum_j^J \frac{d(F_j)_{\text{inj}}}{dV} \hat{c}_{p_j}(T_{\text{inj}}) \right] (T_{\text{inj}} - T_0) \qquad (9.2.34)$$

$$\frac{d\dot{H}}{dV} = \left[\sum_j^J \frac{dF_j}{dV} \hat{c}_{p_j}(T) \right] (T - T_0) + \left[\sum_j^J F_j \hat{c}_{p_j}(T) \right] \frac{dT}{dV} + \sum_m^{n_I} \Delta H_{Rm}(T_0) \frac{d\dot{X}_m}{dV} \qquad (9.2.35)$$

Combining Eqs. 9.2.34 and Eq. 9.2.35 and rearranging,

$$\frac{dT}{dV} = \frac{1}{\displaystyle\sum_j^J F_j \hat{c}_{p_j}(T)} \left[\frac{d\dot{Q}}{dV} + \left(\sum_j^J \frac{d(F_j)_{\text{inj}}}{dV} \hat{c}_{p_j}(T_{\text{inj}}) \right) (T_{\text{inj}} - T_0) - \right.$$

$$\left. \left(\sum_j^J \frac{dF_j}{dV} \hat{c}_{p_j}(T) \right) (T - T_0) - \sum_m^{n_I} \Delta H_{Rm}(T_0) \frac{d\dot{X}_m}{dV} \right] \qquad (9.2.36)$$

Using a material balance over species j,

$$\frac{dF_j}{dV} = \frac{d(F_j)_{\text{inj}}}{dV} + \sum_m^{n_I} (s_j)_m \frac{d\dot{X}_m}{dV} \qquad (9.2.37)$$

and Eq. 9.2.36 reduces to

$$\frac{dT}{dV}=\frac{1}{\sum_j^J F_j \hat{c}_{p_j}}\left[\frac{d\dot{Q}}{dV}+\left(\sum_j^J \frac{d(F_j)_{\text{inj}}}{dV}\hat{c}_{p_j}(T_{\text{inj}})\right)(T_{\text{inj}}-T_0)-\right.$$
$$\left.\left(\sum_j^J \hat{c}_{p_j}\sum_m^{n_I}(s_j)_m\frac{d\dot{X}_m}{dV}\right)(T-T_0)-\sum_m^{n_I}\Delta H_{Rm}(T_0)\frac{d\dot{X}_m}{dV}\right] \quad (9.2.38)$$

We reduce Eq. 9.2.38 to dimensionless form similar to the way we reduced Eq. 9.2.24. When the species heat capacities do *not* vary with temperature, the energy balance equation for gas-phase reactions is

$$\frac{d\theta}{d\tau}=\frac{1}{\text{CF}}\left[\text{HTN}(\theta_F-\theta)+\sum_j^J \frac{d}{d\tau}\left(\frac{(F_j)_{\text{inj}}}{(F_{\text{tot}})_0}\right)\left(\frac{\hat{c}_{p_j}}{\hat{c}_{p_0}}\right)(\theta_{\text{inj}}-\theta)-\right.$$
$$\left.\left(\sum_j^J \frac{\hat{c}_{p_j}}{\hat{c}_{p_0}}\sum_m^{n_I}(s_j)_m\frac{dZ_m}{d\tau}\right)(\theta-1)-\sum_m^{n_I}\Delta H_{Rm}(T_0)\frac{dZ_m}{d\tau}\right] \quad (9.2.39)$$

where CF is the correction factor of the heat capacity of the reacting fluid, defined by

$$\text{CF}\equiv\frac{\sum_j^J F_j \hat{c}_{p_j}}{(F_{\text{tot}})_0 \hat{c}_{p_0}} \quad (9.2.40)$$

and the specific molar heat capacity of the reference stream is

$$\hat{c}_{p_0}=\sum_j^J \frac{(F_j)_{\text{in}}+(F_j)_{\text{inj}}}{(F_{\text{tot}})_0}\hat{c}_{p_j} \quad (9.2.41)$$

When the side injection rate is uniform,

$$\frac{d}{d\tau}\left[\frac{(F_j)_{\text{inj}}}{(F_{\text{tot}})_0}\right]=\frac{1}{\tau_{\text{tot}}}\left[\frac{(F_j)_{\text{inj}}}{(F_{\text{tot}})_0}\right]$$

Equation 9.2.39 reduces to

$$\frac{d\theta}{d\tau}=\frac{1}{\text{CF}}\left[\text{HTN}(\theta_F-\theta)+\frac{1}{\tau_{\text{tot}}}\sum_j^J\left(\frac{(F_j)_{\text{inj}}}{(F_{\text{tot}})_0}\right)\left(\frac{\hat{c}_{p_j}}{\hat{c}_{p_0}}\right)(\theta_{\text{inj}}-\theta)-\right.$$
$$\left.\left(\sum_j^J \frac{\hat{c}_{p_j}}{\hat{c}_{p_0}}\sum_m^{n_I}(s_j)_m\frac{dZ_m}{d\tau}\right)(\theta-1)-\sum_m^{n_I}\Delta H_{Rm}(T_0)\frac{dZ_m}{d\tau}\right] \quad (9.2.42)$$

The correction factor of the heat capacity is

$$\text{CF}(Z_m, \theta) = \frac{1}{\hat{c}_{p_0}} \sum_j^J \left[\frac{(F_j)_{\text{in}}}{(F_{\text{tot}})_0} + \frac{(F_j)_{\text{inj}}}{(F_{\text{tot}})_0} \left(\frac{\tau}{\tau_{\text{tot}}} \right) + \sum_m^{n_I} (s_j)_m Z_m(\tau) \right] \hat{c}_{p_j}(\theta) \qquad (9.2.43)$$

We solve the design equations simultaneously with the energy balance equation, subject to the initial condition that at $\tau = 0$, the extents of all the independent reactions and the dimensionless temperature are specified. Note that we solve these equations for a specified value of τ_{tot} (or reactor volume). The reaction operating curves of plug-flow reactors with side injection are the final value of Z_m's and θ for different values of τ_{tot}.

Example 9.3 Valuable product V is produced in a plug-flow reactor with a distributed feed. The following simultaneous, gas-phase chemical reactions take place in the reactor:

$$\text{Reaction 1:} \quad A + B \longrightarrow V$$
$$\text{Reaction 2:} \quad 2A \longrightarrow W$$

The reaction rates are $r_1 = k_1 C_A C_B$, and $r_2 = k_2 C_A^2$. Two equal streams (50 mol/min each), one of reactant A and one of reactant B, are to be processed in the reactor. Reactant A is injected uniformly along the reactor, and reactant B is fed into the reactor inlet. The reactor is operated at 2 atm.

a. Derive the design equations, and plot the reaction and species operating curves for adiabatic operation with inlet temperature of 300°C and injection temperature of 250°C.

b. Compare the operation to that of an adiabatic plug-flow reactor.

c. For each operation, what is the reactor volume, and what are the species production rates at 90% conversion of reactant A?

d. Examine the effect of the inlet temperature on the performance of the reactor (for $T_{\text{inj}} = 250°C$).

Data: At 300°C, $k_1 = 0.8 \, \text{L/mol sec}^{-1}$ $k_2 = 1.6 \, \text{L/mol sec}^{-1}$

$$\Delta H_{R_1} = -5000 \, \text{cal/mol} \quad \Delta H_{R_2} = -3750 \, \text{cal/mol}$$

$$E_{a_1} = 12,000 \, \text{cal/mol} \quad E_{a_2} = 24,000 \, \text{cal/mol}$$

$$\hat{c}_{P_A} = 30 \, \text{cal/mol K}^{-1} \quad \hat{c}_{P_B} = 10 \, \text{cal/mol K}^{-1}$$

$$\hat{c}_{P_V} = 32 \, \text{cal/mol K}^{-1} \quad \hat{c}_{P_W} = 40 \, \text{cal/mol K}^{-1}$$

Solution   The stoichiometric coefficients of the chemical reactions are

$$s_{A_1} = -1 \qquad s_{B_1} = -1 \qquad s_{V_1} = 1 \qquad w_1 = 0 \qquad \Delta_1 = -1$$
$$s_{A_2} = -2 \qquad s_{B_2} = 0 \qquad s_{V_2} = 0 \qquad s_{W_2} = 1 \qquad \Delta_2 = -1$$

Since each reaction has a species that does not appear in the other, the two reactions are independent, and there is no dependent reaction. We select the total feed into the reactor as the reference stream. Hence, $(F_{\text{tot}})_0 = (F_A)_0 + (F_B)_0 = 100$ mol/h. We also select the reference state at $T_0 = 300°C$ and $P_0 = 2$ atm; hence

$$C_0 = \frac{P_0}{RT_0} = 0.0425 \text{ mol/L}$$

$$v_0 = \frac{(F_{\text{tot}})_0}{C_0} = 2353 \text{ L/h} = 0.653 \text{ L/sec}$$

Since only reactant B is fed into the reactor inlet, $(F_{\text{tot}})_{\text{in}} = (F_B)_{\text{in}} = 50$ mol/h, and $(F_A)_{\text{in}} = 0$. Since only reactant A is injected along the reactor, $(F_A)_{\text{inj}} = (F_{\text{tot}})_{\text{inj}} = 50$ mol/h, and $(F_B)_{\text{inj}} = 0$.

a. We write design equation Eq. 9.2.8 for each independent reaction:

$$\frac{dZ_1}{d\tau} = r_1 \left(\frac{t_{\text{cr}}}{C_0}\right) \qquad 0 \le \tau \le \tau_{\text{tot}} \tag{a}$$

$$\frac{dZ_2}{d\tau} = r_2 \left(\frac{t_{\text{cr}}}{C_0}\right) \qquad 0 \le \tau \le \tau_{\text{tot}} \tag{b}$$

Using Eq. 9.2.19, the concentrations of the reactants at any point in the reactor are

$$C_A = C_0 \frac{0.5(\tau/\tau_{\text{tot}}) - Z_1 - 2Z_2}{[0.5[(\tau_{\text{tot}} + \tau)/\tau_{\text{tot}}] - Z_1 - Z_2]\theta} \tag{c}$$

$$C_B = C_0 \frac{0.5 - Z_1}{[0.5[(\tau_{\text{tot}} + \tau)/\tau_{\text{tot}}] - Z_1 - Z_2]\theta} \tag{d}$$

The rates of the reactions are

$$r_1 = k_1(T_0)C_0^2 \frac{[0.5(\tau/\tau_{\text{tot}}) - Z_1 - 2Z_2](0.5 - Z_1)}{[0.5[(\tau_{\text{tot}} + \tau)/\tau_{\text{tot}}] - Z_1 - Z_2]^2\theta^2} e^{\gamma_1(\theta-1)/\theta} \tag{e}$$

$$r_2 = k_2(T_0)C_0^2 \frac{[0.5(\tau/\tau_{\text{tot}}) - Z_1 - 2Z_2]^2}{[0.5[(\tau_{\text{tot}} + \tau)/\tau_{\text{tot}}] - Z_1 - Z_2]^2\theta^2} e^{\gamma_2(\theta-1)/\theta} \tag{f}$$

We select Reaction 1 as the leading reaction; hence, the characteristic reaction time is

$$t_{cr} = \frac{1}{k_1(T_0)C_0} = 29.4 \text{ s} = 0.49 \text{ min} \tag{g}$$

Substituting (e), (f), and (g) into (a) and (b), the design equations reduce to

$$\frac{dZ_1}{d\tau} = \frac{[0.5(\tau/\tau_{tot}) - Z_1 - 2Z_2](0.5 - Z_1)}{[0.5[(\tau_{tot} + \tau)/\tau_{tot}] - Z_1 - Z_2]^2 \theta^2} e^{\gamma_1(\theta - 1)/\theta} \quad 0 \leq \tau \leq \tau_{tot} \tag{h}$$

$$\frac{dZ_2}{d\tau} = \left(\frac{k_2(T_0)}{k_1(T_0)}\right) \frac{[0.5(\tau/\tau_{tot}) - Z_1 - 2Z_2]^2}{[0.5[(\tau_{tot} + \tau)/\tau_{tot}] - Z_1 - Z_2]^2 \theta^2} e^{\gamma_2(\theta - 1)/\theta} \quad 0 \leq \tau \leq \tau_{tot} \tag{i}$$

To set up the energy balance equation, we have to determine first several related parameters. Using Eq. 9.2.41, the specific molar heat capacity of the reference stream is

$$\hat{c}_{p_0} = (0.5)\hat{c}_{p_A}(T_0) + (0.5)\hat{c}_{p_B}(T_0) = 20 \text{ cal/mol K}$$

The reference temperature is $T_0 = 573$ K, and the dimensionless heats of reaction are

$$\text{DHR}_1 = \frac{\Delta H_{R_1}(T_0)}{T_0 \hat{c}_{p_0}} = -0.436 \quad \text{DHR}_2 = \frac{\Delta H_{R_2}(T_0)}{T_0 \hat{c}_{p_0}} = -0.327$$

The dimensionless activation energies of the two reactions are

$$\gamma_1 = \frac{E_{a_1}}{RT_0} = 10.54 \quad \gamma_2 = \frac{E_{a_2}}{RT_0} = 21.07$$

When $T_{inj} = 250°C$, $\theta_{inj} = 0.8255$. For adiabatic operation, HTN = 0. Using Eq. 9.2.43, the correction factor of the heat capacity is

$$\text{CF}(Z_m, \theta) = \frac{1}{\hat{c}_{p_0}}\left[\left((0.5)\frac{\tau}{\tau_{tot}} - Z_1 - 2Z_2\right)\hat{c}_{p_A} + (0.5 - Z_1)\hat{c}_{p_B} + Z_1\hat{c}_{p_V} + Z_2\hat{c}_{p_W}\right]$$

$$= (0.25) + (0.75)\frac{\tau}{\tau_{tot}} - (0.4)Z_1 - Z_2 \tag{j}$$

Substituting these parameters and (j) into Eq. 9.2.42, the energy balance equation reduces to

$$\frac{d\theta}{d\tau}=\frac{0.75(\tau/\tau_{tot})(\theta_{inj}-\theta)+DHR_1(dZ_1/d\tau)+DHR_2(dZ_2/d\tau)}{(0.25)+(0.75)\dfrac{\tau}{\tau_{tot}}-(0.4)Z_1-Z_2}\ 0\leq\tau\leq\tau_{op}\quad\text{(k)}$$

We solve (k) simultaneously with (h) and (i) subject to the initial conditions (reactor inlet) $Z_1(0)=Z_2(0)=0$, $\theta(0)=1$, for different values of total space time, τ_{tot}, and determine the final extents and θ at each τ_{tot}. Figure E9.3.1 shows the reaction operating curves, and Figure E9.3.2 shows the temperature curve. To obtain the species operating curves, we use Eq. 2.7.8 to express the species molar flow rates at the reactor outlet,

$$\frac{F_A(\tau_{tot})}{(F_{tot})_0}=\frac{F_{A_{in}}+F_{A_{inj}}}{(F_{tot})_0}-Z_1(\tau_{tot})-2\cdot Z_2(\tau_{tot})\qquad\text{(l)}$$

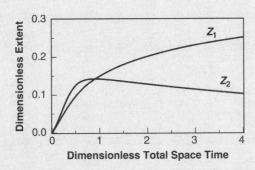

Figure E9.3.1   Reaction operating curves; distributed feed, adiabatic operation.

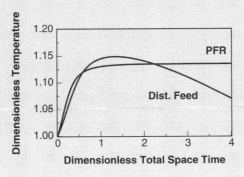

Figure E9.3.2   Temperature curve; adiabatic operation.

$$\frac{F_B(\tau_{tot})}{(F_{tot})_0} = \frac{F_{B_{in}} + F_{B_{inj}}}{(F_{tot})_0} - Z_1(\tau_{tot}) \tag{m}$$

$$\frac{F_V(\tau_{tot})}{(F_{tot})_0} = Z_1(\tau_{tot}) \tag{n}$$

$$\frac{F_W(\tau_{tot})}{(F_{tot})_0} = Z_2(\tau_{tot}) \tag{o}$$

Figure E9.3.3 shows the species operating curves.

b. For plug-flow reactors, the species concentrations at any point in the reactor are

$$C_A = C_0 \frac{0.5 - Z_1 - 2Z_2}{(1 - Z_1 - Z_2)\theta} \tag{p}$$

$$C_B = C_0 \frac{0.5 - Z_1}{(1 - Z_1 - Z_2)\theta} \tag{q}$$

The design equations are

$$\frac{dZ_1}{d\tau} = \frac{(0.5 - Z_1 - 2Z_2)(0.5 - Z_1)}{(1 - Z_1 - Z_2)^2\theta^2} e^{\gamma_1(\theta-1)/\theta} \tag{r}$$

$$\frac{dZ_2}{d\tau} = \left[\frac{k_2(T_0)}{k_1(T_0)}\right] \frac{(0.5 - Z_1 - 2Z_2)^2}{(1 - Z_1 - Z_2)^2\theta^2} e^{\gamma_2(\theta-1)/\theta} \tag{s}$$

The correction factor for the heat capacity is

$$CF(Z_m, \theta) = 1 - (0.4)Z_1 - Z_2 \tag{t}$$

For adiabatic plug-flow operation, HTN $= 0$, and the energy balance equation is

$$\frac{d\theta}{d\tau} = \frac{DHR_1(dZ_1/d\tau) + DHR_2(dZ_2/d\tau)}{1 - (0.4)Z_1 - Z_2} \tag{u}$$

For adiabatic plug-flow reactor, we solve (r), (s), and (u) simultaneously, subject to the initial conditions that $Z_1(0) = Z_2(0) = 0$, $\theta(0) = 1$. Figure E9.3.3 shows a comparison of the V production between the adiabatic distributed-feed reactor and the adiabatic plug flow reactor with $T_{in} = 300°C$. Figure E9.3.4 compares the production of V.

c. For the adiabatic plug-flow reactor with distributed-feed, 90% conversion of Reactant A ($F_A/(F_A)_0 = 0.05$), is reached at $\tau_{tot} = 1.4$ and 3.30. Using 9.2.7 and (g), the reactor volume is

$$V_{R_{tot}} = v_0 \cdot t_{cr} \cdot \tau_{tot} = 1612\,L$$

At $\tau_{tot} = 1.4$, $F_V/(F_{tot})_0 = 0.1748$ and $F_W/(F_{tot})_0 = 0.1372$, and $\theta = 1.149$. Hence, the production rate of Product V is 17.48 mol/min and of product W is 13.72 mol/min. The outlet temperature of the reactor is 385°C. For the adiabatic plug-flow reactor, 90% conversion of Reactant A is reached at $\tau_{tot} = 0.5$. Using 9.2.7 and (g), the reactor volume is 576 L. For $\tau_{tot} = 0.5$, $F_V/(F_{tot})_0 = 0.107$ and $F_W/(F_{tot})_0 = 0.1745$, and $\theta = 1.114$. Hence, the

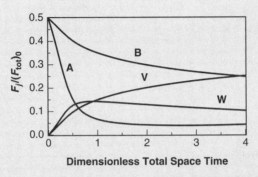

Figure E9.3.3   Species operating curves; distributed feed, adiabatic operation.

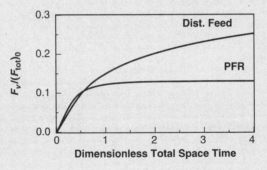

Figure E9.3.4   Comparison of V production; adiabatic operation.

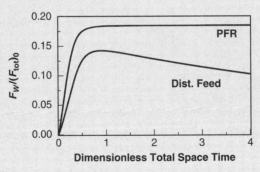

Figure E9.3.5   Comparison of W production; adiabatic operation.

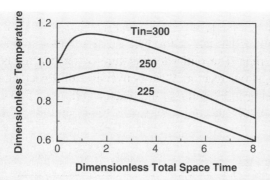

Figure E9.3.6 Effect of inlet temperature on exit temperature; distributed feed, adiabatic operation.

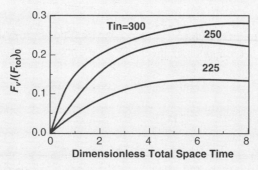

Figure E9.3.7 Effect of inlet temperature on the production of product V; distributed feed, adiabatic operation.

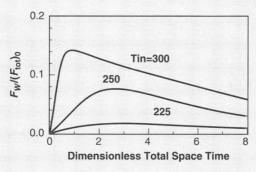

Figure E9.3.8 Effect of inlet temperature on the production of product W; distributed feed, adiabatic operation.

production rate of Product V is 10.7 mol/min and of product W is 17.4 mol/min. The outlet temperature of the reactor is 365°C.

d. To examine the effect of the inlet temperature on the performance of the adiabatic distributed-feed reactor, we solve (h), (i), and (k) simultaneously for

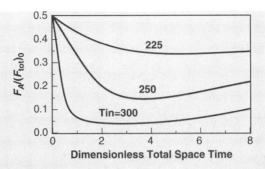

Figure E9.3.9 Effect of inlet temperature on the consumption of reactant A; distributed feed, adiabatic operation.

different values of inlet temperature. Figure E9.3.6 shows a comparison of the exit temperature, Figure E9.3.7 compares the production of product V, Figure E9.3.8 compares the production of product W, and Figure E9.3.9 compares the consumption of reactant A.

9.3 DISTILLATION REACTOR

A distillation reactor is a liquid-phase ideal batch reactor where volatile products are generated and continuously removed from the reactor, as shown schematically in Figure 9.4. Because species are removed, the volume of the reacting fluid reduces during the operation.

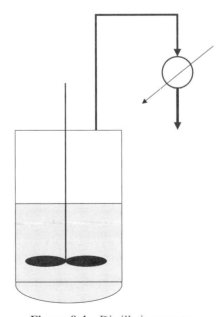

Figure 9.4 Distillation reactor.

To derive a design equation, we write a species balance equation for species j that is *not* removed from the reactor:

$$\frac{dN_j}{dt} = (r_j)V_R(t) \tag{9.3.1}$$

We use Eq. 2.3.3 to relate the moles of species j in the reactor at operating time t to the extents of the independent reactions, $X_m(t)$'s, following the procedure described in Chapter 4 for a batch reactor, and obtain

$$\frac{dX_m}{dt} = \left(r_m + \sum_{k}^{n_D} \alpha_{km} r_k \right) V_R(t) \tag{9.3.2}$$

This is the differential design equation for a distillation reactor, written for the mth-independent chemical reaction. Note that Eq. 9.3.2 is identical to the design equation of an ideal batch reactor. The difference between the two cases is in the variation of the reactor volume and species concentrations during the operation.

To derive a relation for the reactor volume, we write an overall material balance over the reacting fluid. Assuming the mass of the gaseous species inside the reactor is negligible, the reduction in the total mass of the reactor during operating time t is equal to the mass of the volatile species removed,

$$\rho(0)V_R(0) - \rho(t)V_R(t) = \sum_{j}^{n_{\text{evap}}} \text{MW}_j N_j(t) \tag{9.3.3}$$

where MW_j is the molecular weight, and $N_j(t)$ is the mole of gaseous species j formed during the operation. Note that the summation in Eq. 9.3.3 is only over species that evaporate and are removed from the reactor. Assuming the density of the reacting liquid does not vary during the operation, $\rho(t) = \rho(0) = \rho$, differentiating Eq. 9.3.3, with respect to time:

$$\frac{dV_R}{dt} = -\frac{1}{\rho} \sum_{j}^{n_{\text{evap}}} \text{MW}_j \frac{dN_j}{dt} \tag{9.3.4}$$

To relate the formation rates of these species to the extents of the independent reactions, we differentiate stoichiometric relation Eq. 2.3.3,

$$\frac{dN_j}{dt} = \sum_{m}^{n_I} (s_j)_m \frac{dX_m}{dt} \tag{9.3.5}$$

Substituting this into Eq. 9.3.4,

$$\frac{dV_R}{dt} = -\frac{1}{\rho} \sum_j^{n_{\text{evap}}} \text{MW}_j \sum_m^{n_I} (s_j)_m \frac{dX_m}{dt} \tag{9.3.6}$$

Multiplying both sides by dt and integrating,

$$V_R(t) = V_R(0) - \frac{1}{\rho} \left(\sum_j^{n_{\text{evap}}} \text{MW}_j \sum_m^{n_I} (s_j)_m X_m(t) \right) \tag{9.3.7}$$

Equation 9.3.7 provides an expression for the volume of the reactor in terms of the extents of the independent chemical reactions.

To reduce Eqs. 9.3.2 and 9.3.7 to dimensionless form, we select the initial reactor state as the reference state, $(N_{\text{tot}})_0 = N_{\text{tot}}(0)$, $V_{R_0} = V_R(0)$, and define a dimensionless extent of the mth-independent reaction by

$$Z_m(t) \equiv \frac{X_m(t)}{(N_{\text{tot}})_0} \tag{9.3.8}$$

and the reference concentration, C_0, by

$$C_0 \equiv \frac{(N_{\text{tot}})_0}{V_{R_0}} \tag{9.3.9}$$

As for batch reactors, we define the dimensionless time by

$$\tau = \frac{t}{t_{\text{cr}}} \tag{9.3.10}$$

where t_{cr} is the characteristic reaction time, defined by Eq. 3.5.1. Dividing both sides of Eq. 9.3.7 by $V_R(0)$ and $N_{\text{tot}}(0)$, and, using Eqs. 9.3.8 and 9.3.9,

$$\frac{V_R(\tau)}{V_R(0)} = 1 - \frac{C_0}{\rho} \left(\sum_j^{n_{\text{evap}}} \text{MW}_j \sum_m^{n_I} (s_j)_m Z_m(\tau) \right) \tag{9.3.11}$$

Differentiating Eqs. 9.3.8 and 9.3.10,

$$dX_m = (N_{\text{tot}})_0 \, dZ_m$$

$$dt = t_{\text{cr}} \, d\tau$$

Substituting these into Eq. 9.3.2 and using Eq. 9.3.11,

$$\frac{dZ_m}{d\tau} = \left(r_m + \sum_k^{n_D} \alpha_{km} r_k\right)\left[1 - \frac{C_0}{\rho}\left(\sum_j^{n_{evap}} MW_j \sum_m^{n_I} (s_j)_m Z_m(\tau)\right)\right]\left(\frac{t_{cr}}{C_0}\right) \quad (9.3.12)$$

Equation 9.3.12 is the dimensionless, reaction-based design equation for distillation reactors, written for the mth-independent reaction. To describe the operation, we have to write Eq. 9.3.12 for each independent reaction.

To solve the design equations, we express the species concentrations in terms of the extents of the independent reactions. Using stoichiometric relation (Eq. 2.7.4) and accounting for changes in the reactor volume, the concentration of species j is

$$C_j(\tau) = C_0 \left[\frac{V_R(0)}{V_R(\tau)}\right]\left[y_j(0) + \sum_m^{n_I} (s_j)_m Z_m(\tau)\right] \quad (9.3.13)$$

Substituting Eq. 9.3.11 into Eq. 9.3.13, the species concentrations are

$$C_j(\tau) = C_0 \frac{y_j(0) + \sum_m^{n_I} (s_j)_m Z_m(\tau)}{1 - (C_0/\rho)\left(\sum_j^{n_{evap}} MW_j \sum_m^{n_I} (s_j)_m Z_m(\tau)\right)} \quad (9.3.14)$$

Due to evaporation, most distillation reactors operate isothermally. To determine the rate, heat is transferred to (or from) the reactor, we also have to solve the energy balance equation. We modify the energy balance equation (Eq. 5.2.8) by adding a term to account for the enthalpy removed from the reactor by the evaporating species:

$$\Delta H(t) = Q(t) - \int_0^t (\dot{m}_{out} h_{out})\, dt - W_{sh}(t) \quad (9.3.15)$$

where $\dot{m}_{out} h_{out}$ is the rate enthalpy is removed from the reactor by the gaseous species. Differentiating Eq. 9.3.15 with respect to time,

$$\frac{dH}{dt} = \dot{Q} - \dot{m}_{out} h_{out} - \dot{W}_{sh} \quad (9.3.16)$$

Assuming constant specific mass-based heat capacity, the enthalpy change of the reacting liquid in the reactor is

$$H(t) - H(0) = M(t)\bar{c}_p[T(t) - T_0] - M(0)\bar{c}_p[T(0) - T_0] + \sum_m^{n_I} \Delta H_{R_m}(T_0)X_m(t)$$

$$(9.3.17)$$

Differentiating with respect to time,

$$\frac{dH}{dt} = M(t)\bar{c}_p \frac{dT}{dt} + \bar{c}_p[T(0) - T_0]\frac{dM}{dt} + \sum_{m}^{n_I} \Delta H_{R_m}(T_0)\frac{dX_m}{dt} \qquad (9.3.18)$$

Noting that

$$\frac{dM}{dt} = -\dot{m}_{\text{out}} \qquad \bar{c}_p[T(t) - T_0] = h_L$$

where $\bar{c}_p$ is the specific mass-based enthalpy of the liquid phase, we combine Eqs. 9.3.16 and 9.3.18:

$$M(t)\bar{c}_p \frac{dT}{dt} = \dot{Q} - \dot{W}_{\text{sh}} - \dot{m}_{\text{out}}(h_{\text{out}} - h_L) - \sum_{m}^{n_I} \Delta H_{R_m}(T_0)\frac{dX_m}{dt} \qquad (9.3.19)$$

Using Eq. 9.3.5,

$$\dot{m}_{\text{out}}(h_{\text{out}} - h_L) = \sum_{j}^{n_{\text{evap}}} (\Delta \hat{H}_j)_{\text{evap}} \sum_{m}^{n_I} (s_j)_m \frac{dX_m}{dt} \qquad (9.3.20)$$

where $(\Delta \hat{H}_j)_{\text{vap}}$ is the specific molar-based enthalpy of evaporation species j. Also, the rate heat transferred to the reactor is

$$\dot{Q} = U\left(\frac{S}{V}\right)V_R(t)(T_F - T) \qquad (9.3.21)$$

Combining Eqs. 9.3.19, 9.3.20, and 9.3.21, for reactors with negligible shaft work, the energy balance equation becomes

$$\frac{dT}{dt} = \frac{1}{M(t)\bar{c}_p} U\left(\frac{S}{V}\right)V_R(t)(T_F - T) -$$

$$\frac{1}{M(t)\bar{c}_p}\left[\sum_{j}^{n_{\text{evap}}} (\Delta \hat{H}_j)_{\text{evap}} \sum_{m}^{n_I} (s_j)_m \frac{dX_m}{dt} + \sum_{m}^{n_I} \Delta H_{R_m}(T_0)\frac{dX_m}{dt}\right] \qquad (9.3.22)$$

Equation 9.3.22 expresses the temperature changes of the reacting fluid. To reduce it to dimensionless form, we define dimensionless temperature by

$$\theta = \frac{T}{T_0} \qquad (9.3.23)$$

and divide both sides of Eq. 9.3.22 by $(N_{tot})_0 \hat{c}_{p_0} T_0$:

$$\frac{d\theta}{d\tau} = \frac{1}{CF} \left[HTN \left(\frac{V_R(\tau)}{V_R(0)} \right) (\theta_F - \theta) - \sum_{j}^{n_{evap}} \frac{(\Delta \hat{H}_j)_{vap}}{\hat{c}_{p_0} T_0} \sum_{m}^{n_I} (s_j)_m \frac{dZ_m}{d\tau} - \sum_{m}^{n_I} DHR_m \frac{dZ_m}{d\tau} \right]$$

(9.3.24)

where DHR_m is the dimensionless heat of reaction of the mth-independent reaction defined by Eq. 5.2.24, HTN is the heat-transfer number, defined by Eq. 5.2.23, $\hat{c}_{p_0}$ is the specific molar heat capacity of the reference state, and CF is the correction factor of the heat capacity. For *liquid-phase* reactions, the specific molar heat capacity of the reference state is defined by

$$\hat{c}_{p_0} = \frac{M(0)\bar{c}_p}{(N_{tot})_0}$$

(9.3.25)

and, using Eq. 9.3.10, the correction factor is

$$CF = \frac{V_R(\tau)}{V_R(0)} = 1 - \frac{C_0}{\rho} \left[\sum_{j}^{n_{evap}} MW_j \sum_{m}^{n_I} (s_j)_m Z_m(\tau) \right]$$

(9.3.26)

When the reactor temperature does not change, the rate of heat transferred is

$$\frac{d}{d\tau} \left[\frac{Q}{(N_{tot})_0 \hat{c}_{p_0} T_0} \right] = \sum_{j}^{n_{evap}} \frac{(\Delta \hat{H}_j)_{evap}}{\hat{c}_{p_0} T_0} \sum_{m}^{n_I} (s_j)_m \frac{dZ_m}{d\tau} + \sum_{m}^{n_I} DHR_m \frac{dZ_m}{d\tau}$$

(9.3.27)

Example 9.4   The dehydration of two heavy organic species, reactants A and B, is carried out in a distillation reactor where the water, product C, is being removed continuously. The reactor is initially charged with a 60-L organic solution that contains 200 mol of reactant A and 300 mol of reactant B. The reactor is operated isothermally at 200°C, and the following reactions take place in the reactor:

Reaction 1:	A (liquid) $\longrightarrow$ C (g) + D (liquid)
Reaction 2:	B (liquid) $\longrightarrow$ 2 C (g) + E (liquid)

The rates of the chemical reactions are $r_1 = k_1 C_A$ and $r_2 = k_2 C_A$.

a. Derive the design equations and plot the reaction operating curves.
b. Plot the species operating curves.
c. Determine the operating time needed to achieve 80% conversion of B and the reactor volume at that time.

d. Determine the heating load during the operation.

Data: At $200°C$, $k_1 = 0.05 \text{ min}^{-1}$ $\qquad k_2 = 0.1 \text{ min}^{-1}$

$$\Delta H_{R_1} = 30 \text{ kcal/mol} \qquad \Delta H_{R_2} = 40 \text{ kcal/mol}$$

$$\text{MW}_C = 18 \text{ g/mol} \qquad \rho = 800 \text{ g/L} \quad \bar{c}_p = 0.9 \text{ cal/g K}^{-1}$$

Heat of vaporization of species C is 8300 cal/mol

Solution The two chemical reactions are independent and their stoichiometric coefficients are

$$\begin{array}{cccccc} s_{A_1} = -1 & s_{B_1} = 0 & s_{C_1} = 1 & s_{D_1} = 1 & s_{E_1} = 0 & \Delta_1 = 1 \\ s_{A_2} = 0 & s_{B_2} = -1 & s_{C_2} = 2 & s_{D_2} = 0 & s_{E_2} = 1 & \Delta_2 = 2 \end{array}$$

We select the initial state as the reference state. Hence, $V_{R_0} = 60 \text{ L}$,

$$(N_{\text{tot}})_0 = N_A(0) + N_B(0) = 500 \text{ mol}$$

$$C_0 = \frac{(N_{\text{tot}})_0}{V_{R_0}} = 8.33 \text{ mol/L}$$

and $y_A(0) = 0.4$, $y_B(0) = 0.6$. For liquid-phase reactions, using Eq. 5.2.31, the specific molar heat capacity of the reference state is

$$\hat{c}_{p_0} = \frac{(60 \text{ L})(800 \text{ g/L})(0.9 \text{ cal/g K})}{500 \text{ mol}} = 86.4 \text{ cal/mol K}$$

Since only species C is gaseous, Eq. 9.3.10 reduces to

$$\frac{V_R(\tau)}{V_R(0)} = 1 - \frac{C_0}{\rho} \text{MW}_C[Z_1(\tau) + 2Z_2(\tau)] \qquad (a)$$

and, using Eq. 9.3.24, the correction factor of the heat capacity is

$$\text{CF} = \frac{V_R(\tau)}{V_R(0)} = 1 - \frac{C_0}{\rho} \text{MW}_C(Z_1 + 2Z_2) = 1 - 0.187(Z_1 + 2Z_2) \qquad (b)$$

The reference temperature is $T_0 = 473 \text{ K}$, and the dimensionless heats of reaction are

$$\text{DHR}_1 = \frac{\Delta H_{R_1}(T_0)}{T_0 \hat{c}_{p_0}} = 0.734 \qquad \text{DHR}_2 = \frac{\Delta H_{R_2}(T_0)}{T_0 \hat{c}_{p_0}} = 0.978$$

and the dimensionless specific molar enthalpy of vaporization of species C is

$$\frac{(\Delta \hat{H}_C)_{evap}}{\hat{c}_{p_0} T_0} = \frac{8300}{(86.4)(473)} = 0.203 \qquad \text{(c)}$$

a. We write Eq. 9.3.12 for each independent reaction:

$$\frac{dZ_1}{d\tau} = r_1 \left[1 - \frac{C_0}{\rho} MW_C (Z_1 + 2Z_2) \right] \left(\frac{t_{cr}}{C_0} \right) \qquad \text{(d)}$$

$$\frac{dZ_2}{d\tau} = r_2 \left[1 - \frac{C_0}{\rho} MW_C (Z_1 + 2Z_2) \right] \left(\frac{t_{cr}}{C_0} \right) \qquad \text{(e)}$$

We use Eq. 9.3.14 to express the reactant concentrations in terms of the dimensionless extents:

$$C_A(\tau) = C_0 \frac{y_A(0) - Z_1(\tau)}{1 - (C_0/\rho) MW_C [Z_1(\tau) + 2Z_2(\tau)]} \qquad \text{(f)}$$

$$C_B(\tau) = C_0 \frac{y_B(0) - Z_2(\tau)}{1 - (C_0/\rho) MW_C [Z_1(\tau) + 2Z_2(\tau)]} \qquad \text{(g)}$$

We select Reaction 1 as the leading reaction and define a characteristic reaction time by

$$t_{cr} = \frac{1}{k_1} = 20 \text{ min} \qquad \text{(h)}$$

We substitute (f) and (g) in the rate expressions and the latter together with (h) in the design equations and obtain

$$\frac{dZ_1}{d\tau} = (0.4 - Z_1) e^{\gamma_1 (\theta - 1)/\theta} \qquad \text{(i)}$$

$$\frac{dZ_2}{d\tau} = \frac{k_2(T_0)}{k_1(T_0)} (0.6 - Z_2) e^{\gamma_2 (\theta - 1)/\theta} \qquad \text{(j)}$$

b. For isothermal operation, $\theta = 1$, and we solve (i) and (j) subject to the initial conditions that at $\tau = 0$, $Z_1 = Z_2 = 0$. In this case, we obtain analytical solutions:

$$Z_1(\tau) = 0.4(1 - e^{-\tau}) \qquad \text{(k)}$$

$$Z_2(\tau) = 0.6(1 - e^{-2\tau}) \qquad \text{(l)}$$

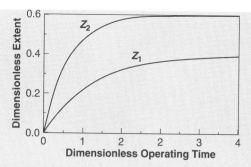

Figure E9.4.1 Reaction operating curves.

The reaction operating curves and the reactor volume curve are shown in Figure E9.4.1.

c. Now that the extents of the independent reactions at any operating time are known, we can determine the amount of each species formed or depleted using relation Eq. 2.5.3:

$$\frac{N_A(\tau)}{(N_{tot})_0} = 0.4 - Z_1(\tau) = 0.4e^{-\tau}$$

$$\frac{N_B(\tau)}{(N_{tot})_0} = 0.6 - Z_2(\tau) = 0.6e^{-2\tau}$$

The species operating curves are shown in Figure E9.4.2.

d. For 80% conversion of B, $N_B = 0.2N_B(0) = 0.2[0.6(N_{tot})_0] = 0.12(N_{tot})_0$, and, using Eq. 2.7.4, $Z_2 = 0.48$. From (1), this value of Z_2 is reached at dimensionless operating time of $\tau = 0.805$, and, at that time, $Z_1 = 0.221$. The operating time is

$$t = \tau t_{cr} = (0.805)(20 \text{ min}) = 16.1 \text{ min}$$

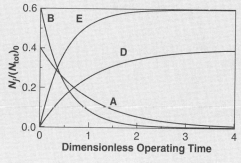

Figure E9.4.2 Species operating curves.

For $\tau = 0.805$, $V_R(t)/V_R(0) = 0.785$. The volume of the reactor when 80% conversion of B is reached is

$$V_R(t) = (0.785)V_R(0) = 46.71 \text{ L}$$

For isothermal operation, using Eq. 9.3.23

$$\frac{d}{d\tau}\left[\frac{Q}{(N_{\text{tot}})_0 \hat{c}_{p_0} T_0}\right] = \frac{(\Delta \hat{H}_C)_{\text{evap}}}{\hat{c}_{p_0} T_0}\left(\frac{dZ_1}{d\tau} + 2\frac{dZ_m}{d\tau}\right) + \text{DHR}_1\frac{dZ_1}{d\tau} + \text{DHR}_2\frac{dZ_2}{d\tau}$$

(m)

which, upon integration, becomes

$$Q = (N_{\text{tot}})_0\left[(\Delta \hat{H}_C)_{\text{evap}}(Z_1 + 2Z_2) + \Delta H_{R_1}Z_1 + \Delta H_{R_2}Z_2\right]$$

(n)

For 80% conversion of B, $Z_1 = 0.221$, $Z_2 = 0.48$, and $Q = 17.82 \times 10^3$ kcal.

9.4 RECYCLE REACTOR

A recycle reactor is a mathematical model describing a steady plug-flow reactor where a portion of the outlet is recycled to the inlet, as shown schematically in Figure 9.5. Although this reactor configuration is rarely used in practice, the recycle reactor model enables us to examine the effect of mixing on the operations of continuous reactors. In some cases, the recycle reactor is one element of a complex reactor model. Below, we analyze the operation of a recycle reactor with multiple chemical reactions, derive its design equations, and discuss how to solve them.

To derive the design equation of a recycle reactor, we consider a differential reactor element, dV, and write a species balance equation over it for species j:

$$dF_j = (r_j)\,dV$$

(9.4.1)

We follow the same procedure as the one used to derive the design equation of a plug-flow reactor (Section 4.4.2) and obtain

$$\frac{d\dot{X}_m}{dV} = r_m + \sum_{k}^{n_D} \alpha_{km} r_k$$

(9.4.2)

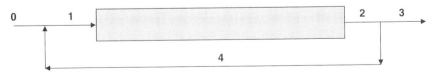

Figure 9.5 Recycle reactor.

where $\dot{X}_m$ is the extent per unit time of the mth-independent reaction from the reactor inlet to a given point in the reactor. This is the differential design equation for a recycle reactor, written for the mth-independent chemical reaction. Note that Eq. 9.4.2 is identical to the design equation of a plug-flow reactor. The main difference between the recycle reactor and a plug-flow reactor is in the way the volumetric flow rate and the concentrations of the species vary along the reactor. Consequently, to solve the design equations, we have to express these quantities in terms of the extents of the independent reactions. Since the reactor inlet is affected by the outlet, let $\dot{X}_{m_{out}}$ denote the extent per unit time of the mth-independent reaction in the *entire reactor*.

To reduce the design equation to a dimensionless form, we select the feed stream to the system (stream 0) as the reference stream and define a dimensionless extent of the mth-independent reaction by

$$Z_m \equiv \frac{\dot{X}_m}{(F_{\text{tot}})_0} \tag{9.4.3}$$

where $(F_{\text{tot}})_0 = (v_0 C_0)$ is the total molar flow rate of the reference system, v_0 is its volumetric flow rate, and C_0 is the reference concentration. We define a dimensionless space time by

$$\tau \equiv \frac{V}{v_0 t_{\text{cr}}} \tag{9.4.4}$$

Differentiating Eqs. 9.4.3 and 9.4.4,

$$d\dot{X}_m = (F_{\text{tot}})_0 \, dZ_m$$

$$dV = (v_0 t_{\text{cr}}) \, d\tau$$

and substituting these in Eq. 9.4.2, the design equation reduces to

$$\frac{dZ_m}{d\tau} = \left(r_m + \sum_{k}^{n_D} \alpha_{km} r_k \right) \left(\frac{t_{\text{cr}}}{C_0} \right) \tag{9.4.5}$$

Equation 9.4.5 is the dimensionless design equation of a recycle reactor, written for the mth-independent reaction. To describe the operation of a recycle reactor with multiple chemical reactions, we have to write Eq. 9.4.5 for each of the independent reactions.

To solve Eq. 9.4.5, we have to express the reaction rates in terms of the extents of the independent reactions. We do so by expressing the local volumetric flow rate and the local molar flow rates of all reactants in terms of Z_m's and calculating the local species concentrations. Using Eqs. 2.7.8 and 2.7.10, the local molar flow rate of species j at any point in the reactor is

$$F_j = F_{j_1} + \sum_{m}^{n_I} (s_j)_m \dot{X}_m \qquad (9.4.6)$$

and the local total molar flow rate is

$$F_{\text{tot}} = (F_{\text{tot}})_1 + \sum_{m}^{n_I} \Delta_m \dot{X}_m \qquad (9.4.7)$$

The local molar flow rate of species j at the reactor inlet (point 1) is

$$F_{j_1} = F_{j_0} + F_{j_4} \qquad (9.4.8)$$

Using the definition of the recycle ratio,

$$R \equiv \frac{F_{j_4}}{F_{j_3}}$$

the molar flow rate at point 4 is related to the molar flow rate at point 2 by

$$F_{j_4} = \frac{R}{1+R} F_{j_2} \qquad (9.4.9)$$

Writing Eq. 9.4.6 for the reactor outlet (point 2 where, $\dot{X}_m = \dot{X}_{m_{\text{out}}}$) and combining it with Eqs. 9.4.8 and 9.4.9,

$$F_{j_2} = (1 + R)\left[F_{j_0} + \sum_{m}^{n_I} (s_j)_m \dot{X}_{m_{\text{out}}} \right] \qquad (9.4.10)$$

Substituting Eqs. 9.4.9 and 9.4.10 into Eq. 9.4.8, the molar flow rate of species j at the reactor inlet (point 1) is

$$F_{j_1} = (1 + R)F_{j_0} + R \sum_{m}^{n_I} (s_j)_m \dot{X}_{m_{\text{out}}} \qquad (9.4.11)$$

Substituting Eq. 9.4.11 into Eq. 9.4.6, and using the definition of dimensionless extent (Eq. 9.4.3), the local molar flow rate of species j is

$$F_j = (F_{\text{tot}})_0 \left[(1 + R)y_{j_0} + R \sum_{m}^{n_I} (s_j)_m Z_{m_{\text{out}}} + \sum_{m}^{n_I} (s_j)_m Z_m \right] \qquad (9.4.12)$$

To obtain the total local molar flow rate, we sum Eq. 9.4.12 over all species and obtain

$$F_{\text{tot}} = (F_{\text{tot}})_0 \left(1 + R + R \sum_{m}^{n_I} \Delta_m Z_{m_{\text{out}}} + \sum_{m}^{n_I} \Delta_m Z_m \right) \qquad (9.4.13)$$

For the reactor inlet and outlet (points 1 and 2), Eq. 9.4.13 reduces, respectively, to

$$(F_{\text{tot}})_1 = (F_{\text{tot}})_0 \left(1 + R + R \sum_{m}^{n_I} \Delta_m Z_{m_{\text{out}}} \right) \qquad (9.4.14)$$

$$(F_{\text{tot}})_2 = (F_{\text{tot}})_0 \left(1 + R + (1 + R) \sum_{m}^{n_I} \Delta_m Z_{m_{\text{out}}} \right) \qquad (9.4.15)$$

Using Eq. 9.4.12, the concentration of species j at any point in the reactor is

$$C_j = \frac{(F_{\text{tot}})_0}{v} \left[(1 + R) y_{j_0} + R \sum_{m}^{n_I} (s_j)_m Z_{m_{\text{out}}} + \sum_{m}^{n_I} (s_j)_m Z_m \right] \qquad (9.4.16)$$

For *liquid-phase* reactions, the density is assumed constant; hence, $v = v_1$, $v_3 = v_0$, and $v_4 = R v_0$. Hence,

$$v_1 = v_0 + v_4 = (1 + R) v_0 \qquad (9.4.17)$$

Substituting Eq. 9.4.17 into Eq. 9.4.16, the local concentration for liquid-phase reactions is

$$C_j = C_0 \left[y_{j_0} + \frac{R}{1 + R} \sum_{m}^{n_I} (s_j)_m Z_{m_{\text{out}}} + \frac{1}{1 + R} \sum_{m}^{n_I} (s_j)_m Z_m \right] \qquad (9.4.18)$$

Note that for $R = 0$ (no recycle), Eq. 9.4.18 reduces to the local concentration in a plug-flow reactor with liquid-phase reactions (Eq. 7.1.11), and for $R \to \infty$ (very high recycle), Eq. 9.4.18 becomes

$$C_j = C_0 \left[y_{j_0} + \sum_{m}^{n_I} (s_j)_m Z_{m_{\text{out}}} \right]$$

which is the outlet concentration of a CSTR. Indeed, at very high recycles, the recycle reactor behaves as a CSTR.

For *gas-phase* reactions, assuming ideal-gas behavior, the local volumetric flow rate at any point in the reactor is

$$v = v_0 \left(\frac{F_{\text{tot}}}{(F_{\text{tot}})_0} \right) \left(\frac{T}{T_0} \right) \left(\frac{P_0}{P} \right) \tag{9.4.19}$$

Using Eq. 9.4.13,

$$v = v_0 \left(1 + R + R \sum_m^{n_I} \Delta_m Z_{m_{\text{out}}} + \sum_m^{n_I} \Delta_m Z_m \right) \theta \left(\frac{P_0}{P} \right) \tag{9.4.20}$$

Substituting Eq. 9.4.20 into Eq. 9.4.16, for isobaric operations, the local concentration of species j is

$$C_j = C_0 \frac{(y_j)_0 + R/(1+R) \sum_m^{n_I} (s_j)_m Z_{m_{\text{out}}} + 1/(1+R) \sum_m^{n_I} (s_j)_m Z_m}{\left[1 + R/(1+R) \sum_m^{n_I} \Delta_m Z_{m_{\text{out}}} + 1/(1+R) \sum_m^{n_I} \Delta_m Z_m \right] \theta} \tag{9.4.21}$$

Note that for $R = 0$ (no recycle), Eq. 9.4.21 reduces to the local concentration in a plug-flow reactor, and for $R \rightarrow \infty$ (very high recycle), it reduces to the concentration of a CSTR.

For nonisothermal recycle reactors, we have to incorporate the energy balance equation to express variation in the reactor temperature. The energy balance equation is the same as that of a plug-flow reactor,

$$\frac{d\theta}{d\tau} = \frac{1}{\text{CF}(Z_m, \theta)} \left[\text{HTN}(\theta_F - \theta) - \sum_m^{n_I} \text{DHR}_m \frac{dZ_m}{d\tau} \right] \tag{9.4.22}$$

The only difference here is that the temperature at the reactor inlet depends on the outlet temperature and the recycle. Hence, we have to solve Eq. 9.4.22 subject to the initial condition that at $\tau = 0$, $\theta(0) = \theta_1$ (the dimensionless temperature at point 1). To determine θ_1, we write an energy balance over the mixing point,

$$(F_{\text{tot}})_1 \hat{H}_1 = (F_{\text{tot}})_0 \hat{H}_0 + (F_{\text{tot}})_4 \hat{H}_4 \tag{9.4.23}$$

where $\hat{H}$ is the specific molar enthalpy of the stream. Taking the feed stream to the system as the reference stream, and assuming no phase change, the enthalpy of a stream at temperature T is

$$(F_{\text{tot}}) \hat{H} = \int_{T_0}^{T} \sum_j^{n_J} (F_j \hat{c}_{p_j}) \, dT \tag{9.4.24}$$

Using the correction factor of the heat capacity (Eq. 5.2.54), Eq. 9.4.23 reduces to

$$\int_1^{\theta_1} CF(Z_m, \theta)_1 \, d\theta = \int_1^{\theta_4} CF(Z_m, \theta)_4 \, d\theta \qquad (9.4.25)$$

which relates to the various process variables by

$$\int_1^{\theta_1} \sum_j^J \left[(1 + R)y_{j_0} + R \sum_m^{n_I} (s_j)_m Z_{m_{\text{out}}} \right] \hat{c}_{p_j}(\theta_1) \, d\theta$$

$$= \int_1^{\theta_4} R \sum_j^J \left[y_{j_0} + \sum_m^{n_I} (s_j)_m Z_{m_{\text{out}}} \right] \hat{c}_{p_j}(\theta_4) \, d\theta \qquad (9.4.26)$$

We solve Eq. 9.4.26 to express θ_1 in terms of θ_4. When the heat capacities of the species are independent of the temperature, Eq. 9.4.25 reduces to

$$\theta_1 = 1 + \frac{CF(Z_m, \theta)_4}{CF(Z_m, \theta)_1} \theta_4 - 1 \qquad (9.4.27)$$

For *liquid-phase* reactions, assuming constant density and a constant specific heat capacity, $CF(Z_m, \theta) = \dot{m}/\dot{m}_0$, and Eq. 9.2.27 becomes, for $\theta_4 = \theta_{\text{out}}$,

$$\theta_1 = 1 + \frac{R}{1 + R} \theta_{\text{out}} - 1 \qquad (9.4.28)$$

Note that, when $R = 0$, $\theta_1 = 1$, and, when $R \to \infty$, $\theta_1 = \theta_{\text{out}}$.

For *gas-phase* reactions with constant species heat capacities, Eq. 9.4.27 becomes, for $\theta_4 = \theta_{\text{out}}$,

$$\theta_1 = 1 + \frac{R \sum_j^J \left[(y_j)_0 + \sum_m^{n_I} (s_j)_m Z_{m_{\text{out}}} \right] \hat{c}_{p_j}}{\sum_j^J \left[(1 + R)(y_j)_0 + R \sum_m^{n_I} (s_j)_m Z_{m_{\text{out}}} \right] \hat{c}_{p_j}} \theta_{\text{out}} - 1 \qquad (9.4.29)$$

With the concentration relations and an expression for θ_1, we can now complete the design formulation of recycle reactors. Substituting the species concentrations and θ in the individual reactions rates, r_m's and r_k's, we obtain a set of first-order, nonlinear differential equations that should be solved simultaneously with the energy balance equation for the initial condition that at $\tau = 0$, Z_m's $= 0$ and $\theta = \theta_1$. Note that we solve these equations for a given value of τ_{tot} corresponding to a given reactor volume. The solutions indicate how the extents and temperature vary along the reactor for the specified reactor. To obtain the operating curves of recycle reactors (for a given recycle ratio), we repeat the calculations for different values of τ_{tot} and for each case determine the final value of Z_m's and θ. Also note that the species concentrations are expressed in terms of the extents at the reactor outlet. Therefore, the solutions are obtained by an iterative procedure. We first guess the outlet extents, $Z_{m_{out}}$'s, solve the set of differential equations, and then check if the calculated outlet extents agree with the assumed values.

Example 9.5 Autocatalytic reactions are chemical reactions where a product of the reaction affects its rate. For such reactions, a recycle reactor provides better performance. This example examines the use of a recycle reactor to carry out an autocatalytic reaction. Consider the liquid-phase chemical reaction

$$A \longrightarrow B + C$$

whose rate expressions is $r = kC_AC_C$. A 450-L tubular reactor is available in the plant, and a stream consisting of 95% of reactant A and 5% of product C (mole %) is to be processed in the reactor. The volumetric flow rate of the stream is 300 L/min and its total molar flow rate is 1500 mol/min. The stream temperature is 60°C.

a. Determine the optimal recycle ratio for isothermal operation.

b. Determine the optimal recycle ratio for adiabatic operation.

c. Compare the production rate in each case to that of a plug-flow reactor and a CSTR operating at the same conditions.

Data: At 60°C, $k = 0.4$ L/mol min, $\Delta H_R = -8$ kcal/mol, $E_a = 6$ kcal/mol

The density of the stream $= 0.9$ kg/L and its mass-based specific heat capacity is 0.85 kcal/kg K.

Solution The stoichiometric coefficients of the chemical reaction are

$$s_A = -1 \qquad s_B = 1 \qquad s_C = 1 \qquad \Delta = 1$$

We select the feed stream as the reference stream; hence, the reference concentration is

$$C_0 = \frac{(F_{tot})_0}{v_0} = 5 \text{ mol/L}$$

The composition of the reference stream is $y_{A_0} = 0.95$, $y_{B_0} = 0$, and $y_{C_0} = 0.05$. For liquid-phase reactions, the specific molar heat capacity of the reference stream is

$$\hat{c}_{p_0} = \frac{v_0 \rho \bar{c}_p}{(F_{tot})_0} = 153 \text{ cal/mol K}$$

The reference temperature is $T_0 = 333.15\,\text{K}$, and the dimensionless heat of reaction is

$$\text{DHR} = \frac{\Delta H_R(T_0)}{T_0 \hat{c}_{p_0}} = -0.16$$

The dimensionless activation energies of reaction is

$$\gamma = \frac{E_a}{RT_0} = 9.06$$

Using Eq. 9.4.5, the design equation of a recycle reactor is

$$\frac{dZ}{d\tau} = r\left(\frac{t_{\text{cr}}}{C_0}\right) \qquad 0 \leq \tau \leq \tau_{\text{tot}} \tag{a}$$

Using Eq. 9.4.19, the local concentrations of reactant A and product C at any point in the reactor are

$$C_A = C_0\left(y_{A_0} - \frac{R}{1+R}Z_{\text{out}} - \frac{1}{1+R}Z\right) \tag{b}$$

$$C_C = C_0\left(y_{C_0} + \frac{R}{1+R}Z_{\text{out}} + \frac{1}{1+R}Z\right) \tag{c}$$

The local rate of the chemical reaction is

$$r = k(T_0)\left(y_{A_0} - \frac{R}{1+R}Z_{\text{out}} - \frac{1}{1+R}Z\right)\left(y_{C_0} + \frac{R}{1+R}Z_{\text{out}} + \frac{1}{1+R}Z\right)e^{\gamma(\theta-1)/\theta} \tag{d}$$

The characteristic reaction time is

$$t_{\text{cr}} = \frac{1}{k(T_0)C_0} = 0.5\,\text{min} \tag{e}$$

Substituting (d) and (e) into (a), the design equation reduces to

$$\frac{dZ}{d\tau} = \left(y_{A_0} - \frac{R}{1+R}Z_{\text{out}} - \frac{1}{1+R}Z\right)\left(y_{C_0} + \frac{R}{1+R}Z_{\text{out}} + \frac{1}{1+R}Z\right)e^{\gamma(\theta-1)/\theta}$$

$$0 \leq \tau \leq \tau_{\text{tot}} \tag{f}$$

Using 9.4.22, the energy balance equation for a recycle reactor is

$$\frac{d\theta}{d\tau} = \frac{1}{\text{CF}(Z_m, \theta)}\left[\text{HTN}(\theta_F - \theta) - \text{DHR}\frac{dZ}{d\tau}\right] \qquad 0 \leq \tau \leq \tau_{\text{tot}} \tag{g}$$

For liquid-phase reactions, assuming a constant specific mass-based heat capacity, the local correction factor is

$$CF(Z_m, \theta) = \frac{v}{v_0} = 1 + R \tag{h}$$

Using Eq. 9.4.28, the relationship between the inlet reactor temperature and the outlet temperature is

$$\theta_1 = 1 + \frac{R}{1+R}(\theta_{out} - 1) \tag{i}$$

For the given reactor,

$$\tau_{tot} = \frac{V_R}{v_0 t_{cr}} = 3 \tag{j}$$

a. For isothermal operation, $\theta = 1$, and we solve (f) for different values of recycle ratio, R, subject to the initial conditions $Z(0) = 0$. Note that each solution involves iterations because Z_{out} is not known a priori. Figure E9.5.1 shows the reaction extent at the reactor outlet as a function of recycle ratio. From the curve, highest production of product B (highest Z_{out}) is achieved at $R = 0.9$, with $Z_{out} = 0.696$. The production rate of product B is

$$F_{B_{out}} = (F_{tot})_0 \left(y_{B_0} + Z_{out} \right) = 1044 \text{ mol/min}$$

For comparison, for isothermal plug-flow reactor ($R = 0$) of the same volume, the outlet extent is $Z_{out} = 0.469$, and the production rate of product B is 703.5 mol/min. For isothermal CSTR ($R = \infty$) of the same volume, the outlet extent is $Z_{out} = 0.641$, and the production rate of product B is 961.5 mol/min.

b. For adiabatic operation, $HTN = 0$, and we solve (f) and (g) simultaneously, subject to the initial conditions $Z(0) = 0$, and $\theta(0) = \theta_1$, where θ_1 is given by

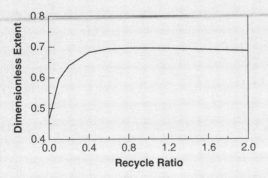

Figure E9.5.1 Effect of recycle ratio—isothermal operation.

(i). Here too, the solution involves iterations because Z_{out} and θ_{out} are not known a priori. Figure E9.5.2 shows the reaction extent at the reactor outlet as a function of recycle ratio, and Figure E9.5.3 shows the exit temperature as a function of the recycle ratio. From Figure E9.5.2, the highest Z_{out} is achieved at $R = 0.2$, with $Z_{out} = 0.890$ and $\theta_{out} = 1.152$. The production rate of product B is

$$F_{B_{out}} = (F_{tot})_0 (y_{B_0} + Z_{out}) = 1335 \text{ mol/min}$$

For comparison, for adiabatic plug-flow reactor ($R = 0$) of the same volume, the outlet extent is $Z_{out} = 0.742$, and $\theta_{out} = 1.135$. The production rate of product B is 1,131 mol/min. For adiabatic CSTR ($R = \infty$) of the same volume, the outlet extent is $Z_{out} = 0.841$, and $\theta_{out} = 1.132$. The production rate of product B is 1260 mol/min. Note that for both isothermal and adiabatic operation, a recycle reactor provides a higher production rate of product B than a corresponding plug-flow reactor and a CSTR.

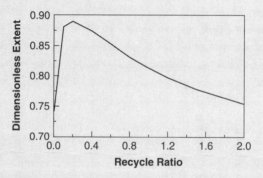

Figure E9.5.2   Effect of recycle ratio—adiabatic operation.

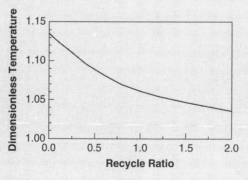

Figure E9.5.3   Effect of recycle ratio on outlet temperature—adiabatic operation.

9.5   SUMMARY

In this chapter, we discussed the design of different reactor configurations. We showed how the reaction-based design formulation is used to describe the operations of these reactors with multiple reactions. We covered in some detail the following reactor models:

a. Semibatch reactor
b. Plug-flow reactor with a distributed feed
c. Distillation reactor
d. Recycle reactor

The reader is challenged to derive design formulations to other reactor configurations and models.

PROBLEMS*

9.1₄ Chlorinations of hydrocarbons are notorious for their undesirable side reactions, where the desired products are further chlorinated to bi- or tri-substitutions. You are on a research team that is assigned to examine the operating mode of a chlorination reactor. The following homogeneous, gas-phase reactions take place in the reactor:

$$A + B \longrightarrow V$$

$$A + V \longrightarrow W$$

$$A + W \longrightarrow P$$

where A is the chlorine, B is the hydrocarbon, and V is the desired product (monochlorinated hydrocarbon). The team was assigned the task to design a continuous reactor. Two members of the team suggest two different approaches. One engineer suggests using a plug-flow reactor, feeding it with a stoichiometric proportion of A and B. The second engineer suggests using a plug-flow reactor feeding the hydrocarbon (B) at the inlet and injecting the chlorine (A) uniformly along the reactor. Here too, A and B are fed in stoichiometric proportion. Each reactant stream is available at 300 K and 1 atm. The flow rate of the hydrocarbon stream is 10 mol/min.

---

*Subscript 1 indicates simple problems that require application of equations provided in the text. Subscript 2 indicates problems whose solutions require some more in-depth analysis and modifications of given equations. Subscript 3 indicates problems whose solutions require more comprehensive analysis and involve application of several concepts. Subscript 4 indicates problems that requires the use of a mathematical software or the writing of a computer code to obtain numerical solutions.

a. Derive the design equations and plot the reaction and species operating curves for each mode.

b. Determine the optimal reactor volume for maximizing the production of product V for each mode.

c. Suggest the preferred operating mode (check $t_{tot} = 10\ t_{cr}$; $t_{tot} = 8\ t_{cr}$; $t_{tot} = 5\ t_{cr}$).

The rate expressions are

$$r_1 = k_1 C_A^{0.5} C_B \qquad r_2 = k_2 C_A C_V \qquad r_3 = k_3 C_A C_W$$

Data: At 300 K: $k_1 C_0^{0.5} = 0.5\,\text{min}^{-1}$; $k_2 C_0 = 1.0\,\text{min}^{-1}$; $k_3 C_0 = 1.2\,\text{min}^{-1}$

9.2₄ You are on a research team that is assigned to examine the operating mode of a chemical reactor. The following liquid-phase reactions take place in the reactor

$$A \longrightarrow 2V$$

$$A + V \longrightarrow W$$

One engineer suggests operating the reactor as a batch reactor, charging it with reactant A. The second engineer suggests operating the reactor as a semibatch reactor with a constant injection rate.

a. Derive the design equations and plot the reaction and species operating curves for isothermal operation of each mode (at 350 K).

b. Derive and plot the heating curve and determine the isothermal HTN for each mode.

c. Determine the optimal operating time for maximizing the production of product V for each mode.

d. Repeat (a) and (b) for adiabatic operation of each mode.

The rate expressions are: $r_1 = k_1 C_A \qquad r_2 = k_2 C_A C_V$

Data: Feed properties: $\rho = 0.85\,\text{kg/L}$ $\qquad C_A = 2\,\text{mol/L}$
At 350 K: $k_1 = 0.2\,\text{min}^{-1}$ $\qquad k_2 C_0 = 0.01\,\text{min}^{-1}$
$$E_{a_1} = 12\,\text{kcal/mol} \qquad E_{a_2} = 6\,\text{kcal/mol}$$
$$\Delta H_{R_1} = 15\,\text{kcal/mol} \qquad \Delta H_{R_2} = -20\,\text{kcal/mol}$$

9.3₄ The elementary gas-phase reactions

$$2A \longrightarrow B$$

$$A + B \longrightarrow C$$

take place in an isothermal recycle reactor. Product B is the desirable product. Derive and plot the reaction curves and the species curves for a recycle reactor when the recycle ratio is 5. Reactant A is fed to the reactor at 137°C and 2 atm at a volumetric flow rate of 10 L/s.

a. Derive and plot the reaction curves and the species curves.

b. What should be the volume of the reactor to obtain the highest production rate of product B?

At the reactor operating temperature, $k_1 = 200$ L/mol s^{-1}, $k_2 = 50$ L/mol s^{-1}.

9.4$_4$ The homogeneous catalytic liquid-phase reaction

$$2A \longrightarrow R$$

is catalyzed by species B (the catalyst), and the reaction rate expression is $r = kC_A^2 C_B$. The reaction is taking place in a semibatch reactor in which 100 L of a solution containing reactant A (2 mol/L) is initially charged to the reactor. A solution containing species B (0.5 mol/L) is injected at a uniform rate of 0.5 L/min. At the operating condition $k = 0.2$ L^2 mol^{-2} min^{-1}.

a. Derive the design equation and plot the reaction and species curves.

b. How many moles of product R are in the reactor after half an hour?

c. Repeat parts (a) and (b) for the case where the reactor is filled initially with 100 L of the B solution, and the solution of reactant A is fed to the reactor at a constant rate of 5 L/min.

d. Which operation is preferable?

9.5$_4$ You are asked to design a semibatch reactor to be used in the production of specialized polymers (ethylene glycol–ethylene oxide co-polymers). The semi-batch operation is used to improve the molecular-weight distribution. Reactant B (EG) and a fixed amount of homogeneous catalyst are charged initially into the reactor (the proportion is 6.75 moles of catalyst per 1000 moles of Reactant B). Reactant A (EO) is injected at a constant rate during the operation. The polymerization reactions are represented by the following liquid-phase chemical reactions:

Reaction 1:	A + B	$\longrightarrow$ C
Reaction 2:	A + C	$\longrightarrow$ D
Reaction 3:	A + D	$\longrightarrow$ E
Reaction 4:	A + E	$\longrightarrow$ F

The rates of the chemical reactions are:

$$r_1 = k_1 \cdot C_{\text{cat}} \cdot C_A \frac{C_B}{C_B + C_C + C_D + C_E + C_F}$$

$$r_2 = k_2 \cdot C_{\text{cat}} \cdot C_A \frac{C_C}{C_B + C_C + C_D + C_E + C_F}$$

$$r_3 = k_3 \cdot C_{\text{cat}} \cdot C_A \frac{C_D}{C_B + C_C + C_D + C_E + C_F}$$

$$r_4 = k_4 \cdot C_{\text{cat}} \cdot C_A \frac{C_E}{C_B + C_C + C_D + C_E + C_F}$$

A batch has to process 4 k-moles of Reactant B (1000 liter) and 8 k-mole of Reactant A (2000 liter). The catalyst (KOH) is dissolved in the solution of Reactant B. The overall objective is to maximize the production of Product D and to minimize the amounts of the other products. Your task is to determine whether semibatch operation is preferable, and to specify the duration of the operation. Consider the following operations:

a. Isothermal operation at 125°C. Derive and plot the reaction and species curves, and compare them to those of batch reactors.

b. What is the required heating load during the semibatch and batch operations in (a)—show the heating/cooling curve, and indicate the total load (in kcal).

c. If the temperature of the jacket is 110°C, what is the isothermal HTN in (a)—show the HTN curve, and estimate the average value.

d. Adiabatic operation. Derive and show the reaction, species, and temperature curves, and compare them to those of adiabatic batch reactor.

e. Adiabatic operation, but Reactant A is injected at 100°C.

f. Nonisothermal operation. Repeat part (e) with HTN = 0.8 of the average isothermal, with Reactant A injected at 100°C.

Data: At 125°C, the reaction rate constants (in $m^3/\text{mol min}^{-1}$) are:

$$k_1 = 1.936 \times 10^{10}; \quad k_2 = k_3 = k_4 = 1.446 \times 10^6$$

The activation energies of the chemical reactions (in kcal/mol) are:

$$E_{a1} = 21.7; \quad E_{a2} = E_{a3} = E_{a4} = 15.8.$$

The heat of reaction of all the chemical reactions (in kcal/mol) is -0.523 (exothermic). The density of the reacting fluid (assumed constant) is 950 g/liter. The heat capacity of the reacting fluid (assumed constant) is 0.5 kcal/kg K^{-1}.

9.6$_4$ The liquid-phase, autocatalytic chemical reaction

$$2A \longrightarrow R$$

Has the rate expression $r = kC_A^2 C_R$ mol/L min^{-1}. An aqueous solution of reactant A ($C_A = 2$ mol/L) is fed into an isothermal recycle reactor. We wish to achieve 85% conversion.

a. What should the recycle ratio be to obtain the highest production rate of product R?

b. What volumetric feed rate can we process if the volume the reactor is 10 L?

10

ECONOMIC-BASED OPTIMIZATION

In the preceding chapters, we described how to design chemical reactors—how to determine the reactor size (or operating time) to obtain a specified production rate, and how to determine the production rate attainable on an existing reactor. However, we have not addressed the following question: Is it desirable to achieve high reaction extents (and use large reactors) or to utilize smaller reactors and recycle unconverted reactants? When more than one reactant is used, we also have to ask what is the most economical proportion of the reactants? The answers to those questions are not straightforward. They depend on the value of the products, the cost of the reactants, the cost of operating the reactor, as well as the cost of recovering unconverted reactants. The underlying motivation to the design and operation of chemical reactors is their economic performance. The dimensionless species operating curves generated from the reactor design relate the species production rates to the reactor size (through dimensionless operating or space time, τ). The dimensionless energy balance equation ties the utilities (heating/cooling) needed for the operation to the extents and τ. Together, they provide the means to conduct an economic-based optimization of reactor operations. In this chapter, we discuss briefly how to apply these relations to optimize the design and operations of chemical reactors.

Principles of Chemical Reactor Analysis and Design, Second Edition. By Uzi Mann
Copyright © 2009 John Wiley & Sons, Inc.

10.1   ECONOMIC-BASED PERFORMANCE OBJECTIVE FUNCTIONS

Optimization of chemical processes is based on the objective of maximizing the profit of the entire process. For processes involving chemical reactions, we can write the objective function as

$$\left\{ \begin{array}{c} \text{Profit} \\ \text{rate} \\ (\$/\text{time}) \end{array} \right\} = \left\{ \begin{array}{c} \text{Rate of} \\ \text{value of} \\ \text{products} \\ (\$/\text{time}) \end{array} \right\} - \left\{ \begin{array}{c} \text{Rate of} \\ \text{cost of} \\ \text{reactants} \\ (\$/\text{time}) \end{array} \right\} - \left\{ \begin{array}{c} \text{Rate of} \\ \text{operating} \\ \text{expense} \\ (\$/\text{time}) \end{array} \right\} \quad (10.1.1)$$

For convenience, the operating expenses are divided into different categories:

$$\left\{ \begin{array}{c} \text{Rate of} \\ \text{operating} \\ \text{expense} \\ (\$/\text{time}) \end{array} \right\} = \left\{ \begin{array}{c} \text{Rate of} \\ \text{cost of} \\ \text{utilities} \\ (\$/\text{time}) \end{array} \right\} + \left\{ \begin{array}{c} \text{Rate of} \\ \text{cost of} \\ \text{labor} \\ (\$/\text{time}) \end{array} \right\} +$$

$$\left\{ \begin{array}{c} \text{Rate of} \\ \text{cost of} \\ \text{maintenance} \\ (\$/\text{time}) \end{array} \right\} + \left\{ \begin{array}{c} \text{Rate of} \\ \text{amortization of} \\ \text{capital equipment} \\ (\$/\text{time}) \end{array} \right\} \quad (10.1.2)$$

The revenue portion of Eq. 10.1.1 depends on the composition of the reactor outlet. A higher purity product is more valuable, and the economics of the process depends on whether unconverted reactants and undesirable by-products are separated from the final product. When species are separated, the separation cost should be incorporated into the analysis of the reactor operation. To maximize the profit of the process, an engineer can adjust the design and several operating parameters:

- The feed rate per reactor volume (expressed in terms of dimensionless operating or space time)
- The proportion of reactants (expressed in terms of y_{j_0}'s)
- The heating (or cooling) rate (by adjusting the temperature of the heating fluid, θ_F)

The profit objective function (Eq. 10.1.1) is then expressed in terms of these (and other) operating parameters, and we determine the optimal values of the parameters by solving the following equations:

$$\frac{\vartheta\{\text{Profit rate}\}}{\partial\tau} = 0 \qquad \frac{\vartheta\{\text{Profit rate}\}}{\partial y_j(0)} = 0 \qquad \frac{\vartheta\{\text{Profit rate}\}}{\partial\theta_F} = 0 \quad (10.1.3)$$

For most operations with single chemical reactions, the profit function is expressed in terms of relatively simple functions of these parameters, and Eq. 10.1.3 can be solved analytically. For operations with multiple chemical reactions, we have to determine the optimal parameters numerically.

Next, we derive the gross revenue function of the process, expressed in terms of the extents of the independent reactions. Let Val_j denote the value of species j

expressed in \$/mol. When all the species in the reactor outlet are separated, using Eq. 2.7.8, the value of the product stream is

$$\sum_{j}^{J} F_{j_{\text{out}}} \text{Val}_j = (F_{\text{tot}})_0 \sum_{j}^{J} \text{Val}_j \left[y_{j_0} + \sum_{m}^{n_I} (s_j)_m Z_m \right] \qquad (10.1.4)$$

The value of the feed stream is

$$\sum_{j}^{J} F_{j_{\text{in}}} \text{Val}_j = (F_{\text{tot}})_{\text{in}} \sum_{j}^{J} \text{Val}_j y_{j_{\text{in}}} \qquad (10.1.5)$$

Hence, when the inlet stream is selected as the reference stream, the gross revenue of the process is

$$\left\{ \begin{array}{c} \text{Rate of} \\ \text{gross} \\ \text{revenue} \\ (\$/\text{time}) \end{array} \right\} = (F_{\text{tot}})_0 \sum_{j}^{J} \text{Val}_j \sum_{m}^{n_I} (s_j)_m Z_m \qquad (10.1.6)$$

In many instances, the separation expense is expressed in terms of cost per mole of product recovered, SC_j (in \$/mol of j); hence, the separation expense rate is

$$\left\{ \begin{array}{c} \text{Rate of} \\ \text{separation} \\ \text{expense} \\ (\$/\text{time}) \end{array} \right\} = (F_{\text{tot}})_0 \sum_{j}^{J} SC_j \left[y_{j_0} + \sum_{m}^{n_I} (s_j)_m Z_m \right] \qquad (10.1.7)$$

Combining Eqs. 10.1.6 and 10.1.7, the gross income rate of the reactor operation (without accounting for reactor operation expense) is

$$\left\{ \begin{array}{c} \text{Rate of} \\ \text{gross} \\ \text{income} \\ (\$/\text{time}) \end{array} \right\} = (F_{\text{tot}})_0 \left[\sum_{j}^{J} (\text{Val}_j - SC_j) \sum_{m}^{n_I} (s_j)_m Z_m - \sum_{j}^{J} (SC_j y_{j_0}) \right] \qquad (10.1.8)$$

When the species in the reactor effluent stream are *not* separated (unconverted reactants are discarded with the product), the gross revenue rate is

$$\left\{ \begin{array}{c} \text{Gross} \\ \text{revenue} \\ \text{rate} \\ (\$/\text{time}) \end{array} \right\} = (F_{\text{tot}})_0 \sum_{j=1}^{J} \text{Val}_j \left(y_{j_0} + \sum_{m}^{n_I} (s_j)_m Z_{m_{\text{out}}} \right)$$

$$- (F_{\text{tot}})_0 \sum_{j=1}^{J} \text{Val}_j y_{j_0} \qquad (10.1.9)$$

Note that the summation in the first term on the right of Eq. 10.1.9 is over the species in the outlet, but the value of products that are not being sold is zero (i.e., unconverted reactants that are discarded with the product). Also, note that polluting species in the product stream may have a negative value (the cost of removing them to meet environmental specifications). To obtain the net profit of the operation, we have to substitute into Eq. 10.1.1 the operating expenses of the reactor, itemized in Eq. 10.1.2.

The example below illustrates an economic-based optimization procedure.

Example 10.1   Product V is produced by reacting reactant A with reactant B where the following liquid-phase, elementary chemical reaction take place:

$$\text{Reaction 1:} \quad A + B \longrightarrow V$$

$$\text{Reaction 2:} \quad V + B \longrightarrow W$$

where product W is undesirable. A stream of reactant A ($C_A = 16\,\text{mol/L}$, $\rho = 800\,\text{g/L}$) and a stream of reactant B ($C_B = 20\,\text{mol/L}$, $\rho = 800\,\text{g/L}$) are available in the plant, and we want to utilize an available 200-L tubular reactor (plug-flow). The reactor is operated isothermally at 190°C. When each stream is fed at the same volumetric flow rate, determine:

a. The total feed rate needed to maximize the yield of product V
b. The total feed rate needed to maximize the profit when the outlet stream of the reactor is not separated
c. The total feed rate when the outlet stream undergoes separation

Data: At 190°C, $k_1 = 0.01\,\text{L/mol h}^{-1}$ $k_2 = 0.02\,\text{L/mol h}^{-1}$
Values of the reactants: A = 1 \$/mol   B = 2 \$/mol
Values of the products: V(raw) = 30 \$/mol   V(pure) = 36 \$/mol   W = −3 \$/mol
Species separation costs: A = 0.02   B = 0.3   V = 0.3   W = 0.03 \$/mol

Solution   The stoichiometric coefficients of the chemical reactions are

$$s_{A_1} = -1 \quad s_{B_1} = -1 \quad s_{V_1} = 1 \quad s_{W_1} = 0 \quad \Delta_1 = -1$$
$$s_{A_2} = 0 \quad s_{B_2} = -1 \quad s_{V_2} = -1 \quad s_{W_2} = 1 \quad \Delta_2 = -1$$

Since each reaction has a species that does not participate in the other, the two reactions are independent and there is no dependent reaction. We select the inlet stream to the reactor as a reference stream and denote its flow rate by v_0, where $v_0 = v_1 + v_2$. The concentration of the reference stream is

$$C_0 = \frac{(F_{\text{tot}})_0}{v_0} = \frac{v_1 C_A + v_2 C_B}{v_1 + v_2} = w_1 C_A + (1 - w_1)C_B \tag{a}$$

where $w_1 = v_1/(v_1 + v_2)$ is the volumetric fraction of stream 1 ($w_1 = 0.5$). The molar fractions of the species in the reference stream are

$$y_{A_0} = w_1 \frac{C_A}{C_0} \qquad y_{B_0} = (1 - w_1)\frac{C_B}{C_0} \qquad \text{(b)}$$

$y_{V_0} = y_{W_0} = 0$, and $Z_{1_{in}} = Z_{2_{in}} = 0$. We write Eq. 7.1.1 for each independent reaction:

$$\frac{dZ_1}{d\tau} = r_1\left(\frac{t_{cr}}{C_0}\right) \qquad \text{(c)}$$

$$\frac{dZ_2}{d\tau} = r_2\left(\frac{t_{cr}}{C_0}\right) \qquad \text{(d)}$$

Using Eq. 2.7.8, the local species molar flow rates are

$$F_A = (F_{tot})_0(y_{A_0} - Z_1) \qquad \text{(e)}$$

$$F_B = (F_{tot})_0(y_{B_0} - Z_1 - Z_2) \qquad \text{(f)}$$

$$F_V = (F_{tot})_0(Z_1 - Z_2) \qquad \text{(g)}$$

$$F_W = (F_{tot})_0 Z_2 \qquad \text{(h)}$$

For liquid-phase reactions, we use Eq. 7.1.11 to express the species concentrations, and the rates are

$$r_1 = k_1 C_0^2 (y_{A_0} - Z_1)(y_{B_0} - Z_1 - Z_2) \qquad \text{(i)}$$

$$r_2 = k_2 C_0^2 (Z_1 - Z_2)(y_{B_0} - Z_1 - Z_2) \qquad \text{(j)}$$

We define the characteristic reaction time on the basis of reaction 1; hence,

$$t_{cr} = \frac{1}{k_1 C_0} \qquad \text{(k)}$$

Substituting (i), (j), and (k) into (c) and (d), the design equations reduce to

$$\frac{dZ_1}{d\tau} = (y_{A_0} - Z_1)(y_{B_0} - Z_1 - Z_2) \qquad \text{(l)}$$

$$\frac{dZ_2}{d\tau} = \left(\frac{k_2}{k_1}\right)(Z_1 - Z_2)(y_{B_0} - Z_1 - Z_2) \qquad \text{(m)}$$

a. We solve (l) and (m) numerically subject to the initial condition that at $\tau = 0$, $Z_1 = Z_2 = 0$. Once we obtain Z_1 and Z_2 as functions of τ, we use (e) through (h) to obtain the species operating curves. Figure E10.1.1 shows the reaction curves, and Figure E10.1.2. shows the species operating curves. From the curve of product V we determine that highest $F_V/(F_{tot})_0 = 0.111$,

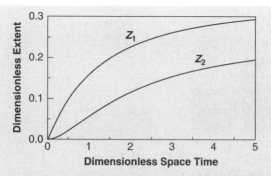

Figure E10.1.1   Reaction operating curves.

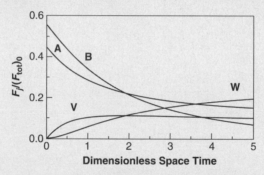

Figure E10.1.2   Species operating curves.

and it is achieved at about $\tau = 1.9$. The feed flow rate that provides the highest profit rate is

$$v_0 = \frac{V_R}{\tau t_{cr}} = 18.9 \text{ L/h}$$

b. When the reactor is operated without a separator, product V is contaminated with unconverted reactants and product W. The profit rate of the operation is calculated by combining Eqs. 10.1.5 and 10.1.6:

$$\{\text{Profit rate}\} = F_V \text{Val}_V + F_W \text{Val}_W - (F_{A_0} \text{Val}_A + F_{B_0} \text{Val}_B) \qquad \text{(n)}$$

where F_V and F_W are given by (g) and (h), respectively. Figure E10.1.3 shows the profit rate as a function of the dimensionless space time. The curve indicates that the highest profit rate is 60.2 $/h, and it is achieved at $\tau = 1.4$. The feed flow rate that provides the highest profit rate is 25.7 L/h. Note that calculation involves iteration since the value of v_0 in not known a priori. Once v_0 is known, we can plot the profit rate as a function of the feed rate, as shown in Figure E10.1.4.

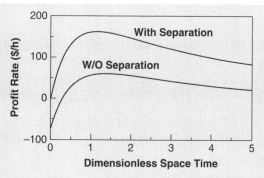

Figure E10.1.3   Profit curve.

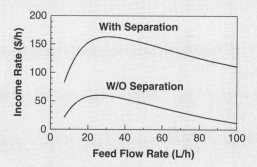

Figure E10.1.4   Profit vs. feed flow rate.

c. When the effluent stream of the reactor is fed to a separator, the profit rate of
the combined operation is calculated by Eq. 10.1.9,

$$\{\text{Profit rate}\} = F_V(36 - 0.03) + F_W(-3 - 0.03) + F_A(0.1 - 0.02)$$

$$+ F_B(0.2 - 0.03) - [F_{A_0}(0.1) + F_{B_0}(0.2)] \tag{o}$$

where F_A, F_B, F_V, and F_W are given by (e), (f), (g), and (h), respectively.
Figure E.3 shows the profit rate of this operation as a function of the dimensionless space time. The curve indicates that the highest profit rate is 162.8 \$/h, and
it is achieved at $\tau = 1.15$. The feed flow rate that provides the highest profit rate
is 31.3 L/h. The plot of the profit rate as a function of the feed rate is shown in
Figure E10.1.4.

   Note that the feed flow rate that provides the maximum profit depends on the
mode of operation (with or without separation) and is different from the feed
flow rate that provides the highest yield of product V. Also note that this example
considered a given proportion of the reactants. The procedure can be repeated
for different reactant proportions to determine the proportion that provides the
highest profit.

10.2 BATCH AND SEMIBATCH REACTORS

The designs of batch and semibatch reactors, covered in Chapters 6 and 9, addressed only the operating time needed to obtain a given conversion (or extent). The size of the batch was not considered. The latter is determined by the required production rate of the unit. The operation of batch reactors usually consists of four steps:

1. Preparation and filling, in duration time of t_f
2. The reaction time (or the operating time), t_r
3. Discharging time, t_d
4. Idle time, t_i

The size of the batch and the duration of the reaction time are determined on the basis of economic consideration. Usually, the desired production rate of a product, F_P, is specified, and the size of the batch, expressed in terms of $(N_{\text{tot}})_0$, is

$$F_P = \frac{(N_{\text{tot}})_0}{t_f + t_r + t_d + t_i} \left[y_{P_0} + \sum_{m}^{n_I} (s_P)_m Z_m \right] \tag{10.2.1}$$

where $(N_{\text{tot}})_0 = C_0 V_0$. The extents of the independent reactions are determined by the design equations for a given operating time. Below, we discuss the determination of the optimal operating time.

Let us denote the respective operating costs per unit time of these steps by W_f, W_r, W_d, and W_i. The total cost of the operation is therefore

$$\text{Cost}_{\text{tot}} = t_f W_f + t_r W_r + t_d W_d + t_i W_i \tag{10.2.2}$$

When the species in the reactor discharge are *not* separated, the gross revenue of a batch is

$$\left\{ \begin{array}{c} \text{Gross} \\ \text{revenue} \\ \text{per batch} \\ (\$) \end{array} \right\} = (N_{\text{tot}})_0 \sum_{j=1}^{J} \text{Val}_j \left(y_{j_0} + \sum_{m}^{n_I} (s_j)_m Z_m \right)$$

$$- (N_{\text{tot}})_0 \sum_{j=1}^{J} \text{Val}_j y_{j_0} \tag{10.2.3}$$

Combining Eqs. 10.1.1 and 10.1.2, the net profit rate is

$$\left\{ \begin{array}{c} \text{Net} \\ \text{profit} \\ \text{rate} \\ (\$/\text{time}) \end{array} \right\} = \frac{(N_{\text{tot}})_0}{t_f + t_r + t_d + t_i} \left[\sum_{j=1}^{J} \overset{\text{sold}}{\text{Val}_j} \left(y_{j_0} + \sum_{m}^{n_I} (s_j)_m Z_m \right) - \sum_{j=1}^{J} \text{Val}_j y_{j_0} \right] -$$

$$\frac{t_f W_f + t_r W_r + t_d W_d + t_i W_i}{t_f + t_r + t_d + t_i} \tag{10.2.4}$$

Note that the reaction time is expressed in terms of the dimensionless operating time, $t_r = t_{cr}\tau$, and relates to the extents by the design equations and to $(N_{tot})_0$ by Eq. 10.2.1. To optimize the operation, we determine the values of the parameters by solving the following equations:

$$\frac{\partial\{\text{profit rate}\}}{\partial\tau} = 0, \quad \frac{\partial\{\text{profit rate}\}}{\partial(y_j)_0} = 0 \qquad (10.2.5)$$

For operations with multiple chemical reactions, the optimal parameters are usually determined numerically. For most operations with single chemical reactions, the profit function is expressed in terms of relatively simple functions of these parameters, and the optimal operating time can be determined analytically. This is illustrated in the example below.

Example 10.2 The liquid-phase, first-order reaction

$$A \longrightarrow B + C$$

is carried out in a batch reactor. The value of reactant A is 0.50 \$/mol, the value of product B is 1 \$/mol, and the value of product C is 1.50 \$/mol. The sum of the feed, discharge, and idle times is 30 mins. The operating cost of the reactor is \$0.10/min. Determine:

a. The optimal operating (reaction) time.
b. The optimal extent.
c. The reactor volume if the desired production rate of product B is 5 mol/min.

Data: $C_A(0) = 2 \text{ mol/L}$ $k = 0.1 \text{ min}^{-1}$

Solution   The stoichiometric coefficients are

$$s_A = -1 \qquad s_B = 1 \qquad s_C = 1$$

We select the initial state as the reference state; hence, $C_0 = C_A(0) = 2 \text{ mol/L}$, and $y_{A_0} = 1$ and $y_{B_0} = y_{C_0} = 0$. For a first-order reaction, the characteristic reaction time is

$$t_{cr} = \frac{1}{k} = 10 \text{ min} \qquad (a)$$

Using Eq. 2.7.6,

$$N_A = (N_{tot})_0(1 - Z) \qquad (b)$$

$$N_B = N_C = (N_{tot})_0 Z \qquad (c)$$

The dimensionless design equation, in this case, is

$$\frac{dZ}{d\tau} = 1 - Z \qquad (d)$$

Solving (d) subject to the initial condition that $Z(0) = 0$, we obtain

$$Z(\tau) = 1 - e^{-\tau} \tag{e}$$

a. Using Eq. 10.2.4, the net profit rate of the operation is

$$\left\{ \begin{array}{c} \text{Rate of} \\ \text{net} \\ \text{profit} \\ (\$/\text{time}) \end{array} \right\} = (N_{\text{tot}})_0 \frac{\text{Val}_B Z + \text{Val}_C Z - \text{Val}_A y_{A_0} - t_r W_r}{t_f + t_r + t_d + t_i}$$

$$= (N_{\text{tot}})_0 \frac{(2.5)Z - 0.5 - t_r 0.1}{30 + t_r} \tag{f}$$

where $t_r = t_{\text{cr}}\tau$ is the operating time to be determined. Substituting (e) into (f) and taking the derivative, we find that the optimal dimensionless time is $\tau = 0.895$. Hence, the optimal operating time is $t_r = t_{\text{cr}}\tau = 8.95$ min, and the duration of a cycle is 38.95 min.

b. Using (e), the optimal extent is $Z = 0.591$.

c. Using Eq. 10.2.1, the specified production rate of product B is

$$5 \text{ mol/min} = \frac{(N_{\text{tot}})_0}{t_f + t_r + t_d + t_i} Z = \frac{(N_{\text{tot}})_0}{38.95}(0.591) \tag{g}$$

Solving (g), $(N_{\text{tot}})_0 = 329.53$ mol, and the required reactor volume is

$$V_R(0) = \frac{(N_{\text{tot}})_0}{C_0} = \frac{329.53 \text{ mol}}{2 \text{ mol/L}} = 164.76 \text{ L}$$

10.3 FLOW REACTORS

The economic-based optimization may be based on other criteria than those discussed in Section 10.1. In some cases, the production rate is specified, and the value of the product is fixed. In such cases, we would like to minimize the total operating cost. This is illustrated in the example below.

Example 10.3 We want to produce 1000 mol of product R per hour from an aqueous solution of reactant A ($C_A = 2$ mol/L) in a CSTR. Product R is formed by the first-order chemical reaction

$$A \longrightarrow 2R$$

The cost of the feed stream is 0.40 $/mol A, and the cost of operating the reactor (labor, recovery of capital cost, overhead, etc.) is 0.10 $/L h. The unconverted reactant A is being discarded. Based on the data below:

a. Determine the optimal operating conditions of V_R, Z_{out}, $(F_{tot})_0$.
b. Calculate the cost of producing R under these conditions.
c. Plot the production cost as a function of the extent.

Data: At the operating conditions $k = 0.125$ h^{-1}.

Solution   The stoichiometric coefficients of the reaction are

$$s_A = -1 \qquad s_R = 2 \qquad \Delta = 1$$

The rate expression is $r = kC_A$. We select the inlet stream as the reference stream; hence, $C_0 = C_{A_0}$, $y_{A_0} = 1$, and $Z_{in} = 0$. The characteristic reaction time is

$$t_{cr} = \frac{1}{k} = 8 \text{ h} \tag{a}$$

Using Eq. 2.7.8,

$$F_A = (F_{tot})_0(1 - Z) \tag{b}$$

$$F_R = (F_{tot})_0 2Z \tag{c}$$

The design equation of a CSTR, for this case, is

$$\tau = \frac{Z_{out}}{1 - Z_{out}} \tag{d}$$

where the dimensionless space time is

$$\tau = \frac{V_R}{v_0 t_{cr}} \tag{e}$$

a. The cost of the operation is

$$\text{Cost}(\$/h) = (0.4)(F_{tot})_0 y_{A_0} + (OC)V_R \tag{f}$$

where OC is the reactor operating cost, OC = 0.10 $/L h. Based on the specified production rate of product R and using (c),

$$(F_{tot})_0 = \frac{1000}{2Z_{out}} \tag{g}$$

From (d) and (e),

$$V_R = v_0 t_{cr} \frac{Z_{out}}{1 - Z_{out}} \tag{h}$$

Substituting (g) and (h) into (f) and noting that $(F_{tot})_0 = v_0 C_0$, the cost of the operation is

$$\text{Cost}(\$/h) = (0.4)\frac{500}{Z_{out}} + (0.1)\left(\frac{500}{C_0} t_{cr} \frac{1}{1 - Z_{out}}\right) \tag{i}$$

To determine the minimum cost for the given production rate of product R, we take the derivative of the cost with respect to Z_{out} and equate it to zero,

$$\frac{d(\text{Cost})}{dZ_{out}} = -\frac{200}{Z_{out}^2} + \left(\frac{50t_{cr}}{C_0}\right)\frac{1}{(1 - Z_{out})^2} = 0 \tag{j}$$

Substituting numerical values, we obtain $Z_{out} = 0.50$, and, from (g), $(F_{tot})_0 = 1000$ mol/h, and

$$v_0 = \frac{(F_{tot})_0}{C_0} = 500 \text{ L/h}$$

From (h), the optimal volume of the reactor is

$$V_R = (500 \text{ L/h})(8 \text{ h})\frac{0.5}{1 - 0.5} = 4000 \text{ L}$$

b. Using (i), the cost of producing R is

$$\text{Cost}(\$/h) = 0.4(1000) + 0.1(4000) = 800/h$$

c. The cost plot is shown in Figure E10.3.1.

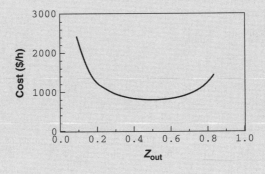

Figure E10.3.1   Profit vs. extent.

10.4   SUMMARY

In this chapter, we discussed briefly the optimization of chemical reactor operations on the basis of economic criteria. We showed how stoichiometric relations are combined with economic data and the design equations to describe the profitability of the operations. We covered the following topics:

1. Economic objective function of reactor operations
2. Economic objective function of reactor and separation operations
3. Economic-based optimization
4. Sizing and optimizing batch reactor operations

The reader is challenged to apply the methods described in this chapter to other applications.

PROBLEMS*

10.1$_2$ The plug-flow reactor is to produce 1000 mol of product R per hour from an aqueous feed of A ($C_{A_0} = 1$ mol/L). The reaction is

$$2A \longrightarrow R$$

and its rate expression is $r = 2kC_A^2$. The cost of reactant stream is 0.50 \$/mol A, and the cost of operating the reactor comes to 0.20 \$/L h. Find V_R, f_A, and F_{A_0} for optimum operations under the following conditions:

a. The unconverted A is discarded.
b. What is the cost of producing R in (a)?
c. The unconverted A is recovered and recycled at a loss of 0.10 \$/mol A.
d. What is the cost of producing R in (c)?

Data: $k = 1$ L/mol h^{-1}.

10.2$_2$ Aqueous feed ($C_A = 1$ mol/L), ($v_{in} = 1000$ L/h) is available at a cost of \$1.00/mol of reactant A. We can produce product R by the second-order chemical reaction

$$2A \longrightarrow R$$

The reaction rate constant is 2 L/mol h^{-1}. The value of product R is \$3.70/mol. The operating cost of a CSTR and a product purification unit is 0.20

*Subscript 1 indicates simple problems that require application of equations provided in the text. Subscript 2 indicates problems whose solutions require some more in-depth analysis and modifications of given equations. Subscript 3 indicates problems whose solutions require more comprehensive analysis and involve application of several concepts. Subscript 4 indicates problems that require the use of a mathematical software or the writing of a computer code to obtain numerical solutions.

$/L of reactor per hour (labor, utilities, value of money). Unconverted reactant A is destroyed when the product is purified. Find the best way to operate such a system (V_R, Z_{out}, F_R, hourly profit rate) and then recommend a course of action.

10.3$_2$ We want to produce 500 mol/h of product R (value $3/mol) by the chemical reaction

$$A + B \longrightarrow R$$

The rate expression is $r = kC_AC_B$. The cost of reactant A is $0.80/mol, and that of reactant B is $0.20/mol. Unconverted reactant A is recovered and recycled to the reactor at a cost of $0.10/mol, and unconverted reactant B is discarded. The cost of operating the reactor is $0.40/L h. Determine:

a. The optimal conversion and feed ratio $F_{B_{in}}/F_{A_{in}}$.

b. The size of the reactor (CSTR).

c. The profit ($/h) of the operation.

Data: $k = 2 \text{ L/mol h}^{-1}$ $C_{A_{in}} = 1 \text{ mol/L}$ $v_{in} = 500 \text{ L/h}$

BIBLIOGRAPHY

More detailed treatment of optimizing the operation of batch reactors can be found in:
R. B. Aris, *An Introduction to Chemical Reactor Analysis*, Prentice-Hall, Englewood Cliffs, NJ, 1960.

A concise review of reactor optimization is provided by:
O. Levenspiel, *Chemical Reaction Engineering*, 3rd ed., Wiley, New York, 1999.

A general treatment of optimizing reactor operations and chemical processes can be found in:
L. T. Biegler, I. E. Grossman, and A. W. Weterberg, *Systematic Methods of Chemical Process Design*, Prentice-Hall, Englewood Cliffs, NJ, 1997.

APPENDIX A

SUMMARY OF KEY RELATIONSHIPS

TABLE A.1 Stoichiometric Relations

General relationships

Definition of stoichiometric coefficients:

$$aA + bB \longrightarrow cC + dD$$

$$s_A = -a \qquad s_B = -b \qquad s_C = c \qquad s_D = d \tag{A}$$

$$\Delta = \sum_{j}^{J} s_j = -a - b + c + d \tag{B}$$

Mathematical condition for a balanced chemical reaction:

$$\sum_{j}^{J} s_j MW_j = 0 \qquad j = A, B, \ldots, \tag{C}$$

Relationship between the kth-dependent reaction and the independent reactions:

$$\sum_{m}^{n_I} \alpha_{km}(s_j)_m = (s_j)_k \qquad j = A, B, \ldots, \tag{D}$$

Definitions

Closed (Batch) Reactor

Extent of the ith chemical reaction:

$$X_i(t) = \frac{(n_j(t) - n_{j0})_i}{(s_j)_i}$$

Dimensionless extent of the ith reaction:

$$Z_i(t) = \frac{X_i(t)}{(N_{tot})_0}$$

Conversion of reactant A:

$$f_A(t) = \frac{N_A(0) - N_A(t)}{N_A(0)}$$

Steady-Flow Reactor

Extent per time of the ith chemical reaction:

$$\dot{X}_i = \frac{1}{(s_j)_i}\frac{d(n_j)_i}{dt} \qquad j = A, B, \ldots, \tag{E}$$

Dimensionless extent of the ith reaction:

$$Z_i = \frac{\dot{X}_i}{(F_{tot})_0} \tag{F}$$

Conversion of reactant A:

$$f_A = \frac{F_{A_{in}} - F_A}{F_{A_{in}}} \tag{G}$$

Stoichiometric relations for multiple reactions

$$F_j = F_{j_{\text{in}}} + \sum_m^{n_I} (s_j)_m \dot{X}_m \qquad j = A, B, \ldots, \tag{H}$$

$$N_j(t) = N_j(0) + \sum_m^{n_I} (s_j)_m X_m(t)$$

$$F_{\text{tot}} = (F_{\text{tot}})_{\text{in}} + \sum_m^{n_I} \Delta_m \dot{X}_m \tag{I}$$

$$N_{\text{tot}}(t) = N_{\text{tot}}(0) + \sum_m^{n_I} \Delta_m X_m(t)$$

$$F_j = (F_{\text{tot}})_0 \left[\frac{(F_{\text{tot}})_{\text{in}}}{(F_{\text{tot}})_0} y_{j_{\text{in}}} + \sum_m^{n_I} (s_j)_m Z_m \right] \tag{J}$$

$$N_j(t) = (N_{\text{tot}})_0 \left[\frac{N_{\text{tot}}(0)}{(N_{\text{tot}})_0} y_j(0) + \sum_m^{n_I} (s_j)_m Z_m(t) \right]$$

$$F_{\text{tot}} = (F_{\text{tot}})_0 \left[\frac{(F_{\text{tot}})_{\text{in}}}{(F_{\text{tot}})_0} + \sum_m^{n_I} \Delta_m Z_m \right] \tag{K}$$

$$N_{\text{tot}}(t) = (N_{\text{tot}})_0 \left[\frac{N_{\text{tot}}(0)}{(N_{\text{tot}})_0} + \sum_m^{n_I} \Delta_m Z_m(t) \right]$$

Stoichiometric relations for single reactions

$$F_j = F_{j_{\text{in}}} + \frac{s_j}{s_A}(F_A - F_{A_{\text{in}}}) \qquad j = B, C, \ldots, \tag{L}$$

$$N_j(t) = N_j(0) + \frac{s_j}{s_A}[N_A(t) - N_A(0)]$$

$$f_A = -\frac{s_A}{F_{A_{\text{in}}}} \dot{X} = -\frac{s_A}{y_{A_{\text{in}}}} Z \tag{M}$$

$$f_A(t) = -\frac{s_A}{N_A(0)} X(t) = -\frac{s_A}{y_A(0)} Z(t)$$

$$F_j = F_{j_{\text{in}}} - \frac{s_j}{s_A} F_{A_{\text{in}}} f_A \qquad j = B, C, \ldots, \tag{N}$$

$$N_j(t) = N_j(0) - \frac{s_j}{s_A} N_A(0) f_A(t)$$

$$F_{\text{tot}} = (F_{\text{tot}})_{\text{in}} - \frac{\Delta}{s_A} F_{A_{\text{in}}} f_A \tag{O}$$

$$N_{\text{tot}} = N_{\text{tot}}(0) - \frac{\Delta}{s_A} N_A(0) f_A(t)$$

TABLE A.2   Summary of Kinetic Relations

Definition of species formation rates:

$$(r_j) \equiv \frac{1}{V}\frac{dN_j}{dt} \qquad (r_j)_S \equiv \frac{1}{S}\frac{dN_j}{dt} \qquad (r_j)_W \equiv \frac{1}{W}\frac{dN_j}{dt} \qquad j = A, B, \ldots, \quad \text{(A)}$$

Relations between the different formation rates:

$$(r_j) = \left(\frac{S}{V}\right)(r_j)_S \qquad (r_j) = \left(\frac{W}{V}\right)(r_j)_W \qquad j = A, B, \ldots, \qquad \text{(B)}$$

Definition of the rate of a chemical reaction:

$$r \equiv \frac{1}{V}\frac{dX}{dt} \qquad r_S \equiv \frac{1}{S}\frac{dX}{dt} \qquad r_W \equiv \frac{1}{W}\frac{dX}{dt} \qquad \text{(C)}$$

Relation between a species formation rate and rates of chemical reactions:

$$(r_j) = \sum_{i=1}^{n_{\text{all}}} (s_j)_i r_i \qquad j = A, B, \ldots, \qquad \text{(D)}$$

Power function rate expression:

$$r = k(T)\prod_{j}^{J} C_j^{\alpha_j} \qquad \text{(E)}$$

where $k(T)$, the reaction rate constant, is expressed by

$$k(T) = k_0 e^{-(E_a/RT)} \qquad \text{(F)}$$

$$k(\theta) = k(T_0)e^{\gamma(\theta-1)/\theta} \qquad \text{(G)}$$

where α_j = order of the jth species
$\quad E_a$ = activation energy
$\quad k_0$ = preexponential factor
$\quad \theta$ = dimensionless temperature, T/T_0
$\quad \gamma$ = dimensionless activation energy, E_a/RT_0

Characteristic reaction time:

$$t_{\text{cr}} \equiv \frac{\text{characteristic concentration}}{\text{characteristic reaction rate}} = \frac{C_0}{r_0} \qquad \text{(H)}$$

For reactions with an overall order of n:

$$t_{\text{cr}} = \frac{1}{k(T_0)C_0^{n-1}} \qquad \text{(I)}$$

TABLE A.3a   Design Equations for Ideal Reactors—Simplified Form[a]

	Batch Reactor	Plug-Flow Reactor	CSTR
Design equation for the mth independent reaction	$\dfrac{dZ_m}{d\tau} = \left(r_m + \displaystyle\sum_k^{n_D}\alpha_{km}r_k\right)\left(\dfrac{t_{cr}}{C_0}\right)\left(\dfrac{V_R}{V_{R_0}}\right)$	$\dfrac{dZ_m}{d\tau} = \left(r_m + \displaystyle\sum_k^{n_D}\alpha_{km}r_k\right)\left(\dfrac{t_{cr}}{C_0}\right)$	$Z_{m_{out}} - Z_{m_{in}} = \left(r_m + \displaystyle\sum_k^{n_D}\alpha_{km}r_k\right)\tau\left(\dfrac{t_{cr}}{C_0}\right)$
Auxiliary relations	Species molar composition: $$N_j = (N_{tot})_0\left(y_{j_0} + \sum_m^{n_I}(s_j)_m Z_m\right)$$ Constant volume: $$V_R = V_{R_0}$$ Variable volume gas phase: $$V_R = V_{R_0}\left(1 + \sum_m^{n_I}\Delta_m Z_m\right)\theta\left(\dfrac{P_0}{P}\right)$$ Species concentration in constant-volume reaction: $$C_j = C_0\left(y_{j_0} + \sum_m^{n_I}(s_j)_m Z_m\right)$$ Concentration in variable-volume gas-phase reactor: $$C_j = C_0\dfrac{y_{j_0} + \sum_m^{n_I}(s_j)_m Z_m}{(1 + \sum_m^{n_I}\Delta_m Z_m)\theta}\left(\dfrac{P}{P_0}\right)$$	Species molar flow rate: $$F_j = (F_{tot})_0\left(y_{j_0} + \sum_m^{n_I}(s_j)_m Z_m\right)$$ Volumetric flow rate for liquid phase: $$v = v_0$$ Volumetric flow rate for gas phase: $$v = v_0\left(1 + \sum_m^{n_I}\Delta_m Z_m\right)\theta\left(\dfrac{P_0}{P}\right)$$ Species concentration in liquid phase: $$C_j = \dfrac{F_j}{v} = C_0\left(y_{j_0} + \sum_m^{n_I}(s_j)_m Z_m\right)$$ Species concentration in gas phase: $$C_j = \dfrac{F_j}{v} = C_0\dfrac{y_{j_0} + \sum_m^{n_I}(s_j)_m Z_m}{(1 + \sum_m^{n_I}\Delta_m Z_m)\theta}\left(\dfrac{P}{P_0}\right)$$	

(Continued)

459

TABLE A.3a *Continued*

	Batch Reactor	Plug-Flow Reactor	CSTR
Definitions	Dimensionless operating time: $$\tau \equiv \frac{t}{t_{cr}}$$ Dimensionless reaction extent: $$Z_m(\tau) \equiv \frac{X_m(\tau)}{(N_{tot})_0}$$ Reference concentration: $$C_0 \equiv \frac{(N_{tot})_0}{V_{R_0}}$$ Composition of the reference state: $$y_{j_0} \equiv \frac{N_{j_0}}{(N_{tot})_0}$$	Dimensionless space time: $$\tau \equiv \frac{t_{sp}}{t_{cr}} = \frac{V_R}{v_0 t_{cr}}$$ Dimensionless reaction extent: $$Z_m \equiv \frac{\dot{X}_m}{(F_{tot})_0}$$ Reference concentration: $$C_0 \equiv \frac{(F_{tot})_0}{v_0}$$ Composition of the reference stream: $$y_{j_0} \equiv \frac{F_{j_0}}{(F_{tot})_0}$$	

[a]Initial state is the reference state; inlet stream is the reference stream.

TABLE A.3b Design Equations for Ideal Reactors—General Form[a]

	Batch Reactor	Plug-Flow Reactor	CSTR
Design equation for the mth independent reaction	$$\frac{dZ_m}{d\tau} = \left(r_m + \sum_k^{n_D} \alpha_{km} r_k\right)\left(\frac{t_{cr}}{C_0}\right)\left(\frac{V_R}{V_{R_0}}\right)$$	$$\frac{dZ_m}{d\tau} = \left(r_m + \sum_k^{n_D} \alpha_{km} r_k\right)\left(\frac{t_{cr}}{C_0}\right)$$	$$Z_{m_{out}} - Z_{m_{in}} = \left(r_m + \sum_k^{n_D} \alpha_{km} r_k\right)\tau\left(\frac{t_{cr}}{C_0}\right)$$
Auxiliary relations	Species molar composition: $$N_j = (N_{tot})_0\left[\frac{N_{tot}(0)}{(N_{tot})_0}y_j(0) + \sum_m^{n_I}(s_j)_m Z_m\right]$$ Constant-volume reaction: $$V_R = V_R(0)$$ Volume of variable-volume gas-phase reactor: $$V_R = V_R(0)\left[\frac{N_{tot}(0)}{(N_{tot})_0} + \sum_m^{n_I}\Delta_m Z_m\right]\left[\frac{\theta}{\theta(0)}\right]\left[\frac{P(0)}{P}\right]$$ Species concentration in constant-volume reactor: $$C_j = C_0\left[\frac{V_{R_0}}{V_R(0)}\right]\left[\frac{N_{tot}(0)}{(N_{tot})_0}y_j(0) + \sum_m^{n_I}(s_j)_m Z_m\right]\left[\frac{\theta(0)}{\theta}\right]\left[\frac{P}{P(0)}\right]$$ Concentration in variable-volume gas-phase reactor: $$C_j = C_0\,\frac{\frac{N_{tot}(0)}{(N_{tot})_0}y_j(0) + \sum_m^{n_I}(s_j)_m Z_m}{\frac{N_{tot}(0)}{(N_{tot})_0} + \sum_m^{n_I}\Delta_m Z_m}\left[\frac{\theta(0)}{\theta}\right]\left[\frac{P}{P(0)}\right]$$	Species molar flow rate: $$F_j = (F_{tot})_0\left[\frac{(F_{tot})_{in}}{(F_{tot})_0}y_{j_{in}} + \sum_m^{n_I}(s_j)_m Z_m\right]$$ Volumetric flow rate of liquid phase: $$v = v_{in} = v_0\left(\frac{v_{in}}{v_0}\right)$$ Volumetric flow rate of gas phase: $$v = v_0\left[\frac{(F_{tot})_{in}}{(F_{tot})_0} + \sum_m^{n_I}\Delta_m Z_m\right]\left[\frac{\theta}{\theta_m}\right]\left[\frac{P_{in}}{P}\right]$$ Species concentration in liquid phase: $$C_j = \frac{F_j}{v} = C_0\left(\frac{v_0}{v_{in}}\right)\left[\frac{(F_{tot})_{in}}{(F_{tot})_0}y_{j_{in}} + \sum_m^{n_I}(s_j)_m Z_m\right]$$ Species concentration in gas phase: $$C_j = \frac{F_j}{v} = C_0\,\frac{\frac{(F_{tot})_{in}}{(F_{tot})_0}y_{j_{in}} + \sum_m^{n_I}(s_j)_m Z_m}{\frac{(F_{tot})_{in}}{(F_{tot})_0} + \sum_m^{n_I}\Delta_m Z_m}\left(\frac{\theta_{in}}{\theta}\right)\left(\frac{P}{P_{in}}\right)$$	

(Continued)

TABLE A.3b *Continued*

Definitions	Batch Reactor	Plug-Flow Reactor	CSTR
	Dimensionless operating time:	Dimensionless space time:	
	$$\tau \equiv \frac{t}{t_{cr}}$$	$$\tau \equiv \frac{t_{sp}}{t_{cr}} = \frac{V_R}{v_0 t_{cr}}$$	
	Dimensionless reaction extent:	Dimensionless reaction extent:	
	$$Z_m(\tau) \equiv \frac{X_m(\tau)}{(N_{tot})_0}$$	$$Z_m \equiv \frac{\dot{X}_m}{(F_{tot})_0}$$	
	Reference concentration:	Reference concentration:	
	$$C_0 \equiv \frac{(N_{tot})_0}{V_{R_0}}$$	$$C_0 \equiv \frac{(F_{tot})_0}{v_0}$$	
	Composition of reference state:	Composition of reference stream:	
	$$y_{j_0} \equiv \frac{N_{j_0}}{(N_{tot})_0}$$	$$y_{j_0} \equiv \frac{F_{j_0}}{(F_{tot})_0}$$	

[a]Initial state is different of the reference state or inlet stream is different of the reference stream.

TABLE A.4 Energy Balance Equations for Ideal Reactors[a]

	Batch Reactor (Constant Volume) / Plug-Flow Reactor	CSTR

Energy balance equation

Batch Reactor (Constant Volume) / Plug-Flow Reactor:

$$\frac{d\theta}{d\tau} = \frac{1}{CF(Z_m, \theta)}\left[HTN(\theta_F - \theta) - \sum_m^{n_I} DHR_m \frac{dZ_m}{d\tau} - \frac{d}{d\tau}\left(\frac{W_{sh}}{(N_{tot})_0 \hat{c}_{p_0} T_0}\right)\right]$$

(For plug-flow reactor, omit the mechanical work term.)

CSTR:

$$HTN\,\tau(\theta_F - \theta) - \frac{\dot{W}_{sh}}{(F_{tot})_0 \hat{c}_{p_0} T_0} = \sum_m^{n_I} DHR_m(Z_{m_{out}} - Z_{m_{in}}) + CF_{out}(\theta_{out} - 1) - CF_{in}(\theta_{in} - 1)$$

Definitions and auxiliary relations

Dimensionless temperature:

$$\theta \equiv \frac{T}{T_0}$$

Specific molar heat capacity of the reference state (Batch) / reference stream (CSTR):

Gas Phase:

$$\hat{c}_{p_0} = \sum_j^J y_{j_0} c_{p_j}(T_0)$$

Liquid Phase (Batch):

$$\hat{c}_{p_0} = \frac{M_0 \bar{c}_p}{(N_{tot})_0}$$

Liquid Phase (CSTR):

$$\hat{c}_{p_0} = \frac{\dot{m}}{(F_{tot})_0}\,\bar{c}_p$$

Dimensionless heat of reaction:

$$DHR_m = \frac{\Delta H_{R_m}(T_0)}{\hat{c}_{p_0} T_0}$$

$$DHR_m = \frac{\Delta H_{R_m}(T_0)}{T_0 \hat{c}_{p_0}}$$

Dimensionless heat-transfer number:

$$HTN = \frac{Ut_{cr}}{C_0 \hat{c}_{p_0}}\left(\frac{S}{V}\right)$$

$$HTN = \frac{Ut_{cr}}{C_0 \hat{c}_{p_0}}\left(\frac{S}{V}\right)$$

Dimensionless heat-transfer rate:

$$\frac{d}{d\tau}\left(\frac{\dot{Q}}{F_{tot_0}\hat{c}_{p_0}T_0}\right) = HTN(\theta_F - \theta)$$

Dimensionless heat-transfer rate for plug-flow reactor:

$$\frac{d}{d\tau}\left(\frac{\dot{Q}}{F_{tot_0}\hat{c}_{p_0}T_0}\right) = HTN(\theta_F - \theta)$$

(Continued)

TABLE A.4 *Continued*

Batch Reactor (Constant Volume)	Plug-Flow Reactor	CSTR
		Dimensionless heat-transfer rate for CSTR: $$\frac{\dot{Q}}{F_{\text{tot}_0}\hat{c}_{p_0}T_0} = \text{HTN}\,\tau(\theta_F - \theta)$$
	Correction factor of heat capacity for liquid phase: $$CF(Z_m, \theta) \equiv \frac{\dot{m}\bar{c}_p}{\dot{m}_0\bar{c}_{p_0}}$$	
	Correction factor of heat capacity for gas phase: $$CF(Z_m, \theta) = \frac{1}{\hat{\bar{c}}_{p_0}}\left[\sum_j^J y_{j_0}\hat{c}_{p_j}(\theta) + \sum_j^J \hat{c}_{p_j}(\theta)\sum_m^{n_I}(s_j)_m Z_m\right]$$	
Correction factor of heat capacity for liquid phase: $$CF(Z_m, \theta) \equiv \frac{M\bar{c}_p}{M_0\bar{c}_{p_0}}$$		
Correction factor of heat capacity for gas phase: $$CF(Z_m, \theta) = \frac{1}{\hat{\bar{c}}_{p_0}}\left[\sum_j^J y_{j_0}\hat{c}_{p_j}(\theta) + \sum_j^J \hat{c}_{p_j}(\theta)\sum_m^{n_I}(s_j)_m Z_m\right]$$		

[a]Initial state is reference state; inlet stream is the reference stream.

APPENDIX B

MICROSCOPIC SPECIES BALANCES—SPECIES CONTINUITY EQUATIONS

Consider a stationary volume element, $\Delta x\, \Delta y\, \Delta z$ shown in Figure B.1, through which species j flows and in which chemical reactions take place. Let J_{jx}, J_{jy}, and J_{jz} be the components of the local molar flux of species j, C_j the local molar concentration of species j, and (r_j) the volume-based formation rate of species j defined by Eq. 3.1.1a. We write a species balance over the element in terms of the molar flux of species j through the six surfaces of the element; each bracket corresponds to a term in Eq. 4.0.1:

$$\left[(J_{jx})_x\, \Delta y\, \Delta z + (J_{jy})_y\, \Delta x\, \Delta z + (J_{jz})_z\, \Delta x\, \Delta y\right] + \left[(r_j)\Delta x\, \Delta y\, \Delta z\right] =$$

$$\left[(J_{jx})_{x+\Delta x}\, \Delta y\, \Delta z + (J_{jy})_{y+\Delta y}\, \Delta x\, \Delta z + (J_{jz})_{z+\Delta z}\, \Delta x\, \Delta y\right] +$$

$$\left[\frac{d}{dt} C_j\, \Delta x\, \Delta y\, \Delta z\right] \tag{B.1}$$

Dividing both sides by $\Delta x\, \Delta y\, \Delta z$ and taking the limit, $\Delta x \to 0$, $\Delta y \to 0$, $\Delta z \to 0$, we obtain

$$\frac{\partial C_j}{\partial t} + \frac{\partial J_{jx}}{\partial x} + \frac{\partial J_{jy}}{\partial y} + \frac{\partial J_{jz}}{\partial z} = (r_j) \tag{B.2}$$

In general, we can write Eq. B.2 in vector notation:

$$\frac{\partial C_j}{\partial t} + \nabla \cdot \mathbf{J}_j = (r_j) \tag{B.3}$$

Principles of Chemical Reactor Analysis and Design, Second Edition. By Uzi Mann
Copyright © 2009 John Wiley & Sons, Inc.

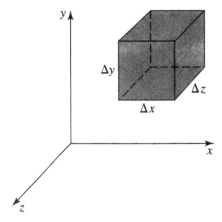

Figure B.1 Diagram of the molar flux components in a Cartesian element.

where ∇ is the divergence operator of the molar flux of species j. Equation B.3 is commonly called the species continuity equation. It provides a relation between the time variations in the species concentration at a fixed point, the local motion of the species, and the rate the species is formed by chemical reactions. The species continuity equations for cylindrical and spherical coordinates are given in Table B.1.

To describe the operation of a chemical reactor, we integrate the species continuity equation over the reactor volume. Multiplying each term in Eq. B.3 by dV and integrating,

$$\int_{V_R} \frac{\partial C_j}{\partial t} dV + \int_{V_R} (\nabla \cdot \mathbf{J}_j) dV = \int_{V_R} (r_j) dV \qquad (B.4)$$

TABLE B.1 Species Continuity Equations

In general vector notation:

$$\frac{\partial C_j}{\partial t} + \nabla \cdot \mathbf{J}_j = (r_j) \qquad (A)$$

For rectangular coordinates:

$$\frac{\partial C_j}{\partial t} + \frac{\partial J_{jx}}{\partial x} + \frac{\partial J_{jy}}{\partial y} + \frac{\partial J_{jz}}{\partial z} = (r_j) \qquad (B)$$

For cylindrical coordinates:

$$\frac{\partial C_j}{\partial t} + \frac{1}{r}\frac{\partial}{\partial r}(rJ_{j_r}) + \frac{1}{r}\frac{\partial J_{j_\theta}}{\partial \theta} + \frac{\partial J_{j_z}}{\partial z} = (r_j) \qquad (C)$$

For spherical coordinates:

$$\frac{\partial C_j}{\partial t} + \frac{1}{r^2}\frac{\partial}{\partial r}(r^2 J_{jr}) + \frac{1}{r\sin\theta}\frac{\partial}{\partial \theta}(J_{j\theta}\sin\theta) + \frac{1}{r\sin\theta}\frac{\partial J_{j\phi}}{\partial \phi} = (r_j) \qquad (D)$$

The first term on the left-hand side reduces to

$$\int_{V_R} \frac{\partial C_j}{\partial t} \, dV = \frac{dN_j}{dt}$$

where N_j is the total number of moles of species j in the reactor. Applying Gauss's divergence theorem, the second term on the left-hand side reduces to

$$\int_{V_R} (\nabla \cdot \mathbf{J}_j) \, dV = \int_{S_R} (\mathbf{J}_j \cdot \mathbf{n}) \, dS = F_{j_{out}} - F_{j_{in}}$$

where $\mathbf{n}$ is the outward unit vector on the boundaries of the reactor. This term provides the net molar flow rate of species j through the boundaries of the reactor. Thus, Eq. B.4 reduces to

$$\frac{dN_j}{dt} + F_{j_{out}} - F_{j_{in}} = \int_{V_R} (r_j) \, dV \qquad \text{(B.5)}$$

which is the integral form of the general species-based design equation of a chemical reactor, written for species j, and is identical to Eq. 4.1.3.

APPENDIX C

SUMMARY OF NUMERICAL DIFFERENTIATION AND INTEGRATION

C.1 NUMERICAL DIFFERENTIATION

For equally spaced points, the first derivative of function $f(x)$ is approximated (to error order of Δx^2) as follows:

Forward differentiation:

$$\left(\frac{df}{dx}\right)_i = \frac{-3f(x_i) + 4f(x_{i+1}) - f(x_{i+2})}{2\,\Delta x} \qquad \text{(C.1)}$$

Central differentiation:

$$\left(\frac{df}{dx}\right)_i = \frac{f(x_{i+1}) - f(x_{i-1})}{2\,\Delta x} \qquad \text{(C.2)}$$

Backward differentiation:

$$\left(\frac{df}{dx}\right)_i = \frac{3f(x_i) - 4f(x_{i-1}) + f(x_{i-2})}{2\,\Delta x} \qquad \text{(C.3)}$$

Principles of Chemical Reactor Analysis and Design, Second Edition. By Uzi Mann
Copyright © 2009 John Wiley & Sons, Inc.

C.2 NUMERICAL INTEGRATION

Trapezoidal Rule The trapezoidal rule provides a first-order approximation of the area of a function between two points:

$$I = \int_{x_1}^{x_2} f(x)\, dx = \frac{x_2 - x_1}{2}[f(x_1) + f(x_2)] \tag{C.4}$$

Simpson's Rule This method is based on a second-order polynomial approximation of the function. For equally spaced points, the integral of the function between x_0 and x_2 is

$$I = \int_{x_0}^{x_2} f(x)\, dx = \frac{\Delta x}{3}[f(x_0) + 4f(x_1) + f(x_2)] \tag{C.5}$$

INDEX

Activation energy 86
    Determination of 87
Activity coefficients 134
Adiabatic operations
    Batch reactor 144, 165, 217, 224
    CSTR 154, 321, 359, 364
    Plug-flow reactor 153, 244,
        284, 289
    PFR with distributed feed 409
    Recycle reactor 431
    Semibatch reactor 387, 394
Adsorption 10
Arrhenius equation 86
Autocatalytic reactions 431

Batch-reactor 3, 29, 159–230
    Constant-volume 167–181
    Variable-volume 181–189
Biological reactions *see* Enzymatic reactions
Bubble column reactor 7

Cascade of CSTRs 4, 336–341
Catalysis 10
Catalytic reactions 10, 257
Characteristic reaction time
    Definition of 92
    Determination of 93

Chemical formula 26
Co-current flow 282
Continuous stirred tank reactor (CSTR)
    see Reactor
Conversion
    Definition of 54
    Relation to extent 54
Counter-current flow 282
Correction factor of heat capacity
    Definition of 139, 150
    for gas-phase reactions 142, 152
    for liquid-phase reactions 142, 152
    for distillation reactor 421
    for PFR with distributed feed 405, 406
    for recycle reactor 430
    for semibatch reactor 386

Damkohler number 13
Dependent reactions 39
    Relation to independent
        reactions 43
Design equation *see* Reactor design
    equation 43
Differential method 190–192
Differential reactor 102
Diffusion coefficient 11
Diffusivity 9

Dimensionless variables, definition of
    Activation energy 88
    Extent 64
    Heat 140
    Heat of reaction 140
    Heat transfer rate 153
    Heat transfer number 140
    Operating time 113
    Space time 115
Distillation reactor 3, 416–425

Economics 441–453
Effectiveness factor 10
Elementary reactions 26
Endothermic reactions 88
Energy of activation *see* Activation energy
Energy balance equation 15,
    135–156
    for batch reactors 136–145
    for plug-flow reactor 150, 243
    for steady flow reactors 147–154
    for CSTR 153, 320
    for semibatch reactors 382–384
    for recycle reactor 429
    for PFR with distributed feed 405
Enzymatic reactions 175, 331
Equilibrium constant 134
Ergun equation 301
Excess reactant 49
Exothermic reactions 88
Extent of reaction
    Definition of 28
Experimental reactors 16

Fixed bed reactor *see* Packed-bed reactor
Fluidized bed reactor 5
Fluidized catalytic cracking 20
Formulation procedure
    of design equations, batch reactor 199
    of design equations, PFR 265
    of design equations, CSTR 341
    of energy balance equation, batch
        reactor 216
    of energy balance equation, PFR 283
    of energy balance equation, CSTR 358
Fractional conversion *see* Conversion
Frequency factor *see* Pre-exponential factor
Friction factor 296

Gas–solid reactions 2, 12
Gaussian elimination 41
Generation rate 31
Global reaction rate 14, 91

Hatta number 13
Heat capacity
    Dependence on composition 142
    Dependence on temperature 143
    of reference state or stream 141, 151
Heat of formation 131
Heat of reaction 131
Heat transfer coefficient 138
Heat transfer number (HTN)
    Definition 140
    Estimation of 165, 244, 321
Heterogeneous reactions 2
Homogeneous reactions 2
Hougen–Watson formulation 10

Independent chemical reactions 39–47
    Determination of number of 40
    Relation to dependent reactions 43
    Selecting a set of 112
Independent species specifications 68–71
Interfacial area 82
Instantaneous HTN 165
Integral method 192
Intrinsic kinetics 91
Isothermal operations
    of batch reactors 166–215
    of plug-flow reactor 245–281
    of CSTR 322–358

Kiln reactor 7
Kinetics 9, 81–97

Laminar flow 239
Langmuir–Hinshelwood formulation 10
Limiting reactant 48
Local HTN 244

Mass transfer 9
Michaelis–Menten rate expression 90
Molar flux 465
Momentum balance equation 15, 296
Moving bed reactor 6
Multiple steady-states 18

Nonideal flow 20
Nonisothermal operation
    of batch reactors 216–230
    of plug-flow reactor 281–295
    of CSTR 358–370
    of PFR with distributed feed 409
    of recycle reactor 431
    of semibatch reactors 387, 393
Numerical differentiation 469

Optimization 441–452
Order of reaction 89–90

Packed-bed reactor 6
Pore diffusion 2, 10
Porous catalyst 2
Power law rate expression 90
Pre-exponential factor 86–90
Pressure drop 296–308
Price of chemicals 443

Rate expression 86
    Determination of 189–193, 261, 333–334
    Forms of 90
Rate of reaction 82–84
Rate of species formation 81–82
Rate law *see* Rate expression
Rate limiting step 10
Reaction intermediates 26
Reaction operating curve 117
Reaction pathway 26
Reaction rate constant 86
Reaction rates 83
Reactor
    Batch 159–230
    Continuous stirred tank 317–370
    Distillation 416–425
    Plug-flow 239–309
    PFR with distributed feed 400–416
    Recycle 425–434
    Semibatch 377–400
Reactor design equation
    Dimensionless forms 113–116
    Reaction-based 107–112
    Species-based 102–107

Reference state 113
Reference stream 114
Resident time 20, 114
Residence time distribution 20
Reynolds number 9

Schmidt number 9
Second law of thermodynamics 300
Selectivity 59
Sherwood number 9
Shrinking core model 12
Shrinking particle model 12
Simpson rule 470
Sound velocity 300
Space time 114
Space velocity 114
Species balance equations 14, 101–107
    Macroscopic form 102–104
    Microscopic form 465–469
Species continuity equations 465
Species formation rate 81
Species operating curves 117
Stirred tank reactor *see* CSTR
Stoichiometric coefficients 27
Stoichiometric proportion 48
Stoichiometric relationships, table of 456
Surface reactions 2

Thiele modulus 10
Transport limitations 9, 91
Trickle-bed reactor 5
Tubular reactor *see* Plug-flow reactor
Turbulent flow 239

Yield, definition of 58